Smart Innovation, Systems and Technologies

Volume 254

The Smart Innovation, Systems and Technologies book series encompasses the topics of knowledge, intelligence, innovation and sustainability. The aim of the series is to make available a platform for the publication of books on all aspects of single and multi-disciplinary research on these themes in order to make the latest results available in a readily-accessible form. Volumes on interdisciplinary research combining two or more of these areas is particularly sought.

The series covers systems and paradigms that employ knowledge and intelligence in a broad sense. Its scope is systems having embedded knowledge and intelligence, which may be applied to the solution of world problems in industry, the environment and the community. It also focusses on the knowledge-transfer methodologies and innovation strategies employed to make this happen effectively. The combination of intelligent systems tools and a broad range of applications introduces a need for a synergy of disciplines from science, technology, business and the humanities. The series will include conference proceedings, edited collections, monographs, handbooks, reference books, and other relevant types of book in areas of science and technology where smart systems and technologies can offer innovative solutions.

High quality content is an essential feature for all book proposals accepted for the series. It is expected that editors of all accepted volumes will ensure that contributions are subjected to an appropriate level of reviewing process and adhere to KES quality principles.

Indexed by SCOPUS, EI Compendex, INSPEC, WTI Frankfurt eG, zbMATH, Japanese Science and Technology Agency (JST), SCImago, DBLP.

All books published in the series are submitted for consideration in Web of Science.

More information about this series at https://www.springer.com/bookseries/8767

Agnessa O. Inshakova · Evgenia E. Frolova
Editors

Smart Technologies for the Digitisation of Industry: Entrepreneurial Environment

Springer

Editors
Agnessa O. Inshakova
Institute of Law
Volgograd State University
Volgograd, Russia

Evgenia E. Frolova
Department of Civil Law and Procedure
and International Private Law
Peoples' Friendship University of Russia
(RUDN University)
Moscow, Russia

Vladivostok State University of Economics
and Service
Vladivostok, Russia

ISSN 2190-3018 ISSN 2190-3026 (electronic)
Smart Innovation, Systems and Technologies
ISBN 978-981-16-4620-1 ISBN 978-981-16-4621-8 (eBook)
https://doi.org/10.1007/978-981-16-4621-8

This Springer imprint is published by the registered company Springer Nature Singapore Pte Ltd.
The registered company address is: 152 Beach Road, #21-01/04 Gateway East, Singapore 189721, Singapore

Preface

The book is designed to fuse technology and a body of knowledge through the elaboration of theoretical concepts and conceptual frameworks to ensure the economic growth of the Russian Federation by utilizing the huge potential for innovative and environmental entrepreneurship in Russia. The research is intended to solve the most challenging problems facing digitalization in the field of environmental entrepreneurship in the country: the need for specialized personnel training; the considerable financial resources needed for the maintenance of digital technologies; how to market environmentally focused enterprises and organizations, as well as the environmental goods, works, and services that they provide; the greening of consumer preferences; and the emergence and development of a green economy and green investment and financial instruments designed to support greener industrial development and agricultural projects. The proposed results will create the conditions for a systemic approach to tilting the Russian economy toward supporting new eco-businesses through an improved regulatory framework—currently virtually absent in the field of environmental entrepreneurship at the national level.

The book will make it possible to identify new common points of law and economics and to define the prospects for investment in renewable energy sources, circulation of energy resources, and energy efficiency improvements to gain positive economic effects from the introduction of new eco-friendly technologies in Russia. The aim is to create conditions and incentives for energy efficiency improvements and support the energy and environmental security of the Russian Federation by balancing private and public interests in renewable energy sources (RES), circulation of energy resources and energy efficiency improvements, and to define the prospects for improvement of the Russian legislation.

To solve particular problems that have been identified in the Strategy for the scientific and technological development of the Russian Federation, it is necessary to collect empirical evidence on the current interactions between society and nature, interdisciplinary studies on the impact of sophisticated technologies in the field of renewable energy sources (RES), information on the circulation of energy resources and energy efficiency improvements on the state of the environment, the health and safety of citizens, and the current state of public and state institutions designed to contribute to environmental protection. The study of the dynamics of interactions between economic and legal institutions in the field of RES development, the circulation of energy resources, and energy efficiency improvements make it possible to formulate some specific proposals to enable the consistent expansion of RES as one of the green sectors of the economy.

The rapid technological progress being achieved by contemporary society is manifested in the wide use of digital technologies in various spheres of production, business management, health care, provision of services in the banking sector, and the social sphere. Higher living standards for individuals, security for processes through qualitatively new regulatory approaches, the provision of State security (including cybersecurity), overcoming the adverse effects of natural disasters, the prevention of man-made disasters are some of the dozens of goals and problems that cannot be achieved or solved without the use of digital information technologies. As a result, the wide range of issues related to the development and support of digital infrastructure in various areas of the economy cannot be ignored: virtual reality technologies, the creation of innovative entities using, inter alia, augmented reality technologies, cross-industry solutions, resource management, and utilization mechanisms both within the Common Market and in the institutions of the Eurasian Economic Union (EEU) and BRICS, cross-cutting technologies used to control the transborder movement of capital both at the global level and at the domestic level, integration of new technologies and new mechanisms of communication between parties to judicial proceedings in the judiciary system.

This book brings together collected empirical evidence, important pilot projects run by the author team, as well as proposals for the modernization of Russian legislation and existing economic regulations that can be used to train highly qualified personnel in law and economics at the most important higher educational establishments in the Russian Federation.

Russia is currently actively investing in the digital sector of the economy. However, digital innovations technology presents new problems. Many questions are left regarding the legally justified implementation of the digitization strategy. Digital transformation affects data protection, information technology security, contract development, settlement of liability issues and gives rise to other problems of legal regulation. In addition, the use of digital innovation has an impact on the environment. A study conducted in France in 2019 and presented at the International Weather and Climate Forum (IWF) showed that only 40% of the survey participants are aware of the strong link between digital and climate change. It turned out that currently most of the energy is consumed mainly by small items: smartphones, tablets, connected objects, computers. Scientific research has confirmed that the digital sector creates several different types of pollution: the pollution from manufacturing IT equipment; the pollution from electronic waste, that is, used electrical and electronic equipment; the pollution from our daily digital use. Thus, the digital sector may well be invisible, but it has very real consequences for our environment.

Sustainable development is largely associated with the preservation of the biosphere and natural capital together with the technosphere and sociosphere. However, in real life, environmental aspects are overlooked. Essential changes are needed in the mechanism of economic and regulatory legal regulation of entrepreneurship, in particular, strengthening its environmental component through the use of neo-industrial technologies. In this regard, the search for effective "common grounds" between digital technologies and environmental entrepreneurship is particularly significant and promising. Digital technologies and environmental entrepreneurship should not be mutually exclusive. The environmental effect of digitization will primarily be achieved through the digital environment, which will manifest itself in ensuring the efficient use of resources. One of the tools contributing to the ecologization of the economy and law can be a system of a "green" economy. In addition, technological change is directly related to the use of big data, artificial intelligence, and robotics, which are generally transforming global manufacturing, industrial processes and as a result changing labor power. The digital economy serves the effective automated management of urban and rural production in the context of advanced information technologies. The digital economy serves the effective automated management of urban and rural production in the context of advanced information technologies. All this makes it possible to carry out both scientific–technological progress and the development of environmental entrepreneurship, also for the purpose of a safe environmental state. The two pillars—digital technology and environmental sustainability—should complement and reinforce each other as mutually conditioning components. The unity of digitalization and industrial production should strategically coexist in the interests of the society, in the interests of the ecological environment safety, and the sustainability of the ecosystem.

There are many problems in establishing the relationship between the results and the research methodology in the economic and legal sciences. Many of the findings obtained as a result of economic research cannot be applied in practice, since there is no mechanism for interaction between representatives of these two branches of scientific knowledge. On the contrary, the expansion of the "green" sectors of the economy (the production of environmentally friendly products, the development of "green" energy, the development of eco-tourism, insurance and audit, etc.), effective counteraction to the global environmental threats (the climate change, the need for the development of renewable energy sources) requires not only the development of specific legal norms but also taking into account modern economic concepts (e.g., the concept of The Circular Economy by Ken Webster), which effectively counteract environment degradation state in Russia. In addition, in order to preserve and restore the environment and protect natural resources, there is a necessity for the integrated use of the end-to-end digital technologies, including neo-industrial technologies, which ensure the accelerated development of the "green" economy in the Russian Federation and the environmental entrepreneurship.

The actuality of solving this problem is due to the fact that the use of digital technologies in modern society, as well as the development of technical capabilities, forms a new digital environment, as well as in the system of environmental entrepreneurship. The digital environment plays a definite and increasingly important role in the relationship between the technosphere and the natural environment. The impact of digitalization on the economic and legal regulation of environmental entrepreneurship is unavoidable.

The scientific problem, which the project is aimed to solve, is to develop balanced mechanisms for the impact of digital technologies on the effective production regulation and non-production areas of environmental entrepreneurship at the present stage of Russian development using the scientific methods of economics and law interaction. It is also planned to develop economic and legal mechanisms for the interaction of scientific methods of jurisprudence and economics concerning the cross-sectoral problem of the renewable energy sector development, the turnover of energy resources, and energy efficiency.

The scientific and practical relevance of solving the indicated problem is that it will allow creating scientific premises in the form of the theoretical concepts' development and conceptual foundations in the stated field of research, at a level close to the world level, ensuring the economic growth of the Russian Federation through the intensive involvement of the Russian environmental entrepreneurship potential. The study assumes the solution of the most serious challenges facing digitalization in the field of environmental entrepreneurship in modern Russia—training of specialized personnel, significant financial costs for operation digital technologies, marketing of the environmentally friendly enterprises and organizations, as well as environmental goods, works, and services, and the formation of the ecological component of consumer preferences. The proposed results will create conditions for a systematic approach to changing the vector of the Russian economy toward the eco-market and eco-business and identify promising areas for improving the Russian regulatory framework, which is currently practically absent in the field of environmental entrepreneurship at the state level.

The proposed scientific research will reveal new "common grounds" of jurisprudence and economics, determine promising areas for investment in the renewable energy sources' development, the energy resources turnover and improve energy efficiency, get a positive economic effect from the introduction of new environmentally friendly technologies, create conditions for the implementation of the international obligations in the field of environmental protection by the Russian Federation. The proposed solutions will create conditions and incentives for increasing energy efficiency and ensuring the energy and environmental security of the Russian Federation, will balance private and public interests in the development of the renewable energy sources (RES) sector, the energy resources turnover, and increase energy efficiency, and will identify promising areas for improving Russian legislation.

The proposed project is closely related to the formation of the scientific–technological premise for economic growth and social development of the Russian Federation. Solving specific tasks set in the Strategy for Scientific and Technological Development of the Russian Federation assumes the collection of empirical data on the current situation in the field of interaction between society and nature, cross-sectoral research on the impact of modern technologies in the field of renewable energy sources, energy turnover and energy efficiency on the environment, life and the health of citizens, the state of public and governmental institutions aimed to participate in environmental protection. The dynamic study of the interaction between economic and legal institutions in the field of RES development, energy turnover, and energy efficiency will make it possible to formulate a number of specific suggestions to the authorities, allowing to gradual expanding RES as one of the "green" sectors of the economy. The collected empirical materials, doctrinal developments of the authors' team, suggestions for the modernization of Russian legislation, and existing economic regulators can be used to train highly qualified personnel in higher educational institutions of the Russian Federation, in law and economics faculties.

In the modern world, there is a staged greening of industrial, agricultural, and other products, which manifests itself in a gradual decrease of the negative impact on the environment. This work is carried out to achieve several goals related to reducing greenhouse gas emissions (which should help slow down the process of global climate change), with the transition to sustainable development standards, with a reduction in waste production and consumption, etc. However, to achieve the set goals, typically the narrow-industry approaches are used, limited either to a separate approach (instrumental, university, according to the social belonging of the consumption of the goods, according to the objects of research) within the framework of economic science, or to any separate branch of law (environmental law, civil law, and financial law).

The proposed project is aimed at developing an integrated approach to the study of economic and legal mechanisms for the development of digital technologies in the regulation of environmental entrepreneurship; will allow attracting both the results of the latest researches in the field of economics and the latest achievements of the several branches of Russian and foreign legal science. The theoretical standards development for cross-sectoral interaction between economics and law will help to build a balanced system of mechanisms for the economic and legal regulation of the production and non-production areas of environmental entrepreneurship. The development of an economic and legal strategy will allow a more complete reflection of the economic science achievements and to a more complete reflection of the main elements of economic incentives for environmental protection and support for environmental entrepreneurship in the proposed draft Federal Laws.

The methodological basis of the research is a systematic approach, which will be applied within the framework of the materialist view of nature and the general method of research—dialectic materialism. The systematic approach will be implemented through the use of a number of interrelated methods, namely analysis, synthesis, system-element, system-structural, system-functional, system-communication, system-integrative, system-historical methods. The study will use an interdisciplinary approach that allows combining the achievements of economic and legal sciences. Using the systemic–structural method of scientific knowledge, the place of environmental entrepreneurship in the system of types of entrepreneurial activity and the green economy as a whole will be identified, as well as its importance for achieving the sustainable development goals (SDGs). This method will also make it possible to formulate a list of forms and methods of economic incentives within the framework of state support for Russian businesses that use neo-industrial technologies to expand the "green" sectors of the economy. This will allow substantiating the measures' targeting of state support for environmental entrepreneurship at the federal and regional levels, the reasoning for the delimitation of environmental entrepreneurship from other areas of entrepreneurial activity, and, as a result, proposing a program document—the Strategy for the Development of Economic and Legal Regulation of Environmental Entrepreneurship in the Russian Federation.

General scientific methods used in the project are divided into three groups. The group of theoretical methods includes the method of rising from the abstract to the concrete, hypothetical-deductive method, axiomatic method, and the formalization method in building models for the formation of new social relations within the framework of the "green" economy and its part—the environmental entrepreneurship.

The group of empirical methods provides the collection of data (including the composing of the dataset and the formation of a database), comparison, description, measurement, and assessment of the impact of digital technologies and environmental entrepreneurship on modern social development.

The group of general logical methods includes abstraction, generalization, idealization, induction, analogy, modeling, probabilistic, and statistical methods, which will make it possible to achieve consistent scientific conclusions in exact accordance with the objectives of the project.

As a part of private scientific legal methods, the project will use the dogmatic, legal, and technical method, the method of interpreting the law, state legal modeling, comparative legal method, which allows forming a scientifically based structure of legal regulation of the rights and obligations of entities whose economic activities are directly related to production and commercialization of environmental goods, works, and services. As a part of private scientific economic methods, the project uses SWOT analysis, mathematical, graphical, comparative, formalization method, which will allow assessing the effectiveness of the use of regulatory mechanisms and neo-industrial end-to-end technologies for the development of environmental entrepreneurship as an element of a resource-saving and socially inclusive model of a green economy.

Using the political and cultural method, the experience of state support for environmental entrepreneurship in European and Asian countries will be investigated that will reveal the influence of political traditions on the content of measures for the development of environmental entrepreneurship and determine the measures applicable in the specific conditions of the Russian Federation. In addition, this method will make it possible to identify the effectiveness's conditionality of state support for environmental entrepreneurship by the status of the social–environmental culture, as well as to determine measures to increase its level through a series of educational, educative, and awareness-raising activities to popularize the socially inclusive model of the "green" economy and the effective use of digital technologies by people in everyday life.

Using a comparative method, the laws of the EAEU and BRICS countries, aimed at supporting various types of entrepreneurial activity, including environmental entrepreneurship, will be investigated. An analysis of the provisions of such laws and arising from them regulatory enforcement will make it possible to determine which of the measures provided by them are effective and deserve support in the model Strategy for the Development of Economic and Legal Regulation of Environmental Entrepreneurship in the Russian Federation, proposed by the authors' team, and which are contrary to the current legislation or standards of legal technology and economic laws.

The scientific works of representatives of foreign economic and legal science, dedicated to the development of state strategies and other political and legal documents in the field of environmental protection and the development of environmental entrepreneurship, as well as existing scientific approaches to the economic and legal regulation of environmental entrepreneurship, including the use of digital technologies, in Russia and foreign countries, will be investigated using the comparative method.

The uniqueness of the book, according to its editors, is not the study of end-to-end technologies themselves, but the economic and legal regulation of environmental entrepreneurship and innovation with the use of end-to-end technologies.

The book is aimed at developing a comprehensive approach to the study of economic and legal mechanisms for the development of digital technologies in the regulation of environmental entrepreneurship. This combined to attract the results of the latest research in the field of economics, as well as the latest achievements of several branches of Russian and foreign legal science. The theoretical development of standards for the cross-sectoral combination of economics and law allowed us to build a balanced system of mechanisms for the economic and legal regulation of the production and non-production spheres of environmental entrepreneurship. The development of an economic and legal strategy will make it possible to fully reflect the achievements of economic science and to more adequately reflect in the proposed draft federal laws the main elements of economic incentives for environmental protection and support for environmental and innovative entrepreneurship.

Agnessa O. Inshakova
Doctor of Law, Professor, Head of the Department of Civil and Private International Law (The Basic Department of the Southern Scientific Center of the Russian Academy of Sciences)
Volgograd State University
Volgograd, Russia
gimchp@volsu.ru
ainshakova@list.ru

Evgenia E. Frolova
Doctor of Law, Professor, Head of the Department of Civil Law and Procedure and International Private Law
Peoples' Friendship University of Russia (RUDN University)
Moscow, Russia
frolova_ee@rudn.ru

Professor of the Department of Theory and History of State and Law
Far Eastern Federal University
Vladivostok, Russia

Leading Researcher
Vladivostok State University of Economics and Service
Vladivostok, Russia

Contents

Editors and Contributors

About the Editors

Professor Doctor of Law Agnessa O. Inshakova is Honorary Worker spheres of education of the Russian Federation and is Head of the master's degree program "Civil law, business law, private international law". She is Head of the scientific school "Law of economic entities in the context of technological and socio-economic transformations" and also Head of the scientific and educational center "Modernization of the legal system of modern Russia" of the Volgograd state University, Institute of law. She is Head of the basic Department of the Southern Scientific Center of the Russian Academy of Sciences. She is Author of more than 400 scientific, educational and methodological publications, including more than 30 monographs, including leading foreign publications indexed in the international scientific-metric databases Scopus/Web of Science Core Collection in English (as a leading author and editor); 36 textbooks, workshops and training manuals (including more than 10 publications with the stamp recommendations of the Ministry of science and higher education, as well as leading Federal Publishers, including Yurayt, Zertsalo, etc.). She is Author of more than 200 articles in journals in the List Higher attestation Commission of the Ministry of education and science of the Russian Federation; 66 articles in journals indexed in the international database Scopus, including Q1 and Q2; 41 articles in journals indexed in the international database Web of Science Core Collection, including Q1 and Q2. It ranks second in the country in the number of publications in Scopus and Web of Science Core Collection in the specialty—Law. She has high scientific-metric indicators: citation index in RSCI-1580, Scopus—165, Web of Science Core Collection—49; h-index in RSCI—16; Scopus—7; Web of Science Core Collection—4.

Professor Evgenia E. Frolova holds Doctor of Law and is Professor, Head of the Department of civil law and procedure and international private law of Peoples' Friendship University of Russia (RUDN University); she is Honored Lawyer of the Russian Federation, Honorary Worker of higher professional education of the Russian

Federation, Honorary Professor of Shanghai University of Political Science and Law (China) and Research Supervisor (consultant) of the department of financial law and legal regulation of economic activity of the Belarusian State University. She is Author of more than 200 scientific, educational and methodological publications, including 10 monographs; 20 textbooks and teaching AIDS; more than 50 articles in journals reviewed by the higher attestation Commission of the Russian Federation; 60 articles in journals indexed in the international citation and analytical database Scopus; 15 articles in journals indexed in the international citation and analytical database Web of Science Core Collection; more than 20 publications in publications included in the database Russian Science Citation Index (RSCI) on the Web of Science platform. She is also Member of the Editorial Boards of the Higher attestation Commission of the Ministry of science and higher education of the Russian Federation journals, such as Eurasian Law Journal, problems of economy and legal practice.

Contributors

Aslan Kh. Abashidze Department of International Law, Peoples' Friendship University of Russia (RUDN University), Moscow, Russia

Aleksey P. Anisimov Volgograd State University, Volgograd, Russia

Natalia V. Antonova Department of Social Legislation, Institute of Legislation and Comparative Law Under the Government of the Russian Federation, Moscow, Russia

Alexander V. Begichev Department of Notaries, Kutafin Moscow State Law University (MSAL), Moscow, Russia

Anastasia A. Belousova Peoples' Friendship University of Russia (RUDN University), Moscow, Russia

Vitaliy V. Bezbakh Department of Civil Law and Procedure and International Private Law, Peoples' Friendship University of Russia (RUDN University), Moscow, Russia

Marghoob Saleem Butt Independent Permanent Human Rights, Commission (IPHRC) of the Organization of Islamic Cooperation, Department of International Law, Peoples' Friendship University of Russia (RUDN University), Moscow, Russia

Tatyana I. Chursina Constitutional Law Department, Institute of Legislation and Comparative Law, Government of the Russian Federation, Moscow, Russia

Andrey A. Dementev RUDN University, Moscow, Russia

Tatiana V. Deryugina Moscow State University of the Ministry of Internal Affairs of Russia Named After V.Ya. Kikot, Moscow, Russia

Maria O. Diakonova Department of Civil Legislation and Procedure, Institute of Legislation and Comparative Law, Government of the Russian Federation, Moscow, Russia

Natalia N. Emelianova RUDN University, Moscow, Russia

Elena P. Ermakova Department of Civil Law and Procedure and International Private Law, Peoples' Friendship, University of Russia (RUDN University), Moscow, Russia

Evgenia E. Frolova Department of Civil Law and Procedure and International Private Law, Peoples' Friendship University of Russia (RUDN University), Moscow, Russia;
Vladivostok State University of Economics and Service, Vladivostok, Russia

Olga K. Goncharenko Moscow State Institute of International Relations (MGIMO University), Moscow, Russia

Alexander I. Goncharov Volgograd State University, Volgograd, Russian Federation

Marina V. Goncharova Plekhanov Russian University of Economics, Volgograd, Russia

Marianna Ilyashevich Peoples' Friendship University of Russia (RUDN University), Moscow, Russia

Agnessa O. Inshakova Volgograd State University, Volgograd, Russia

Dmitry V. Ivanov Moscow State Institute of International Relations (MGIMO University), Moscow, Russia

Yuliya N. Kashevarova Department of Civil Law and Process, Institute of Legislation and Comparative Law Under the Government of the Russian Federation, Moscow, Russia

Alexey Yu. Mamychev Vladivostok State University of Economics and Service, Vladivostok, Russia;
Head of the Laboratory of Political and Legal Studies, Deputy Dean of the Faculty of Political Science, Lomonosov Moscow State University, Moscow, Russia;
Department of Theory and History of State and Law, Far Eastern Federal University, Vladivostok, Russia

Denis E. Matytsin Department of Business Law, Arbitration and Civil Procedure, Institute of Law, Volgograd State University, Volgograd, Russia;
Department of Civil Law and Procedure, Volga Branch of the International Law Institute, Volgograd, Russia

Maryia V. Miashchanava Department of Civil Law, Belarusian State University, Minsk, Belarus

Tatsiana N. Mikhaliova Law Faculty of the Belarusian State University, Minsk, Belarus

Valentina G. Mikrina Odintsovo Branch of the Moscow State Institute of International Relations of the Ministry of Foreign Affairs of the Russian Federation (MGIMO University), Moscow, Russia

Olga I. Miroshnichenko Head of the Department of Theory and History of State and Law, Far Eastern Federal University, Vladivostok, Russia

Nikita Yu. Molchakov Moscow State Institute of International Relations (MGIMO University), Moscow, Russia

Dmitry A. Pashentsev Department of Theory Law and Interdisciplinary Studies of Legislation, Institute of Legislation and Comparative Law Under the Government of the Russian Federation, Moscow, Russia

Natalia V. Putilo Deputy Head of the Department of Social Legislation Institute of Legislation and Comparative Law Under the Government of the Russian Federation, Moscow, Russia

Ekaterina P. Rusakova Department of Civil Law and Procedure and International Private Law, Peoples' Friendship University of Russia (RUDN University), Moscow, Russia

Sergei A. Sinitsyn Department of Civil Legislation and Procedure, Institute of Legislation and Comparative Law, Government of the Russian Federation, Moscow, Russia

Ekaterina A. Tsepova Law Firm "PharmConsulting", Moscow, Russia;
Vladivostok State University of Economics and Service, Vladivostok, Russia

Albert V. Tumakov Moscow State University of the Ministry of Internal Affairs of Russia Named After V.Ya. Kikot, Moscow, Russia

Grigory A. Vasilevich Department of Constitutional Law, Belarusian State University, Minsk, Belarus

Natalia S. Volkova Deputy Head of the Department of Social Legislation Institute of Legislation and Comparative Law Under the Government of the Russian Federation, Moscow, Russia

Part I
Digital Platforms and Cloud Technologies in the Context of Neo-industrial Transformation of Economic and Legal Systems

Chapter 1
Digital Platforms and Cloud Technologies in Post-capitalist Discourse: A Doctrinal and Legal Perspective

Olga I. Miroshnichenko and Alexey Yu. Mamychev

Abstract The ongoing shift from the industrial to the digital era dramatically affects the institutional structures, public and private communication in society, the dynamics of the search for a balance of interests, etc. Data is becoming a new driving force for economic and social development, shaping the post-capitalistic era. The authors discuss one of the most rapidly developing technological trends in the modern world—digital platforms, cloud technologies, and artificial intelligence systems in general. At the same time, the research notes that the existing legal regulation tries to "fit" the changes taking place within the existing concepts, models, legal structures, and terms. And, these models, constructions, and terms often come into conflict with newly emerging specific phenomena and the practice of their implementation. In the context of the collision of the state and society with the global challenges, there is a strong necessity to create institutions and mechanisms that provide social and legal control, algorithmization of political processes based on digital platforms, etc. It is also noted that it is reasonable to start the solution of these problems from the formulation of the concept of advanced legal modeling, the creation of a common doctrine, methodology, and standardization of the processes of developing autonomous vehicles, software, forecasting the relationship between man and machine, and so on. And only then is it possible to implement a comprehensive program of legislative changes, the development of relevant sub-sectors, and individual branches of law.

O. I. Miroshnichenko (✉)
Head of the Department of Theory and History of State and Law, Far Eastern Federal University, 10 Ajax Bay, Russky Island, 690922 Vladivostok, Russia

A. Yu. Mamychev
Vladivostok State University of Economics and Service, 41 Gogol St., 690014 Vladivostok, Russia

Head of the Laboratory of Political and Legal Studies, Deputy Dean of the Faculty of Political Science, Lomonosov Moscow State University, 27 b.4 Lomonosovsky Prospekt, 119234 Moscow, Russia

Department of Theory and History of State and Law, Far Eastern Federal University, 10 Ajax Bay, Russky Island, 690922 Vladivostok, Russia

A. O. Inshakova and E. E. Frolova (eds.), *Smart Technologies for the Digitisation of Industry: Entrepreneurial Environment*, Smart Innovation, Systems and Technologies 254, https://doi.org/10.1007/978-981-16-4621-8_1

Keywords Digital technologies · Cloud technologies · Digitalization · Digital platforms · Artificial intelligence · Legal regulation of digital technologies · Trends in the development of digital platforms · Risks and threats of digital technologies

JEL Codes F01 · O21 · O32 · O33 · K39

1.1 Introduction

It is difficult to predict the dynamics of modern socioeconomic and political–legal systems transformation since the speed and nature of changes in various sectors of public life destroy the traditional socio-cultural dominants of stability. Today, any forecast and any strategy will be inadequate if, apart from behavioral and social factors, they do not include modeling the development of digital forms and technologies (Abashidze et al., 2020). In other words, modern political–legal modeling and public–legal governance no longer assume only the "social" as a fundamental element and dominant trend. Traditional social organization forms, socio-political management technologies, the development of the social relations normative mediation, without any doubt, remain significant but no longer the only factors in the modern social systems dynamics.

Today, social and political thinking and research practices are also transforming the basic definitions used to characterize unfolding events and processes, the penetration of concepts and notions from different descriptive systems, and research protocols into others. There is an intensified search/formulation of a new "vocabulary" which, *on the one hand*, can adequately describe the current dramatic changes in society, politics, the law under the impact of digital transformation, unfolding long-term viral pandemic, and accelerated climate change and *on the other hand* present a possibility to consider various systems (social, environmental, biological, physical, digital) as interrelated and equivalent, without giving priority to any one dimension.

This transformation is already tangible, as the importance of digital systems, technical needs of digital infrastructure, biological and ecological dominants become equal in status and often lead people's interests and needs and different types of communities (Matytsin & Rusakova, 2021). Today, other orientations are coming to the fore, driven by the needs of systems (technical, informational, digital, biological, climatic, etc.) that "nominally serve us," but for which social value and normative codes and standards are not leading in the trajectory of their development. These trends are associated with the loss of anthropocentric character of all unfolding events and processes and the realization of multifactorial and complexity of the existing reality, which unites visible and hidden, intangible, and indistinguishable factors. The so-called Fourth Industrial Revolution originated long before the present moment was gradually prepared throughout the twentieth century, bringing into play more and more new elements, causes, and conditions of social, biological, technological, ecological, and other processes and events unfolding.

The latter has led to the fact that nowadays social activity (both at the level of the individual and at the level of collectives) "neighbors" with the activity of "non-human" elements (Inshakova et al., 2020a, 2020b). Today, the dynamics of social interaction are increasingly being defined by digital forms, which generate specific events in socio-cultural and digital reality. It is important to consider that social reality also modifies digital development trajectories and shapes the development of digital algorithms in machine learning. Specific social data and cultural artifacts mediate the latter. As a result, the development of these algorithmic systems gets a specific socio-cultural coloring. All this determines a rather unique direction or trajectory of further development of digital algorithmic systems, artificial intelligence systems. Thus, the introduction of an autonomous digital system or a certain algorithmic solution in different communities, trained on a specific "socio-cultural material," leads to completely different trajectories of their development, to the formation of so-called digital system biases, digital forms of discrimination, digital forms of differentiation, etc.

The latter applies, in particular, to the development of digital platforms based on artificial intelligence systems, which act as a fundamental driver of the transformation of traditional connections and relationships in the social system and as a public arena, forming and representing social interests and needs, new forms of identity, etc. Today, digital platforms are the key solution capable of summarizing the contradictory dynamics of social and technical systems, extracting and summarizing a huge array of data, and organizing and controlling the latter's exchange. Overall, data, rather than capital, becomes the new driving force of economic and social development, shaping the post-capitalist era.

1.2 Methods and Materials

The study is based on such methods as the logical, which allows using the methods of analysis and classification to methodize ethical problems of the technological revolution; the system–structural analysis, which allows applying differentiation and detailing of socio-political processes influenced by digitalization and which allows to elaborate ethical aspects in relations related to the use of AI and to highlight ethical and legal problems of these relations; formal–legal method, which analyzes the normative legal acts regulating relations with the use of digital technologies; predictive method and modeling method, which were used to identify trends in the development of Russian society under the influence of digital technologies shortly.

1.3 Results

1.3.1 Development of Digital Platforms and Artificial Intelligence Systems: Social and Legal Aspect

The development of digital platforms and artificial intelligence systems (as the "algorithmic content" of these platforms' functioning) has become one of the most actively discussed technological trends of our time. The last decade has witnessed the rapid development of information technology. The global community has already given this process a name—"Uberization" (from Uber). We are talking about profound economic and societal transformation levels in how supply and demand interact (it is not the Uber platform itself that has revolutionized, but the new business model) (Kuprevich, 2018, p. 313). There are currently no univocal definitions of the terms "platform"/"digital platform." A definition of the platform includes a virtual trading platform, its users, software, hardware and networking complexes, business model, and the firm that implements it (Kuprevich, 2018, p. 313). In other words, the concept of digital platforms can be approached in different ways: as a purely technical phenomenon, socio-technical or economic. Note that new phenomena do involve alternative interpretations due to internal, including technological, complexity. However, experts agree that a "platform" itself is a *complex infrastructure* mediating the relationship between users (Srnichek, 2019), while a "digital platform," unlike a non-digital platform, is firstly constituted through software code, and secondly, contains such components as devices operating systems, databases, etc. (Ryzhkova, 2019, p. 50). Even today, platforms are called the future of the market economy. Another promising technology that has already been introduced in several areas (economy, medicine, manufacturing, transport, and others), and at the same time, one of the most ambiguous in terms of public attitudes is artificial intelligence technology—computer or cyber-physical systems with anthropomorphic (humanoid) "intelligence." Philosophically, and legally, we are moving away from that utilitarian view of the world in which man is the only subject, reasonable, rational, and everything else is a world of objects, a world appropriated by man or subject to inevitable appropriation as soon as technical possibilities permit. As the researchers correctly point out, today, "few people are confused by a view of the world in which man is not the only subject" (Gabov, 2018, p. 105).

Simultaneously, an adaptation of existing categories and legal and technical "fitting" of existing normative acts to a fundamentally new sphere of social relations still seems to be not very effective.

We believe that the solution of this problem should begin with the formation of a general doctrine, methodology, and standardization of autonomous vehicles development processes, software, prediction of human–robot relations, and only then it is possible to implement a comprehensive program of legislative changes, develop relevant sub-sectors and separate branches of law.

For example, Nicholas Petit identifies the following main approaches to regulating the use of artificial intelligence and robots:

- formal approach, according to which we should proceed from the peculiarities of the relevant legal system and its institutions and ensure that all institutions that may affect the use of artificial intelligence, e.g., liability, privacy, cybersecurity applicable to all types of artificial intelligence by establishing universal rules;
- Technological approach, which implies solving individual issues concerning each category of technological application of artificial intelligence technologies and units: unmanned vehicles, social robots, etc. (Petit, 2019).

Modern researchers, when considering the application of law to artificial intelligence systems, as Hugo Pagallo notes, mainly follow three main approaches to their understanding, interpretation, and consideration:

- determining exactly how the development and application of artificial intelligence technologies and units affects and changes existing legal concepts and principles;
- determining how exactly the development and application of artificial intelligence technologies and units determines the creation of new legal concepts and principles;
- seeking to adapt those legal issues that arise when using AI technologies and units within already existing legal approaches and concepts without changing them (Morhat, 2018, p. 99–100).

Meanwhile, legal regulation in the context of technological advances involves complex trade-offs (Petit, 2019). First of all, regulation can nip technical progress in the bud. A historically famous example is the Red Flag Act passed in Britain in 1895, which sought to regulate the introduction of the automobile into society. Not only did the law limit out-of-town speed to four miles per hour (6 km/h) and in the city to two miles per hour (3 km/h), but it also required a signalman with a red flag to walk 60 yards in front of the car in the city limits. Such a restriction made the operation of cars absurd and delayed British car manufacturing development for 30 years (Petit, 2019).

Drawing analogies with artificial intelligence systems, for example, there is a discussion of imposing restrictions on the ability of drones to fly beyond-line-of-sight, which could constrain the development of goods delivery services.

In addition, ineffective over-regulation can occur in response to risks, incidents, and accidents. However, it should be taken into account that "over-regulation" of potential risks may be ineffective since it will also prevent further development of this area and use the favorable positive potential of artificial intelligence to the fullest extent (Petit, 2019).

Summarizing the analysis, we should note that the creation and use of artificial intelligence affect various branches of legal regulation. The difficulty of this regulation is due to the rapid pace of technological development, which does not quite correspond to the pace of development of legal concepts and norms; thus, the law naturally lags behind technological development. In addition, the field is interdisciplinary.

The legal regulation should, on the one hand, provide a balance between the interests of society, which is to use the greatest possible potential of new technologies that can provide positive solutions to a variety of significant problems of both individuals and society as a whole and on the other hand—the need to minimize the negative consequences of the use of innovative technologies.

The difficulty for the legislator in creating an effective system of regulation of the development and use of artificial intelligence technologies is also the lack of completeness of the necessary information about such technologies and their special characteristics.

The most obvious feature of artificial intelligence that allows it to be distinguished from earlier technologies is the ability of artificial intelligence to act autonomously. Artificial intelligence systems can already perform complex tasks such as driving and portfolio creation without active human control. The complexity and volume of tasks to be performed by artificial intelligence will undoubtedly continue to grow in the coming years. Just as the industrial revolution has caused socioeconomic disruptions such as a reduction in the need for human manual labor in manufacturing and agriculture, artificial intelligence and related technological advances reduce the demand for human labor in services.

One of the most important characteristics of artificial intelligence that challenges legal regulation is unpredictability, compared to natural intelligence. There are already numerous examples of artificial intelligence actions, which can manifest non-standard thinking. Thus, computer chess programs make moves contrary to the basic principles of human chess strategy. Another example is the C-Path software developed for diagnostic pathologists. Instead of focusing on the tumor cells themselves, C-Path found that the most predictive signs should be looked for in the cells surrounding the tumor—in an area called the stroma—which was contrary to prevailing medical thinking in cancer diagnosis (Scherer, 2016).

Probably, such examples should not be considered exactly as "creative" steps of artificial intelligence adepts; they are just a consequence of the huge amount of computing resources available to the machine, combined with the absence of cognitive biases (let us call them stereotypes) that affect humans. Thus, when discussing chess programs, Nate Silver notes, "it probably shouldn't be described as 'creative' for finding moves, it's more likely to be at the expense of its computational speed. But it also has the advantage of having no idea how to play chess correctly in these particular circumstances. For humans, however, such a move requires creativity and confidence to transcend traditional thinking" (Scherer, 2016).

The above shows the fundamental difference between human consciousness and the operating principles of artificial intelligence—differences that can lead a machine to create solutions that humans would not expect.

It is this ability to generate unique solutions that make the use of AI units so attractive. Until recently, the unexpectedness of these machines has been severely limited by the domain in use: For example, a computer chess program can make an unexpected move, but it still does nothing more than play chess. Meanwhile, the development of more universal artificial intelligence systems will lead to an increase in unexpected behavior, whereby, even with the most careful system design and programming, the actions of artificial intelligence cannot be predicted. Thus, the issues of foreseeability and causation pose a challenge to legal regulation to ensure the protection of persons who the actions of artificial intelligence may harm.

Based on the foreign countries and Russian Federation adopted acts analysis, we believe it is possible to allocate the following main stages of artificial regulation intelligence development:

The first stage is the development and adoption of doctrinal acts—programs, strategies, plans, concepts of development of artificial intelligence and robotics technologies.
The second stage is the development and adoption of a framework for legislation in this area. Thus, V. V. Arkhipov and V. B. Naumov speak about the option of a special federal law, "On Robotics in the Russian Federation", which could become a basic document, based on and following subsequent changes in other legal acts could be introduced (Arkhipov & Naumov, 2017, p. 58).
The third stage is the introduction of amendments to legislative acts and bylaws. Based on the developed legal regulation strategy, special acts should be developed aimed at:

- establishing the legal regime of robots in the structure of legal relations;
- regulating how data is processed and accessed for research purposes;
- formation of unified systems of standardization, certification, and evaluation of technological solutions based on artificial intelligence;
- development of ethical rules of human interaction with artificial intelligence;
- introduction of amendments to sectoral legislation providing for the possibility of performing activities with the help of artificial intelligence (e.g., transport, medical, tax legislation, etc.).

The above is supported by the findings of authoritative researchers in the field. Thus, A. V. Neznamov speaks of the need for three types of regulatory acts on artificial intelligence: First is a state strategy for developing artificial intelligence and robotics; second is a legislative framework in this area; third is regulation of applied aspects of artificial intelligence (Agapov, 2018, p. 12).

A separate complex branch of law should be formed—robotics law—which has its subject and method of legal regulation. The subject of this law branch should be a set of legal norms regulating social relations in the field of development and production of software, the use of robotic technologies, and autonomous vehicles based on AI. This complex branch of law should include: first, the norms of public law—constitutional, criminal, administrative, land, maritime, water, air, etc., which are based on imperative legal regulation and ensure the protection of rights, freedoms, and legitimate interests of all participants of legal relations, as well as ensure the national security of the Russian state and sustainable national development; second, legal norms of private law branches, which form specific legal regimes of interaction.

1.3.2 Specific Trajectories of Digital Platforms Under the Impact of the Global Pandemic

The pandemic associated with the spread of coronavirus infection has initiated a particular trajectory of digital technology, political and legal practices. Thus, the spread of COVID-19 has significantly influenced the development of digital platforms, the sophistication, and mass adoption of digital surveillance systems, registration, and identification of both social actors and the movement of infection across the social network. Mobile applications and online services have been introduced into human life and society, shaping new modes of citizen interaction, control procedures, differentiated oversight, and disciplinary practices.

In this aspect, the proliferation of mass data collection and processing systems of public (various activities and movements in the public environment) and private (monitoring the state and location of subjects in private space) nature require provision:

- *firstly*, an adequate level of data protection from unlawful transfer to third parties;
- *secondly*, social and legal control over their storage, processing, and use, which requires the formation of fundamentally new socio-political institutions and practices.

Besides, new areas of extra-legal state power activity have emerged, related to public-power activity not regulated by legal norms but necessary to prevent infestations, restrictions related to movement, self-isolation regime, etc. New forms of state-legal coercion (e.g., to the regime of self-isolation), types of adverse legal consequences (e.g., sanctions related to the violation of the mass regime of personal protective equipment use) have also emerged. In this aspect, it is necessary to intensify legal modeling, focused on the formation of legal principles and legal mechanisms to regulate the above-mentioned public activity of state power institutions, as well as to update the tools of legal technique for quick and effective response to non-standard situations and their resolution within the law, not allowing the possible discretionary violation of the rights and freedoms of man and citizen.

Socio-political processes during the pandemic showed the absence of real institutional and non-institutional mechanisms of influence on decision making by the authorities and efficient mechanisms of control over the collection, processing, and use of personal data.

The focus has now shifted to administrative decision-making. For example, all "unpopular" management decisions as part of the pandemic response (mask regime, self-isolation, and observance of socio-medical distance), and the use of information technology (surveillance cameras, social monitoring, and digital passes) are introduced based on decisions of regional authorities, while there are quite legitimate law-making procedures, such as declaring a state of emergency following Federal Law No. 68-FZ of 21 December 1994 "On the Protection of Population and Territories from Emergency Situations of Natural and Man-made Origin." However, the legal policy has followed the path of establishing an alternative legal regime of restrictive measures, including the use of information technology. There was a threat of loss of consistency in the law. The introduction of alternative legal regimes and power arrangements requires harmonization with existing law.

At present, there is a policy aimed more at legitimizing the decisions taken. However, such permissive rules are not always consistent with general principles of law or law principles in force in a particular field. We consider it advisable to use the mechanism established by the Constitution for responding to emergencies to preserve the systemic nature of the right. It is necessary to develop and substantiate specific proposals on the key trends affecting the development of the public organization and relations in strategic planning and public administration, as well as to adopt doctrinal–political and regulatory models of an adequate and effective state response to new digital challenges, biological threats, and civilizational risks.

1.3.3 Medium- and Long-Term Changes Related to the Introduction of Digital Platforms and Artificial Intelligence Systems: A Socio-political Aspect

In modern society, in addition to traditional actors (political actors and subjects of law), there are innovative actors (digital actors, digital personalities, autonomous robotic algorithms, and technologies). The latter act not only as digital aggregators (through which most public relations are currently organized and implemented, a whole variety of public interaction unfolds, and modern systems of political communication function) but also as significant, active elements (digital and virtual actors) of the environment, acting according to their digital strategies and machine learning trajectories.

At the same time, in many spheres, the latter implement not only advisory/expert functions but also the functions of administrative nature (e.g., in the sphere of public order, the digital algorithms not only mark traditional political subjects, giving them criminogenic indices or social trustworthiness rating but also present a list of adequate response measures and actions of power structures or making independent decisions—blocking access, changing the presented list of powers, possible action options, etc.). New actors, on the one hand, actively influence (sometimes even more seriously than the established actors) the decision-making of traditional political actors and the nature of interaction and directions of development of public–power relations in the system person–society–state and on the other hand even more importantly, the development of strategies in both the public and individual (private) spheres.

The ongoing shift between the industrial and post-industrial eras is becoming a catalyst for the next kind of change:

- *Firstly*, the "public arena" of interests confrontation is being replaced by digital platform solutions;

- *Secondly*, institutional structures and public and private communication in society are changing, and under the impact of digitalization, they are losing their traditional resource of legitimacy and social significance. Most interactions are implemented through digital intermediaries (digital platforms and online services hosted on them). At the same time, the real centers of management decision making are shifting from the public space to a new spatial sphere—"digital laboratories" (which develop, implement, and operate complex digital platform systems);
- *Thirdly*, the forms and technologies of civil (digital activism, digital civil society, etc.) and military resistance (remote, distant projection of power, force, authority) are changing. Today's end-to-end digital technologies make it possible to broadcast and project both tangible and physical and symbolic impacts of a meaningful and defining nature in various spheres. In doing so, digital platforms and cloud technologies are acting as new centers of influence;
- *Fourthly*, the impact of digital platforms restructures key centers of mobility, forms, and technologies of social communication, key resources of social organization change as well, the most important becoming the data produced by population, organizations, mechanisms, algorithms. Data becomes the basis for the constant circulation of information and content, the basis of the modern "digital formation."

Digital platforms are a new technological and organizational element capable of changing the dynamics of consolidated socioeconomic and socio-political models. Thus, non-institutional initiatives on digital platforms influence the prospects of socioeconomic, political, cultural, and other dynamics, shape democratic or authoritarian modes of development of social interaction (Inshakova et al., 2020a, 2020b).

Non-institutional platforms influence the situation by overcoming the "democratic deficit," "party identification," "social mobilization," etc., to some extent also aim at transforming the country's politics by creating new mechanisms for political participation and promoting a change in political culture. The latter can change the domestic political landscape, creating new spaces for socio-political mobilization through technology, even if they are built on the digital services of a global oligopoly. The proliferation of digital technologies in developed countries leads to the involvement of a much larger number of individuals in the political agenda. In addition, there are so-called entertainment politics, where serious political issues are discussed on digital platforms in a form that appeals to the broad majority (memes, short humor videos, dipfeeds, etc.). Studies in countries with a high degree of information technology penetration in public life show that the trend toward an increasing share of digital entertainment political content will continue. Shortly, the political agenda will involve strata that have not previously been involved. This can lead to further mobilization of young people political activity who has not previously participated in political processes, increase responsiveness to political events, and the degree of participation in electoral events at various levels.

In turn, when the state and society face global challenges, the population needs to get information about the situation in the country and the world in near real-time mode. Information should be provided in all accessible forms (TV and radio broadcasting, newspapers, Internet, etc.), taking into account the needs of all citizens needs, based on their age, social, and material status, which affect the way of obtaining information. Simultaneously, the disseminated information requires special control, as it may contain personal data of individuals, the use of which is possible only in strict compliance with legal regulations based on human rights. Therefore, the state should take additional measures to create a safe environment for collecting and processing personal data. A side effect of widespread digitalization as the state and society collide is a significant increase in cybercrime and its refocusing mainly on individuals. Combating cybercrime requires a comprehensive cybersecurity system capable of countering these threats. Large-scale consolidated measures by the government, the business community, and society as a whole are needed to ensure security in cyberspace.

The key risks and threats nowadays are related to the fact that the intensive development of digital platform systems can devalue the meaning and value of public-power interaction between the individual, society, and the state. Moreover, the development of autonomous expert systems, automatic collection of information, machine processing of social inquiries, and formation of responses may not only question the need for specialized professional knowledge and skills of civil servants but also form a rather large distance between the administration and the population, reduce the potential of the legitimacy of power and management structures in general in the eyes of the public.

In the medium and long term, the development of digital platforms and artificial intelligence systems will lead to dramatic changes in the public-power organization of society. Thus, in our view, the following scenarios will unfold in the medium term:

(a) renewal of economic and political elites, whose competence will be dominated not by the desire for innovative technological breakthroughs, but by a strategic vision of possible scenarios of "harmonization" of social, digital, biological, and natural factors and development dominants, and the ability to predict their possible interaction and mutual influence;
(b) the formation of new institutions that will ensure "intergenerational mobility" oriented to reconcile the existing interests of the individual, society, and the state;
(c) doctrinal formalization and development of effective mechanisms for the preservation and reproduction of the "humanitarian core" of social dynamics, the preservation of the cultural and anthropological basic characteristics of a man threatened by biotechnological and bio-digital influences, and mechanisms for the identification of man as a biological species and a subject of political history;
(d) establishment of institutions and mechanisms that provide social and legal oversight and set standards for the development, deployment, and exploitation of digital technologies and biotechnological developments.

1.4 Conclusion

Some trends can be identified in the long term.

First, the political agenda, as well as the development of future doctrinal and programmatic provisions of political parties, public organizations, and movements, will predictably include not only the interests of society and specific social groups but also the fundamental trajectories of digital technological elements, artificial intelligence systems. Simultaneously, the efficiency of forecasting and modeling social dynamics will increase if, in addition to behavioral, cultural, institutional, and other social factors, they include specific trajectories of the development of digital forms and technologies. Besides, the "projective future" as a fundamental "driver" of socio-political and socioeconomic transformation, a special matrix in assessing ongoing events and processes, leads to asynchrony of temporal perspectives, to the reduction of time, changing its understanding. In this regard, the chronotype of politics will change; the past, present, and future will "converge" on the political agenda. And, the latter will act as the key to understanding the past and the present. The interests of the future generation and future states (social, biological, digital, physical) will dominate the current interests.

Second, one of the key factors in developing socio-political dynamics is algorithmization based on digital platforms of political processes, particularly the processes of managerial decision-making. Digital platforms and algorithmic solutions do not just collect and systematize data; their grouping and interpretation (based on the original codes and generated indicators) ultimately transform the political process, forming specific trajectories and an established frame of reference. Otherwise, the algorithm itself creates a certain socio-political reality and specific orientation of socio-political dynamics. Here, collection by algorithmic systems is not reduced to a simple reproduction of aggregated data; in autonomous digital systems, data collection and representation are irreversibly transformed in the latter's algorithmic aggregation. In turn, the collected and aggregated data itself transforms the original algorithm itself in machine learning.

Thirdly, the centers of managerial decision making, centers of socio-political expertise, expert and analytical communities are now shifting from the public space to a new hidden spatial sphere—"digital laboratories" (which develop, implement, and operate complex digital algorithmic systems). We are talking about the shadow space of the development of source codes and initial algorithmic solutions to create specific information and communication systems, robotic technologies, autonomous digital programs, digital technologies, systems of weak and strong artificial intelligence, etc. All these activities are "beyond the space" of normative legal regulation, social control and are not regulated by ethical codes and moral standards or other value and ideological regulators. This is a sphere of total voluntarism and arbitrariness. At the same time, it is necessary to consider the fact that the intensive development of the above-mentioned end-to-end digital technologies is implemented in the shadow space, and today, neither public institutions nor political centers have effective mechanisms of control and influence on this process.

Acknowledgements This work was financially supported by the Grant of the President of the Russian Federation No. NSH-2668-2020.6 "National-Cultural and Digital Trends in the Socio-Economic, Political and Legal Development of the Russian Federation in the 21st Century."

References

Abashidze, A. K., Inshakova, A. O., Solntsev, A. M., & Gugunskiy, D. A. (2020). Socio-economic inequality as a barrier on the path of sustainable development: "Institutional traps" and perspectives of overcoming them. *International Journal of Sociology and Social Policy*. https://doi.org/10.1108/IJSSP-03-2020-0069

Agapov, I. (2018). Artificial intelligence in the law. *Standard, 3*, 10–14.

Arkhipov, V. V., & Naumov, V. B. (2017). Artificial intelligence and autonomous devices in the context of law: On the development of the first law on robotics in Russia. *Proceedings of SPIIRAS, 55*, 46–58.

Gabov, A. V. (2018). Legal personality: The traditional category of law in the modern era. *Bulletin of the SSLA, 2*(121), 96–113.

Inshakova, A. O., Frolova, E. E., Rusakova, E. P., & Kovalev, S. I. (2020a). The model of distribution of human and machine labor at intellectual products in industry 4.0. *Journal of Intellectual Capital, 21*(4), 601–622.

Inshakova, A. O., Goncharov, A. I., & Ershova, I. V. (2020b). Digital blockchain registration of legally significant stages of complex good's export-import supplies by business entities of the EAEU and BRICS Jurisdictions. In A. Inshakova & E. Inshakova (Eds.), *Competitive Russia: Foresight model of economic and legal development in the digital age*. CRFMELD 2019. Lecture Notes in Networks and Systems (Vol. 110, pp. 328–336). Cham: Springer. https://doi.org/10.1007/978-3-030-45913-0_37

Intel's Platform Approach. [Electronic resourse]. IT Analytics. https://www.bytemag.ru/articles/detail.php?ID=8655 Accessed: 03.03.2021.

Kuprevich, T. S. (2018). Digital platforms in the global economy: Current trends and directions of development. *Economic Bulletin of the University. Collection of Scientific Works of Scientists and Graduate Students, 37-1*(313), 311–318.

Matytsin, D. E., & Rusakova, E. P. (2021). Strategy of quality management in industry 4.0 and formation of cognitive economy based on industrial and manufacturing engineering in the Russian Federation and countries of the EU. *International Journal for Quality Research, 15*(4). https://doi.org/10.24874/IJQR15.04-03

Morhat, P. M. (2018). *Legal personality of artificial intelligence in the field of intellectual property law: Civil law problems*. Moscow. 420.

Petit, N. (2019). Law and regulation of artificial intelligence and robots: Conceptual framework and normative implications [Electronic resourse]. https://papers.ssrn.com/sol3/papers.cfm?abstract_id=2931339 Accessed: 15.09.2019.

Ryzhkova, M. V. (2019). Conceptualizing the digital platform phenomenon: Market or business? *Bulletin of Tomsk State University. Economy, 47*(50), 48–66.

Scherer, M. U. (2016). Regulating artificial intelligence systems: Risks, challenges, competencies, and strategies [Electronic resourse]. *Harvard Journal of Law & Technology, 29*(2). https://papers.ssrn.com/sol3/papers.cfm?abstract_id=2609777 Accessed: 12.12.2019.

Srnichek, N. (2019). Platforms capitalism. *Economic Sociology, 20*(1), 75, 72–12. Intel platform approach [Electronic resource]. IT Analytics. https://www.bytemag.ru/articles/detail.php?ID=8655 Accessed: 03.03.2021.

Chapter 2
Utilizing Artificial Intelligence in Legal Practice

Evgenia E. Frolova and Elena P. Ermakova

Abstract The authors investigate the experience of using artificial intelligence (AI) in law firms' activities globally. During the research, it was revealed that: (1) Legal Tech is a branch of business specializing in information technology services for professional legal activities and providing consumers with legal services using information technologies; (2) there is no single list of Legal Tech; lawyers, theorists, and practitioners present Legal Tech classifications based on various criteria; (3) according to the authors of this monograph, currently the primary technologies are (a) "predict courts' decisions" or "prediction technology" and (b) "predictive coding"; (4) the recognized advantage of using AI tools in legal practice is efficiency-increasing capacity. The future of AI technology will give legal practitioners a competitive edge in litigation, enabling them to serve their clients better. Law firms that use AI will be more in demand, and firms unable to automate their activities may lose clients due to higher prices for the same services.

Keywords Artificial intelligence · Civil proceedings · Arbitration mediation · Legal tech · Predictive coding

JEL Codes F36 · G15 · G21 · K24 · O16

E. E. Frolova (✉) · E. P. Ermakova
Department of Civil Law and Procedure and International Private Law, Peoples' Friendship University of Russia (RUDN University), 6 Miklukho-Maklay Str., 117198 Moscow, Russia
e-mail: frolova_ee@rudn.ru

E. P. Ermakova
e-mail: ermakova_ep@pfur.ru

E. E. Frolova
Leading Researcher, Vladivostok State University of Economics and Service, 41 Gogol St., 690014 Vladivostok, Russia

A. O. Inshakova and E. E. Frolova (eds.), *Smart Technologies for the Digitisation of Industry: Entrepreneurial Environment*, Smart Innovation, Systems and Technologies 254, https://doi.org/10.1007/978-981-16-4621-8_2

2.1 Introduction

Technology for law firms and lawyers is growing steadily and hit a record $1 billion in investments in 2019. In terms of business technology and innovation, legal technology follows the world's most influential trends (Matytsin & Rusakova, 2021), with artificial intelligence and workplace automation (KeyVision, 2020). Technology provides new, more efficient, and faster tools that help legal services enter the market with more excellent reliability, consistency, and faster speed. Technological advances can be expected to create stress and job losses but also create new opportunities (Inshakova et al., 2020a, 2020b).

The authors fully share the views of Heather Sutty, a market strategist and law firm management consultant, who, in her 2020 article "The Next Decade of Legal Services: Embracing a Change in the Law Market," stated that "If you thought the legal services industry had radically changed in the last decade, then you are badly mistaken. Hurricane changes will mark the next decade. The legal market is becoming tougher, denser, and more fragmented, and competition in various forms continues to intensify. For lawyers, the coming era is turbulent, confusing, and disturbing. The key to survival is flexibility" (Suttie, 2020).

Since the beginning of 2019, several exciting events have taken place that, from our point of view, have put the problem of artificial intelligence in the first row of dispute resolution problems that must be taken into account as practitioners (judges, lawyers, and businessmen) and legislators. For the first time, in February 2019, in the High Court of England and Wales, in mediation, a decision was made by an AI—a "robot mediator." An English mediator, Graham Ross (Hilborne, 2019) spoke about this in his interview with "Legal Futures." The dispute concerned approximately £2000, representing the outstanding payments requested by the client's coach after completing a personal counseling course. Graham Ross stated that the parties to the dispute, whom he named E and D, decided to use the online court at the stage of public testing in the High Court of England and Wales (Ermakova, 2018). After initial attempts to resolve the dispute failed, a hearing was scheduled. The controversy continued for three months when Mr. Ross encouraged the parties to use "Smartsettle ONE," an AI-based Internet dispute resolution (ODR) tool. The system, developed by the Canadian company—iCan Systems, allows both parties to make private offers and uses algorithms to bring both parties closer to an acceptable settlement. As a result, the system allowed the parties to settle the dispute in less than an hour (Bailey, 2019).

In an increasingly competitive environment, forward-thinking law firms are focusing on the technology of the future. In the future, advanced legal technologies support their lawyers and increase their professional potential, allowing them to provide their clients with services of the maximum value (Ermakova & Frolova, 2020).

2.2 Materials

The scientific base of the article is formed based on the scientific works of Russian and foreign scientists. Among the works of Russian authors studying the legal regulation of artificial intelligence in the provision of legal services, it should be noted the monograph Ermakova E. Reforms of civil proceedings, arbitration and mediation in foreign countries, 2014–2018 (Australia, England, Germany, Canada, USA and France) (2018); and scientific articles: Biryukov (2019), Matytsin and Rusakova (2021), Gololobovm (2020), Ermakova and Frolova (2020), Kupchina (2020), and Rozhkova (2020).

General issues of regulation of artificial intelligence in the field of legal services in foreign countries are disclosed in the works of foreign researchers: Apostolova (2020), Bailey (2019), Chen (2019), Dalton (2020), De Westgaver and Turner (2020), Engstrom and Gelbach (2020), Hilborne (2019), Moran (2020), Morgan and Reed (2019), Pasquale and Cashwell (2018), Patton (2018), Suttie (2020), and Trasberg (2020).

The empirical base is provided by the judicial practice, reflecting the characteristic aspects of the legal regulation of artificial intelligence, inn particular the precedent in the case "Da Silva Moore v. Publicis Groupe et al." (2012).

2.3 Methodology

The scientific development of the content of this chapter of the monograph is carried out based on the general scientific method of historical materialism. General scientific methods of cognition are used: dialectical, hypothetical-deductive method, generalization, induction, and deduction, analysis and synthesis, empirical description. The study also used private science methods: juridical-dogmatic, statistical method, method of comparative legal analysis, and others.

2.4 Results

2.4.1 On Legal Tech

Legal Tech is a branch of business specializing in information technology services for professional legal activities, and since the late 2000s, it provides legal services to consumers using information technology. Russian scholar M. A. Rozhkova pointed out that "Legal Tech" (short for legal technology) is a variety of platforms, programs, products, and tools, specially designed to simplify and optimize the processes that make up the professional activities of lawyers. Legal Tech is a technological solution created for professional lawyers and legal businesses to increase the efficiency of the provision of legal services or legal support for businesses. The most prominent domestic example of Legal Tech, undoubtedly, "is the services of well-known reference and legal systems that offer verification of counterparties, drafting of contracts, selection of judicial practice in a specific case, and others" (Rozhkova, 2020).

The provision of legal services to consumers using information technology is implemented through online mediation between the customer and the law firm or the provision of legal self-service tools, eliminating the need to contact professional lawyers. The USA became a pioneer in implementing Legal Tech solutions where, in the early 2000s, startups began to appear, actively introducing information technologies in solving legal problems. Among the first such companies are Rocket Lawyer and LegalZoom, which provide services for creating dynamic documents, smart contracts, and legal advice (Inshakova et al., 2020a, 2020b).

It is also necessary to note the Legal Tech movement, which is to revise the traditional views on resolving legal issues by introducing modern information technologies in legal services. In Russia, an independent magazine, "Legal Insight," created in 2011, by Ph.D. in Law, M. Gaskarova, is devoted to Legal Tech technologies. There are many Legal Tech Web sites in Russia. For example, the information blog "Pravo Tech," oriented to young lawyers, which in addition to the section "Legal Tech," contains sections such as (a) automation cases; (b) conferences and forums; (c) chatbots[1]; (d) "dashboards" for legal practice,[2] and others. The blog also contains links to lawyers' services: Casebook for checking contractors; Caselook for search and analysis of judicial practice; Case.one for automation of the lawyer's work; Form.one is a constructor for chatbots; Doc.one is document constructor; File.one for knowledge base and eDiscovery.

Legal Tech's list does not exist; the list of services is constantly updated and changed, some services replace others, and new findings and directions emerge. However, without any exception, all technological solutions of Legal Tech use AI. Computer technology development follows Moore's law, which states that the number of transistors in an integrated circuit doubles every one and a half or two years (Moore's Law, 2014). This exponential growth has continued for over forty years of the computer era. At the end of the twentieth century—the beginning of the twenty-first century—computers played the same role as steam engines in the nineteenth century—they are the first engines and a symbol of progress. Scientists recently reaffirmed that Moore's law holds and noted the same exponential growth in telecommunications and information storage (McGinnis & Pierce, 2019).

In 2020, Freshfields lawyer K. Apostolova argued that many AI-driven tools have appeared on the legal market. Recently, peer reviewing of documents has already

[1] A chatbot is a software application used to conduct an online chat conversation via text or text-to-speech, instead of providing direct contact with a live human agent. A chatbot is a type of software that can automate conversations and interact with people through messaging platforms.

[2] "Dashboard" is an information panel that displays the key performance indicators ("KPI") of the department or the company as a whole. Thus, a dashboard's main function as a business intelligence tool is to collect the necessary information from various programs (cloud services, databases, and others) and clearly show it in one place for the user. If necessary, one can generate a report as soon as possible.

been done with technology, and now AI technologies are ubiquitous to improve predictive coding and minimize viewing time. Legal research platforms use AI to help identify the most relevant and authoritative case law. Also, unique resources for international arbitration began to emerge, including AI tools that help in the process of choosing an arbitrator and in the study of arbitral awards (Apostolova, 2020).

Throughout the trial, lawyers constantly have to make crucial decisions. Which particular court should the claim be filed with? How to react to the statement of the opposing party? Do we need to seek an amicable settlement of the dispute? What should the client say about the final costs of the litigation? Until recently, lawyers made such decisions based on their own experience and knowledge. Today, there is a revolution in how litigants decide the most critical issues of litigation strategy and tactics (Kupchina, 2020).

The legal profession is belatedly entering a phase based on Big Data. With the advent of forensic analytics, lawyers can scientifically approach success or the amount of risk for almost every option for decisions about the fate of the process (Dalton, 2020).

However, noting the prospects and directions of using Legal Tech, one should not discount the opposing views on this problem. Let us turn to the opinion of D. Gololobov (University of Westminster), expressed on the Pravo.ru portal in April 2020: "Legal Tech is a modern god (or idol), to whom many earnestly pray, but no one has yet seen unique miracles of it. Apologetics predict its instant and imminent coming. Nevertheless, in Russia, it is still visible only for experts who are very keen on the topic. For down-to-earth realists, the issue of 'smart contracts' is not as relevant as the question of the need for a modern law firm to have its office" (Gololobovm, 2020). From our perspective, D. Gololobov's statement is a rather challenging and pragmatic view of the legal profession's problems in modern conditions.

The Russian lawyer, Nikolay Teterev, Senior Manager in Forensic Practice, the Head of eDiscovery and Computer Forensics, Deloitte (Pravo.ru., 2017), tells why to automate the work of the legal department. He gave the example of a Russian company that was in suing in the USA. The American court ordered it to provide documents on particular topics within 90 days since 2009, with the participation of more than 30 employees, some of whom were no longer working there. It was correspondence, contracts, and acts, information from mail and mobile devices, and others. "We started work when there were 60 days left, and the company's lawyers could not postpone the deadline," said Mr. Teterev. A conventional "keyword" search returned 87,485 documents, which would have taken 175 person-days to analyze. The AI program coped on time and produced 99% accurate results—the court received 189 documents on the case and on time (the algorithm included machine learning analysis). Besides, the program has useful analytical functions; for example, it searches for similar or consistent documents and determines connections (who, how, and with whom corresponded)," Teterev added.

2.4.2 Categories of AI Application in Legal Practice

Experts name various categories of AI applications in legal practice. John McGinnis, Professor at Northwestern University Law School, Chicago, and Russell J. Pearce, Professor at Fordham University Law School, New York, presented their perspective on the categories of AI in legal practice in their 2019 article "The Great Disruption: How Artificial Intelligence is changing the role of lawyers in the provision of legal services." They noted: "we describe five areas in which artificial intelligence will provide services or inputs as long as provided by lawyers: (1) disclosure of the facts of the case, (2) investigation of eligibility, (3) drafting of documents, (4) preparation of the case materials, and (5) forecasting of results of cases" (McGinnis & Pierce, 2019).

In their article "Dispute Resolution in the Age of Big Data and AI," Herbert Smith Freehills' lawyers Charlie Morgan and Rebecca Reed identified the following seven categories of using AI in dispute resolution: (1) at the stage of contracting, (2) at the stage of violation or failure to fulfill an obligation, (3) at the stage of developing strategies for preparing a statement of claim or response to a claim, (4) at the stage of collecting factual data and evidence, (5) at the stage of preparing a set of documents for the court, and performing administrative tasks, (6) at the stage of disclosing evidence, and (7) at the stage court session (Morgan & Reed, 2019).

In their 2020 article titled "Legal technologies, civil procedures and the future of American contestability," American scientists David Engstrom and Iona Gelbach have already identified nine categories of AI application in jurisprudence (Engstrom & Gelbach, 2020):

1. Lawyer marketplace and matching are AI tools that allow clients to facilitate the assessment and selection of a potential lawyer. This category also includes tools that help lawyers manage their business (AI applications—Avvo, Ravel, Lexicata, and Atticus).
2. Legal (Re) Search is a tool that helps lawyers find and collect relevant materials on the case (case law, regulations, and rules) (AI applications—CaseText, Judicata, and ROSS Intelligence).
3. Outcome prediction is AI tool that predicts the outcome of specific cases. Forecasts can relate to individual judges, courts and also be used to compare court forums (sites). Predictions may include estimates of the time taken to process the case from the time the documents were filed to the decision on the case in various courts (i.e., forum shopping). Predictions can include assessing how a particular judge will respond to specific motions (e.g., to dismiss) (AI Apps—Colossus, Ravel, Case Crunch, Lex Machina, Gavelytics, and Blue J Legal). Meanwhile, the forecasting accuracy can reach from 70 to 90% (in the latter case, the Canadian company Blue J Legal carries out forecasts on labor disputes in the USA) (Moran, 2020).
4. Legal analytics are tools that perform other analytical tasks, in addition to legal search and forecasting the results, including displaying citations, analytics of court decisions in the text (judge-level analytics) (e.g., adapting arguments

to a particular judge), and document analytics (document-level analytics), or brief document assessment (AI applications—Ravel, FastCase, Gavelytics, Premonition, and CaseText).

5. Discovery tools support or supplant the process of identifying relevant documents and marking privileged documents (AI applications—Everlaw, Relativity, OpenText, and Exterro).
6. Document assembly and creation are AI tools that allow drafting legal documents from simple petitions (responses to petitions) to more complex claims and documents (requests for discovery, petitions, and even simple notes). (AI applications—Legalmation and, RockerLawyer).
7. Practice management is law firm management tool which include dashboards that manage client admission, organize critical facts and documents, and support billing or other administrative tasks (AI applications—Needles).
8. Contract management and analysis are tools that store, analyze, create, and control contracts' execution (AI applications—Kira Systems, Ravn, eBrevia, LexCheck, KMStandards, and UnitedLex).
9. DIY dispute resolution and online legal advice are tools that facilitate alternative dispute resolution and tools that provide automated (often online) legal advice or assistance to parties without a lawyer with litigation (AI applications—LegalZoom, Rocket Lawyer, Modria, Intraspexion, and Nolo).

As seen from the above material, there is no single list of Legal Tech. The criteria for constructing such classifications also differ. In our opinion, currently, there are two main AI technologies used in jurisprudence. The first is "**predict courts' decisions**" or "prediction technology" or "data-driven justice." Moreover, the second technology is "**predictive coding**," which uses artificial intelligence at the stage of document verification to study and identify relevant documents. Predictive coding is used at the stage of electronic evidence disclosure in countries of the English legal tradition. The two technologies are actively used in civil proceedings and arbitration and mediation (De Westgaver & Turner, 2020).

2.4.3 *Predict Courts' Decisions*

"Predict courts' decisions" are the intriguing ability of legal analytics algorithms to conclude court cases, for example, to determine specific models of judges' behavior and the consequences that these algorithms cause (Trasberg, 2020). As P. N. Biryukov states, "now there are programs that foresee the outcome of lawsuits, including potential compensation. Thus, a new concept—"predicted justice"—has become firmly established in everyday life of European countries. While traditional justice is trying to fulfill its mission of deciding within a reasonable time frame, the predicted justice goes further. It provides algorithms for analyzing a vast number of situations in a short time that allow predicting the outcome of a dispute or at least estimate the chances for success. The predicted justice allows choosing the correct way of defense, choosing the most reasonable arguments, evaluating the estimated amount of compensation, and others. Thus, we are not talking about justice itself, but only about analytical tools that would make it possible to predict future decisions in disputes similar to those analyzed" (Biryukov, 2019).

According to Daniel Chen (University of Toulouse, France), "predict courts' decisions" promise to improve the efficiency and fairness of the law (Chen, 2019). Forensic analysts can assess extra-legal factors that influence decisions. To date, with the advent of high-performance AI, legal analytics has started to generate quite a stir in the legal practice of the USA and Europe, as it began to turn into a helpful tool. Legal analytics algorithms are provided with an extensive set of data consisting of the results of previous court cases and specific information about the features of these cases, such as the text of the claim or case metadata (name of the judge, court, subject of dispute, and others). The algorithm then examines the correlations between the data and the outcome of these cases, based on which it either predicts the outcome of a new case, analyzing the features of the case, or identifies patterns as to which arguments, jurisprudence, or evidence used in previous cases are most significantly correlated with a positive outcome. These correlations can be determined for specific regions, courts, judges, or juries. It is noteworthy that this technology makes it possible to effectively conclude what a particular judge may or may not like in a trial.

In 2018, American authors Frank Pasquale (Brooklyn Law School) and Glyn Cashwell presented a critical analysis of the concept of "predicted justice" (Pasquale & Cashwell, 2018). Remarkably, they argued that a new answer is starting to emerge in the twenty-first century: deploying natural language processing (NLP) and machine learning (ML) techniques to predict whether judges will hear a case, and if so, how they will decide it, as well as metadata about the ideological obligations of judges, past voting and many other variables. By processing case-related data and the text of judges' opinions, these AI systems are designed to predict how judges will decide cases, how individual judges will vote, and how to optimize the lawyers' speech and arguments' presentation before those judges. The authors emphasized that this form of forecasting is like forecasters using Big Data to predict storms' movement (instead of understanding atmospheric dynamics). Algorithmically, analyzing a database of, say, 10,000 past cumulus raindrops sweeping over Lake Ontario might be a better predictor of the next cumulus rain trail than an experienced meteorologist who does not have access to such a database. However, despite individual authors' critical assessments, the technology of predicted justice is receiving more and more rave reviews from law firms and individual lawyers, especially in countries of the Anglo-Saxon legal tradition.

Examples of platforms and software in this category include:

(a) Document search and forecasting system—"Intraspexion," USA;
(b) Determination of the results based on the relevant case law by the system "Ravel Law's Judge Analytics," USA;
(c) Lex Machina Legal Analytics Platform based on LexisNexis Company;
(d) Database of trials—"Premonition," USA.

2.4.4 Predictive Coding

Predictive coding is an AI technology that assists document reviewers during the review phase. Typically, machine learning technology is used to denote documents after reviewers have completed some reviews by creating a set of encoded (or tagged) documents. It improves the efficiency of reviewers by showing how they mark documents. The machine then applies these criteria to create labels for documents that have not yet been reviewed (Patton, 2018).

In legal practice, predictive coding is used primarily in the eDiscovery procedure, which is typical for the Anglo-Saxon legal tradition countries. In eDiscovery, AI is needed to navigate massive datasets in search of relevant legal documents. This process, called technology-assisted review (TAR) in the USA, begins by humanlike examining a few documents and coding them as relevant or unrelated.

Besides, several predictive coding applications are already in use in arbitration practice: "ArbiLex," "Solomonic," "Lex Machina's Legal Analytics Platform," and "Ravel Law."

However, several obstacles prevent the introduction of this AI technology into the legal industry.

First, predictive coding is complex. It relies on technology that includes advanced data science and statistical sampling, which requires specialized skills.

Second, predictive coding is expensive to implement. It requires a significant amount of time and money to develop software and the proper "training" before predictive coding reaches its full potential as a valuable tool for electronic evidence detection.

Finally, court approval for predictive coding is still new. Moreover, lawyers fear the risk of adverse consequences if the court rejects documents that were found using predictive coding. Currently, only one US Federal district judge, Andrew Peck, had consistently approved predictive coding technology in several decisions in 2012–2016. A case in point is "Da Silva Moore v. Publicis Groupe & MSL Group" (2012), "Rio Tinto v. Vale" (2015), and "Hyles v. City of New York" (2016).

2.5 Conclusion

All the above technologies are low-level AI. At first glance, it seems that using a low-level AI in resolving disputes is generally a joyous process. The most recognized benefit of using AI tools in legal practice is efficiency gains. AI software uses algorithms that speed up document processing when errors and other problems are detected. However, lawyers often need more time to complete an assignment or prepare a document since a lawyer is paid according to the "paid hours" scheme, and the use of AI to speed up the work of a lawyer cannot be justified. According to many researchers, law firms' pressure to adopt AI is likely to come from competing companies. Law firms that use AI will be more in demand, and firms unable to automate their activities may lose clients due to higher prices for the same services. There is no doubt that AI technology's future use will give legal practitioners a competitive edge in litigation, enabling them to serve their clients better ultimately. It is imperative that the Russian legal sector does not get left behind and can take advantage of these competitive advantages that artificial intelligence can offer.

Acknowledgements This work was financially supported by the Grant of the President of the Russian Federation no. NSH-2668-2020.6 "National-Cultural and Digital Trends in the Socio-Economic, Political and Legal Development of the Russian Federation in the 21st Century."

References

Apostolova, K. (2020). *The dawn of a new era: Arbitration in the age of AI and digitalization*. Freshfields Bruckhaus Deringer LLP. https://www.freshfields.com/495e22/contentassets/ Accessed: 03.08.2020.

Bailey, K. (2019). *Robot mediator settles court case for the first time*. Karl Bayer. Apr 11, 2019. https://www.disputingblog.com/robot-mediator-settles-court-case-for-the-first-time/ Accessed: 18.06.2020.

Biryukov, P. (2019). Artificial intelligence and "predicted justice": Foreign experience. *Lex Russica, 11*(156), 79–87. https://cyberleninka.ru/article/n/iskusstvennyy-intellekt-i-predskazannoe-pravosudie-zarubezhnyy-opyt Accessed: 15.09.2020.

Chen, D. (2019). Judicial analytics and the great transformation of American Law. *Artifical Intelligence and Law, 27*, 15–42. https://doi.org/10.1007/s10506-018-9237-x

Da Silva Moore v. Publicis Groupe et al., No. 1:2011cv01279—Document 96 (S.D.N.Y. 2012). https://law.justia.com/cases/federal/district-courts/new-york/nysdce/1:2011cv01279/375665/96/ Accessed: 16.11.2020.

Dalton, B. (2020). *Big Data and the litigation analytics revolution*. Breaking Media, Inc. https://www.lawtechnologytoday.org/2020/01/data-analytics-to-track-legal-insights/ Accessed: 11.08.2020.

De Westgaver, C., & Turner, O. (2020, February 23). *Artificial intelligence, a driver for efficiency in international arbitration—How predictive coding can change document production*. Kluwer Arbitration Blog.. http://arbitrationblog.kluwerarbitration.com/2020/02/23/artificial-intelligence-a-driver-for-efficiency-in-international-arbitration-how-predictive-coding-can-change-document-production/?doing_wp_cron=1596445691.7806251049041748046875 Accessed: 11.11.2020.

Engstrom, D., & Gelbach, J. (2020). *Legal tech, civil procedure, and the future of American Adversarialism*. University of Pennsylvania Law Review. March 9, 2020. https://papers.ssrn.com/sol3/papers.cfm?abstract_id=3551589 Accessed: 08.08.2020.

Ermakova, E. (2018). *Reforms of civil proceedings, arbitration, and mediation in foreign countries 2014–2018 (Australia, England, Germany, Canada, USA, France)* (p. 177). Infotropic.

Ermakova, E., & Frolova, E. (2020). Dispute resolution and artificial intelligence: Pros and cons. Prospects and problems of the development of arbitration proceedings in Russia. Collection of articles based on the materials of the scientific-practical conference with international participation. Under the general editorship of L. V. Tumanova. Tver, 8 July 2020, 29–36.

Gololobovm, D. (2020). Metro 2021, or virtual lawyers in the era of the early apocalypse, Pravo.ru. 04.20.2020. https://pravo.ru/opinion/220822/ Accessed: 08.29.2020.

Hilborne, N. (2019, February 19). Robot mediator settles first ever court case. Legal Futures. https://www.legalfutures.co.uk/latest-news/robot-mediator-settles-first-ever-court-case Accessed: 15.06.2020.

Inshakova, A. O., Goncharov, A. I., & Salikov, D. A. (2020a). *Electronic-digital smart contracts: Modernization of legal tools for foreign economic activity*. Lecture Notes in Networks and Systems, Digital Economy.Complexity and Variety vs. Rationality. Springer Science.

Inshakova, A. O., Frolova, E. E., Rusakova, E. P., & Kovalev, S. I. (2020b). The model of distribution of human and machine labor at intellectual production in Industry 4.0. *Journal of Intellectual Capital, 21*(4), 601–622.

KeyVision. (2020). Top legal tech trends to follow in 2020. Keyvision. https://www.keyvision.eu/top-legal-tech-trends-to-follow-in-2020/ Accessed: 22.07.2020.

Kupchina, E. (2020). Some issues of the application of artificial intelligence technology in the civil procedure of Japan. *Eurasian Law Journal, 8*, 136–138.

Matytsin, D. E., & Rusakova, E. P. (2021). Strategy of quality management in Industry 4.0 and formation of cognitive economy based on industrial and manufacturing engineering in the Russian Federation and countries of the EU. *International Journal for Quality Research, 15*(4). https://doi.org/10.24874/IJQR15.04-03

McGinnis D., & Pierce R. (2019). The great undermining: How artificial intelligence changes the role of lawyers in the provision of legal services. *Actual Problems of Economics and Law, 13*(2), 1230–1250. https://doi.org/10.21202/1993-047X.13.2019.2.1230-1250

Moore's Law. (2014). Intel Corp. http://www.intel.com/content/www/us/en/silicon-innovations/moores-law-technology.html?wapkw=moores+law Accessed: 26.09.2020.

Moran, L. (2020, August 25). Law firm teams up with Canadian legal tech company on AI-powered case prediction tool. *ABA Journal*. https://www.abajournal.com/lawscribbler/article/before-lawyers-can-ethically-adopt-and-integrate-ai-into-their-practices-they-must-first-be-competent Accessed: 06.09.2020.

Morgan, C., & Reed, R. (2019, September 18). *Dispute resolution in the era of Big Data and AI*. Herbert Smith Freehills. https://www.herbertsmithfreehills.com/latest-thinking/dispute-resolution-in-the-era-of-big-data-and-ai Accessed: 09.06.2020.

Pasquale, F., & Cashwell, G. (2018). Prediction, persuasion, and the jurisprudence of behaviorism. *University of Toronto Law Journal, 63*. https://core.ac.uk/download/pdf/212819507.pdf Accessed: 27.11.2020.

Patton, N. (2018). *The basics of E-discovery: Predictive coding and AI*. Exterro. 14.09.2018. https://www.exterro.com/blog/the-basics-of-e-discovery-predictive-coding-and-ai Accessed: 27.11.2020

Pravo.ru. (2017). Investments in litigation and "smart" contracts: What's new in LegalTech today. Pravo.ru. 25.09.2017. https://pravo.ru/court_report/view/144621/ Accessed: 11.26.2020.

Rozhkova, M. (2020, February 14). LegalTech and LawTech—What are they and what is their significance for law? Zakon.ru. 2020. https://zakon.ru/blog/2020/02/14/legaltech_i_lawtech_-%C2%A0chto_eto_takoe_i_v_chem_ih_znachimost_dlya_prava Accessed: 08.15.2020.

Suttie, H. (2020, January 15). *The next decade of legal services: Embracing law market change*. Attorney at Work. https://www.attorneyatwork.com/next-decade-legal-services-part-1-law-market-change/ Accessed: 12.08.2020.

Trasberg, H. (2020). The advent of legal analytics. RAILS-blog. 30/03/2020. https://ai-laws.org/en/2020/03/the-advent-of-legal-analytics/ Accessed: 27.11.2020.

Chapter 3
Augmented Reality and Civil Law Regulation of Business Relations

Vitaliy V. Bezbakh and Evgenia E. Frolova

Abstract Digital technologies are swiftly getting into the material relations area, inflating it with innovative means of doing business. However, due to the high rates of digital technologies development, its legal regulation remains behind, so the use of selected digital technologies causes unpredictable business flow and diverse law enforcement policies. One such sphere originated from digitalization development is augmented, virtual, and mixed reality. Due to the ability of legal doctrine to react to social and legal changes quicker than a legislator, the array of Russian does and foreign publications about augmented reality, in general, its technologies and platforms, in particular, indicated that the widest and the most important area of affected social relations enforcement is intellectual property. In this article, the authors discover terminology origins of the augmented, virtual, and mixed reality sphere and reveal specific aspects of its practical use, demanding prior legal regulation.

Keywords Digital technologies · Augmented reality · Civil law regulation · Business relations · Intellectual property

JEL Codes O33 · K15 · K29

3.1 Introduction

The area of reasons, indicating the origin of new digital events worth legal regulation, is growing steadily. The tendency toward such growth had come from long ago and was caused by no means imperfection or incompleteness of theoretical study of

---

V. V. Bezbakh (✉) · E. E. Frolova
Department of Civil Law and Procedure and International Private Law, Peoples' Friendship University of Russia (RUDN University), 6 Miklukho-Maklaya Str., 117198 Moscow, Russia

E. E. Frolova
Leading Researcher, Vladivostok State University of Economics and Service, 41 Gogol St., 690014 Vladivostok, Russia

A. O. Inshakova and E. E. Frolova (eds.), *Smart Technologies for the Digitisation of Industry: Entrepreneurial Environment*, Smart Innovation, Systems and Technologies 254, https://doi.org/10.1007/978-981-16-4621-8_3

civil rights subject matter. Its appearance appeared to be a result of swift penetration of technologies into ownership relations, providing it with revolutionary new possibilities of doing business.

Such technologies had already created a wide variety of new categories, becoming subjects of legal regulation or to one degree or another worthy of being regulated—"big data," "blockchain," "cryptocurrency," "tokens," "bitcoins," "digital assets," "digital rights," "virtual reality," etc. (Rusakova et al., 2020). The growth rate of data, whose aspects of keeping and granting access to it also need legal regulation, makes itself unforgettable. An increase in data flow in the digital area happens exponentially: daily increase estimates 2.5 quintillion bytes, which led in 2017 to an assertion that 90% of all world data existing has been created during the last two years (Kartskhia, 2017).

Moreover, widespread digital technology implementation led to the appearance of such unexpected software products that their existence defies the imagination of uninformed ("non-digitized") humans. It also includes technologies widening and increasing human abilities to percept, self-express, and communicate. The prime example of it appears to be the technologies named "augmented reality." They are as difficult to be described as to be determined.

The striking example of this technology's practical use is "AR Smart Glasses" (synonyms: AR Smart Glasses, smart glasses, or data glasses)—computer-based glasses, showing user augmented information in a form of a 3D image, animation, or video additional to real-world surroundings. In other words, a built-in eyeglass-frame computer imposes its generated data on the user's image of the real world. Millions of potential customers got to know from ads that "smart glasses with augmented reality (AR) are wearable AR devices, which combine virtual information with physical surroundings in customer's field of view while being used as ordinary glasses." Smart glasses first appeared in 2013 and were forecasted as a recent breakthrough in the popular gadgets' world. However, by means, not caused by software errors, this product left the market almost as fast as entered (Gvora, 2020), though without losing illustration value for possibilities, granted to humans by AR technologies.

And such possibilities, indeed, are significantly broad and worthy. Rather than constantly studying instructions and manuals about usage and technical maintenance of complicated machinery and devices, stated on hundreds and hundreds of pages, the user became able to AR-directions to see corresponding illustrations just for the inspected part of the equipment. Customs officials got a chance to see detailed information on design features of transport containers, evaluate the risk of smuggling and make a decision on whether to conduct an inspection or not. Rescue services, working in a car accident, became able to see and determine a lancing point for extracting an injured person from either car type much faster than before (Cook et al., 2019).

3.2 Materials

Under the widespread implementation of digital technologies and augmented reality as their essential part across the world, the authors form the study basis not only on domestic but also foreign doctrinal sources, dealing with augmented, virtual, and mixed reality terminology origins, investigating the role of virtual reality on business relations and evaluating ways to develop legal regulation of digital technologies in a specified area (Inshakova A., Goncharov A., and Matytsin D.).

The analysis of business regulation distinctive features, arising from the implementation of augmented reality technologies, is based on research of regulatory legal acts, also with strategic and policy nature, adopted in the Russian Federation.

3.3 Methods

Methodological bases for the study consist of general learning practices: dialectical, consolidation methods, theoretic-empirical, induction, deduction, abstraction, and other means, among which the method of complex learning plays the most valuable role and allows investigating of the theoretical and practical basis of digitalization and its law enforcement in conjunction.

Moreover, distinctive features of the research object required wider use of specific learning practices: comparative legal study, legal modeling, law, and historical analysis, legal formality method, and others, which contributed to the complex evaluation of legal regulation development trends on implementing digital technologies in business flow, with particular catch on business relations and intellectual property spheres.

3.4 Results

3.4.1 Reality and Existence: Definitions and Terminology

It is not annoying for lawyers that information technologies development rates are taking the lead over legal theory development. Law is dealing with social relations in the regulatory sphere in general, advance in taking account of changes, happening under the revolutionary influence of information technologies (Inshakova et al., 2017).

The definition of augmented reality is studied significantly in popular publications, scientific articles, and monographs, regulatory acts, though the range of features characterizing this concept remains excessively wide.

Generally speaking, the method of augmenting reality with any constructions or elements in it is not solely caused by the origin of digital technologies. In particular, another way was mentioned by V. P. Malakhov concerning the anthropological methodology of the modern theory of state and law (Malakhov, 2010). Indeed, the appearance of such group in itself is not necessarily determined by the results of information technologies development and application—as, for example, know-how, turned out to be recognized by our legislation in due time under the influence of other factors, that were developing quite rapidly, for its existing designation not to require Russian translation.

Sometimes augmented reality technology is determined alongside virtual reality technology as a synonym. For example, according to the Action Plan ("road map") of the National Technology Policy "Autonet," "Augmented and virtual reality are the technologies for bringing any sensory data into the field of perception to complete information about the environment and to improve data perception."

Augmented and virtual reality terms are not assimilated among themselves but are determined together as a subject to study and enforcement in separate legal acts (e.g., in Russian Federation Government Order "On approval of the Concept for transport sector personnel training till 2035" dated February 06, 2021, No. 255-p):

- Indicating priorities and goals of government policy (e.g., Russian Federation Government Decree "On approval of the 'Information Society' government program" dated April 15, 2014, No. 313);
- Stipulating budget allocation (e.g., Russian Federation Government Decree "On approval of Rules of assigning federal budget grants for advanced training of higher and secondary vocational education teachers on new programs for IT specialties and various subject areas and securing the achievement of certain results of the federal project "Personnel for the Digital Economy" hosted by the autonomous non-profit organization of higher education "Innopolis University" dated December 15, 2020, No. 2110);
- Approving the technologies range (e.g., Russian Federation Government Decree "On approval of the list of technologies, used in the experimental legal framework in digital innovations area" dated October 28, 2020, No. 1750); and others.

Meanwhile, differences between "augmented reality" and "virtual reality" concepts are not difficult to distinguish even though clear distinctions between them are usually raised not by legal scholars, but by IT specialists. Talking about the concept of virtual reality, it is hardly possible to recognize a unified definition. It is possible, for example, to meet its interpretation as "experience imitation, which may be similar to the experience achieved in the real world or maybe completely different and unique." However, the "freshest" trend should be recognized as the use of the so-called umbrella terms, covering technologies of augmented, virtual, and mixed reality (Goode, 2019). The positive side of the "umbrella" method is shifting from the pedantic differentiation of concepts expressing various information technology achievements, without prejudice to keeping these concepts in focus by those researchers who are aimed at finding an effective legal regulation of relations arising from the use of such achievements.

Indeed, the meaning of virtual and augmented reality terms is not getting opposed, since both have significant prospects for use and embody important tools for digitalizing economics (Kalinina et al., 2019). It can be noted, however, that some researchers have not only utilitarian but also emotional assessments of the virtual reality value: "It is the widespread use of VR (virtual reality) that will change the modern world beyond recognition. This technology allows creating of the worlds that, according to some researchers, people will choose as a cheaper alternative to the real world" (Dremlyuga, 2020).

Attendant terms are also showing up. These include, for example, an "augmented reality platform," defined as "a multimedia product or software, whose toolkit allows creating of virtual content" (Koroleva, 2020).

The need for such accounting has already been reflected in many by-law acts—it is the level that allows securing the definitions of the events caused by the use of these technologies.

The fact that definitions of augmented reality are often supplemented or replaced by definitions of augmented reality raises the question of whether it is important to choose one of them. From a utilitarian point of view, definitions that interpret augmented reality as some kind of technology can be considered productive and therefore deserving of preference, although one can see the same connection between augmented reality technology and augmented reality itself as the connection that exists between a tool and a thing created with its help. In the context of intellectual property relations, as it will be shown below, this is precisely the relationship that is assumed.

The dynamics of the processes currently taking place in the economic and social spheres are associated with their digitalization, which generates new technologies in the information field. It became quite obvious that legal regulation is designed to affect many aspects of digitalization, in particular, neurotechnology and artificial intelligence, distributed ledger systems and quantum technologies, the industrial Internet and robotics components, virtual and augmented reality technologies (Vaipan, 2017).

As applied, for example, to the assessment of the digital transformation of education as "an indisputable global trend in educational system modernization" (*Order of the Ministry of Education of Russian Federation "On approval of methodological recommendations for the implementation of modern digital technologies into the main educational programs"*), the definition of AR (augmented reality) as "a technology for introducing information into the field of human perception, which is perceived by a person as a part of the surrounding world" (Ibid.). It should be noted that the definition mentioned above is about the introduction of information called virtual into the human perception field, although the general context excludes the assumption of a possible synonymy of the augmented and virtual reality concepts. On the contrary, it removes doubts that information meant in it allows a person to perceive data as a part of the real world around him.

The modernization potential of this technology rises from its ability to significantly enrich the educational process by visualizing necessary elements and the possibility of completing those (Tarakanov et al., 2020). At the disposal of trainers and trainees appears to be a technology that functions as a kind of constructor, or creativity material supplier, or a tool for revealing and stimulating the creativity of the perception of the studied materials. Methodologically, the value of augmented reality technology is revealed through its suitability to make the educational process visual, interactive, interesting, increase students‘ motivation and thereby have a positive impact on academic results.

3.4.2 On the Spheres of Commercial Use of Augmented Reality Technology

The areas of such application are quite extensive. Some of them have already been mentioned above. The newest begin to be used in vehicles. For example, the system in the Audi Q4 e-Tron electric car is a screen positioned at the driver's eye level in a way that the images displayed appear to be at a distance of 10 m. It allows the driver not to separate the focus of his vision between the road and the screen. The information on such screen is varied: the movement speed, the speed limit on a given section of the roadway, the safe distance to the vehicle in front, the illumination of the road markings, that appear when the car starts to leave its lane, etc. (*Audi demonstrated an augmented reality system for cars*).

Generally speaking, in the long term the range of possible applications of augmented reality technology can hardly be limited to some exhaustive list. After all, the range of solutions that, in combination with other information methods, are becoming applicable with augmented reality technology is the entire space of goal-directed human activity. Virtual museums, planetariums, lecture halls, laboratories that do not require significant expenses for experimental equipment, the Internet of Things, etc., all this constitutes an area where changes in economic and social processes can exclude, for example, the need for human participation while performing certain actions. Attention to this circumstance has already been drawn in regulatory enactments (*Order of the Ministry of Education of Russian Federation "On approval of methodological recommendations for the implementation of modern digital technologies into the main educational programs"*).

3.4.3 The Results of Creative Activity, Obtained with the Use of Augmented Reality Technologies

Among the questions that require an answer and arise in connection with the use of augmented reality technologies, attention is drawn to the question of who should be recognized as the owner of the right to the creative activity result, obtained based on one or another software platform. If a platform is created by a programmer and augmented reality is created by the person who is using such platform ("the end-user"), then are the benefits delivered by the commercial use of augmented reality subject to distribution among them, as among the co-authors? Is such distribution conflict-prone? Are potential conflicts predictable and preventable?

Answers can probably be found out by defining the legal status and the subject range of people recognized as owners of intellectual rights to virtual and augmented reality technologies and objects generated by them (Grin, 2019; Koroleva, 2020).

But in the course of their search, apparently, the issue of defining augmented reality technology as an environment for the software (Sw) functioning to allow this technology to appear, should be recognized as an even more important question. If the purpose of such an environment is to empower the user of the software with the ability to express various creative ideas, then how can the involvement of the software creator in the image created by the software itself, as a phenomenon of augmented reality, be traced? After all, software users are not indifferent to the type of rights they will achieve for the result created with the help of the software in the new "supplemented" world.

It is possible that augmented reality technologies would be considered and evaluated differently from other types of digital technologies. For comparison, it is possible to address the specific example of them (though not the newest one)—the Word text editor. The revolutionary nature of this software can, of course, be felt only by those users who remember the pre-digital era—the era of typewriters—and the torment of correcting typos, formatting footnotes, typing text that requires the simultaneous use of a Cyrillic and Latin font, and so on. The benefits of digital typing technology over analog, of course, are quite obvious even to those who have no experience with typewriting. However, even for authors with the freshest impressions of such advantages, whose literary works were textually captured as Word files, there was never a need to find out whether the copyright in their work, to any extent, is also extended to the creators of Word editor.

Another, no less vivid example of the same kind, is the three-dimensional computer-aided design programs, in particular, Autocad. Their widespread use in architecture, construction, design, and mechanical and other engineering, etc., speeds up and facilitates the creation of objects worthy of copyright protection. But does the architect's reference to the fact that the object was designed by him with the help of such software means that he recognizes the software developer as a co-author?

Using these examples to find an answer to the question of copyright for a certain augmented reality, it should, of course, be borne in mind that both of the above examples refer to the range of technologies that are not AR technologies since neither Word nor Autocad can create a combination of the physical and real world. But is not the use of AR technology as a way of creative expression the result of creating an appropriate software product, an AR platform? Doesn't such a platform act as a method for an augmented reality creator, a tool for creating a copyrighted result of creative activity?

3.5 Conclusion

Questions of this kind encourage noting the growing relevance of the study that analyzes objects of copyright protection in the field of creation and application of augmented reality technologies results. Noteworthy are concerns expressed about "difficulty and ineffectiveness of detecting potential copyright infringers on the sheer volume of works basis because it can be created using AR, and considering the ways of their storage, since many works may never become available publicly, remaining only in the memory of the AR software product instead" (Afoaku, 2017).

Moreover, the array of Russian and foreign publications devoted to the augmented reality, as well as to the technologies and platforms of augmented reality, confirms the assumption that the most extensive and important area of legal regulation of public relations affected by these categories is beginning to be the sphere of intellectual property (Matytsin & Rusakova, 2021).

Investments in businesses that use augmented reality technologies have already reached enormous levels. Business relations that have entered the era of digital economics cannot exist without the use of such technologies. They need formulated and effectively enforced rules establishing liability for copyright infringement. The provision of reliable protection to such rights will be determined not only by drawing a line between relations regarding the creation of AR programs, platforms, and relations associated with the creation of AR products but also by the development of fairly clear criteria for their protection. Such criteria should not only take into account the differences between digitalized reality, virtual, and augmented but also allow assessing the level of creativity that led to the creation of each. If in a virtual reality environment all objects possible are generated by a computer, then the legal protection criteria cannot but differ from those that are required concerning the spectrum of everything created in an augmented reality environment, where everything generated by a computer is superimposed on the existing physical reality.

The development of reliable and/or effective criteria will certainly be determined by the joint efforts of civil law scientists and IT specialists since the efforts disunity of both remains largely counterproductive.

Acknowledgements This work was financially supported by the Grant of the President of the Russian Federation No. HSH-2668-2020.6 "National-Cultural and Digital Trends in the Socio-Economic, Political and Legal Development of the Russian Federation in the 21st Century".

References

Afoaku, M. (2017). The reality of augmented reality and copyright law. *Northwestern Journal of Technology and Intellectual Property, 15*(2), 110–128.

Audi Demonstrated an Augmented Reality System for Cars. (2021). https://pulse.mail.ru/article/kompaniya-audi-pokazala-sistemu-dopolnennoj-realnosti-dlya-avtomobilej-3447636169423809151-4821624069279329947/?utm_content=lenta_main_mail_ru&utm_campaign=

main&utm_referrer=https%3A%2F%2Fpulse.mail.ru&utm_source=pulse_mail_ru. Accessed: 21.04.2021.

Cook, A. V., Mariani, J., Kishnani, P. K., & Harr, C. (2019). How to begin regulating a digital reality world. Businesses and governments should guide augmented reality development. https://www2.deloitte.com/us/en/insights/industry/public-sector/regulating-digital-reality-augmented-spaces.html. Accessed: 21.04.2021.

Dremlyuga, R. I. (2020). Virtual reality: Common problems of legal regulation. *Actual Problems of Russian Law, 9*, 39–49.

Goode, L. (2019). Get ready to hear a lot more about "XR". https://www.wired.com/story/what-is-xr/ Accessed: 21.04.2021.

Grin, E. S. (2019). The concept of virtual and augmented reality in the field of intellectual property. Russian legal system in the context of the fourth industrial revolution. In XVI International Research-to-Practice Conference (Kutafin Readings): Conference Materials: in 3 vol. Moscow: RG-Press, 2 (pp. 243–246).

Gvora, J. (2020). Google glass: What happened to the futuristic smart glasses? https://screenrant.com/google-glass-smart-glasses-what-happened-explained/ Accessed: 21.04.2021.

Inshakova, A. O., Goncharov, A. I., Mineev, O. A., & Sevostyanov, M. V. (2017). Amendments to the Civil Code of the Russian Federation: Contradictions of theory and practice. In *Contributions to Economics. Russia and the European Union: Development and Perspectives Number XIII* (pp. 147–153).

Kalinina, A. E., Inshakova, A. O., & Goncharov, A. I. (2019). Polysubject jurisdictional blockchain: Electronic registration of facts to reduce economic conflicts. *Studies in Computational Intelligence, 826*, 205–213.

Kartskhia, A. A. (2017). Digital imperative: New technologies create a new reality. *IP. Copyright and Related Rights, 8*, 17–26.

Koroleva, A. G. (2020). Subjects of intellectual property rights in the field of virtual and augmented reality technologies. *Actual Problems of Russian Law, 6*, 133–142.

Malakhov, V. P. (2010). Variety of methodologies of modern theory of state and law. *History of State and Law, 6*, 2–17.

Matytsin, D. E., & Rusakova, E. P. (2021). Strategy of quality management in Industry 4.0 and formation of cognitive economy based on industrial and manufacturing engineering in the Russian Federation and countries of the EU. *International Journal for Quality Research, 15*(4). https://doi.org/10.24874/IJQR15.04-03

Order of the Ministry of Education of Russian Federation "On approval of methodological recommendations for the implementation of modern digital technologies into the main educational programs" dated May 18, 2020 No. R-44. SPS "ConsultantPlus". Accessed: 21.04.2021.

Rusakova, E. P., Frolova, E. E., & Gorbacheva, A. I. (2020). Digital rights as a new object of civil rights: Issues of substantive and procedural law. *Advances in Intelligent Systems and Computing, 1100*, 665–673.

Tarakanov, V. V., Inshakova, A. O., & Goncharov, A. I. (2020). Interactive demo on-line maintenance foreign trade deals of business entities of BRICS jurisdictions. In: A. Inshakova & E. Inshakova (Eds.), *Competitive Russia: Foresight model of economic and legal development in the digital age. CRFMELD 2019*. Lecture Notes in Networks and Systems (Vol. 110, pp. 319–327). Springer. https://doi.org/10.1007/978-3-030-45913-0_36

Vaipan, V. A. (2017). Fundamentals of legal regulation of the digital economy. *Law and Economics, 11*(357), 5–18.

Chapter 4
A Typology of Risks and Threats Associated with the Digital Transformation of Economic and Legal Systems

Alexey Yu. Mamychev and Olga I. Miroshnichenko

Abstract The research is devoted to the analysis of potential risks and threats provoked by the fourth industrial revolution. The authors note that the digital transformation of society leads to new assemblies of digital technologies and people, generating a chain of visible and hidden changes that, in their interconnected totality, change the usual political, legal, and economic landscape. The concept of the "driver of change" is introduced, which is proposed to be understood as a certain technological agent and digital actant. The main threats are categorically differentiated; within each category, conclusions and proposals are proposed for the possible leveling of existing risks.

Keywords The 4th industrial revolution · Digitalization · Digital actant · Technology · Legal regulation · Doctrinal act · Information society · Threats and risks

JEL Codes O33 · K24 · K39 · D81

4.1 Introduction

The Fourth Industrial Revolution is radically transforming the socio-cultural and axiological-normative foundations of society. We are already witnessing dramatic changes in the development of economic, political, legal, and other types of social relations. Yet there is nothing unexpected and unpredictable about these arguably equivocal and ambiguous processes associated with the digital transformation of

---

A. Yu. Mamychev (✉)
Vladivostok State University of Economics and Service, 41 Gogol St., 690014 Vladivostok, Russia

Laboratory of Political and Legal Studies, Faculty of Political Science, Lomonosov Moscow State University, 27 b.4 Lomonosovsky Prospekt, 119234 Moscow, Russia

A. Yu. Mamychev · O. I. Miroshnichenko
Department of Theory and History of State and Law, Far Eastern Federal University, 10 Ajax Bay, Russky Island, 690922 Vladivostok, Russia

A. O. Inshakova and E. E. Frolova (eds.), *Smart Technologies for the Digitisation of Industry: Entrepreneurial Environment*, Smart Innovation, Systems and Technologies 254, https://doi.org/10.1007/978-981-16-4621-8_4

society, which is actively taking place throughout the world. Each industrial revolution is known to have led to radical changes in the existing political, legal, economic, and socio-cultural paradigms and altered the ideological and axiological-normative foundations of society. The emergence of new tools and technologies tends to bring about new socio-technological configurations and assemblages, which alter the existing forms and nature of labor, social communication, power/hierarchical organization, etc. Just about every new "assemblage" of this kind alters, gradually or intermittently, the systemic setup of social relations and existing political, economic, or legal practices.

In this context, according to Gilles Deleuze and Félix Guattari, "the material or machinic aspect of an assemblage relates not to the production of goods but rather to a precise state of the intermingling of bodies in society, including all the attractions and repulsions, sympathies and antipathies, alterations, amalgamations, penetrations, and expansions that affect bodies of all kinds in their relations to one another… Even technology makes the mistake of considering tools in isolation: tools exist only concerning the interminglings they make possible or that make them possible. The stirrup entails a new man-horse symbiosis that at the same time entails new weapons and new instruments. Tools are inseparable from symbioses or amalgamations defining a Nature-Society machinic assemblage. They presuppose a social machine that selects them and takes them into its 'phylum': a society is defined by its amalgamations, not by its tools" (Deleuze & Guattari, 2010, p. 28).

The Fourth Industrial Revolution, likewise, has led to new assemblages of digital technologies and humans, which have given rise to a chain of visible and hidden interrelated changes, which, in turn, have combined to alter the existing political, legal, and economic landscape (Inshakova et al., 2020; Tarakanov et al., 2019). Besides, any revolution also gives birth to a new subject of political, legal, and economic history. A case in point is the proletariat, which has long served as a driver of political-legal and socio-economic changes and a foundational element in ideological systems. While its role and significance have been subject to different interpretations, the proletariat has been recognized as a significant factor in the transformation of society. A similar kind of "the subject of history" is being engendered by the current societal transformations too. At present, there is no accurate definitive term for it, with terms such as "technological agent," "digital actant," and "unit" variously employed. The one obvious thing is that the essence and conceptual component of the new "driver of change" are being reflected in the current socio-technological landscape in increasingly distinct ways.

Moreover, the Fourth Industrial Revolution is distinguished in the following major way: compared with its "predecessors," it forms a whole new dimension, a whole new reality for socio-economic and political-legal processes to take place in. The three preceding industrial revolutions altered substantially the way of interaction and the nature of relationships between and the significance of the three principal "realities"–physical, biological, and socio-cultural (or intercommunicating, i.e., created (constructed) and collectively developed by people). The fourth revolution, in turn, has given birth to a new type of reality–digital reality, which with each passing year becomes increasingly significant as it interacts with the basic realities of human existence.

4.2 Methods and Materials

The study is based on such methods as the logical, which allows using the methods of analysis and classification to methodize ethical problems of the technological revolution; the system-structural analysis, which allows applying differentiation and detailing of socio-political processes influenced by digitalization and which allows to elaborate ethical aspects in relations related to the use of AI and to highlight ethical and legal problems of these relations; predictive method and modeling method, which was used to identify trends in the development of Russian society under the influence of digital technologies soon.

4.3 Results

The present work explores some of the key risks and threats associated with the development of digital reality, and its influence on economic and legal systems.

4.3.1 New "Drivers of History"

Today, it is no longer questioned, especially in light of the rapid digitalization of society and the global challenges caused by the COVID-19 pandemic, that at the present social activity (at both the individual and collective levels) is "neighboring" or unfolding in conjunction with the activity of certain nonhuman elements–"actants," as they are termed in the contemporary literature (e.g., digital actants, like standalone digital algorithms or standalone robotized apparatuses, and biological actants, like viruses and bacteria).

According to Bruno Latour, an *actant* is an active principle or a source of action that, *on one hand*, can be human (e.g., conscious acts of will, joint actions, mental structures, and social institutions) or nonhuman (e.g., speaking biologically–a virus, technologically–a standalone digital technology, and environmentally–a natural phenomenon) (Bruno,), and, *on the other*, can be a combination or an assemblage (Bruno, 2018)[1] of the two, i.e., a fusion of the social and the material, the human and the nonhuman.

---

[1] If the term "system" is mainly used to describe interconnected social objects (people, created institutions), then "assemblage" is used to represent other specific relationships and mutual influences of both human and non-human (material, biological, digital, and other) elements. Assemblage is "the combination or assembly of something together, or the result of such a combination or assembly" (Bruno, 2018, p. 9). It is important to note that assemblage focuses on the close relationship of various elements (social, digital, biological, physical), however, "relationships between them (unlike the system-*auth.*) are not logically necessary but contingent (that have a predominantly random character of combination/assembly—*auth.*) mandatory as the historical result of their co-evolution (joint movement and development—*auth.*)" (Bruno, 2018, p. 21).

An example of the latter may be a person's orientation expressed through digital forms, which can generate specific events in socio-cultural and digital reality or cities, as "assemblages of people, networks, organizations, as well as of a variety of infrastructural components" (DeLanda, 2012, p. 13).

With that said, "an actant never really acts alone. Its efficacy or agency always depends on the collaboration, cooperation, or interactive interference of many bodies and forces" (Bennett, 2018, p. 45). In this theoretical-methodological model, the actant "is neither subject nor object but a 'mode' of what Spinoza calls 'Deus sive Natura' (God or Nature)... every mode is itself a mosaic or assemblage of many simple bodies... What it means to be a 'mode,' then, is to form alliances and enter assemblages: it is to modify and be modified by others. The process of modification is not under the control of anyone mode–no mode is an agent in the hierarchical sense" (Bennett, 2018, p. 47).

Processes of this kind include, for instance, modifications to digital algorithms in the machine learning process. Machine learning is implemented on specific social data and cultural artifacts; as part of the above process, the development of the digital algorithm produces a specific socio-cultural component, as well as quite a unique path for the future development of digital algorithmic systems.

This effect has already been captured in the development of various digital technologies and is quite common nowadays (Brockman, 2017; Kelly, 2017). Thus, the implementation of a standalone digital system that learns based on particular "socio-cultural material" or a certain algorithmic solution in different communities may lead to completely different paths for their development and to "bias in digital systems" or "digital discrimination" (Greenfield, 2018; Kaku, 2018; Stuart, 2019).

Another example of modification changes in people's mental activity and their volitional and emotional characteristics is based on the inclusion of digital technology in everyday life, which has been described in the contemporary literature quite substantively. Or take the effect of a virus pandemic—it may result in special trends in the development of digital systems for the identification, differentiation, and keeping track of social processes, modify political, and legal systems, alter practices related to bio-protection, social and medical distancing, and stay-at-home restrictions, transform people's axiological-normative orientations, etc. (Goodbye, Covid, 2020; Digital Agenda and Digital Initiatives During COVID-19, 2020, p. 19; The end of the familiar world, 2021, 380p.).

In our mental activity and exploratory practices, there, also, takes place a transformation of a sort of "dictionary" as we try to describe various events and processes and concepts and notions from some descriptive systems and exploratory protocols penetrate others. There occurs an intense search for new terminology that, *on one hand*, can make it possible to properly describe the latest radical changes in society, politics, and law taking place in light of the digital transformation of society and the spread of a virus pandemic, as well as global climate changes, and, *on the other*, can make it possible to view the various systems (social, environmental, biological, physical, and digital) as interrelated and equally significant (Bennett, 2018, p. 26).

In general, it is quite conventional for concepts and exploratory practices to interpenetrate from one area to another. For instance, using the discourse on the political body or the analogy with social diseases and viruses that destroy the social organism, which traces their roots back to ancient philosophy, has already become a well-established trend these days. Today, the use of this kind of analogies is becoming a widespread scholarly trend again.

By way of example, it is fitting to share here a quote from a recently published book devoted to the philosophical-political analysis of the relationship between virus contaminations, digital evolution, and the transformation of political and legal practices. The author, Eugene Thacker, suggests, "we have two separate fields, each of which integrates informatics and materiality differently through a network paradigm—this last part is crucial. If information security tells us that certain kinds of computer behavior can be understood through the lens of epidemiology, then it is equally important to note that modern epidemiology tells us that infectious disease can be understood through the lens of mathematics, statistics, and informatics. In one the basic idea is that we can understand particular types of computer behavior through the lens of biology, while in the other the basic idea is that we can understand infectious disease through the paradigm of informatics... The view of contagion presumes a condition of biological materiality, that can then be abstracted into metaphor (computer 'virus') when contagion is considered within epidemiology, it also implicitly links contagion with material and biological processes of the rate of infection, logistic growth, and epidemic thresholds" (Thacker, 2020, p. 74).

4.3.2 Artificial Intelligence, New Technological Formate of Transformation of Socio-economic and Political-Legal Relations: Main Risks

The term "intelligence" derives from the Latin "intellectual" understanding, reason, mind, and it generally means the thinking ability, the mental principle of a person (Ozhegov, 2014, p. 315). According to Patrick Henry Winston, an exhaustively precise and comprehensive definition of natural intelligence in its ordinary meaning seems impossible since intelligence is a "complicated mixture of a significant number of diverse skills in the field of information processing and presentation" (Morkhat, 2018, p. 59).

George Luger points out that today the concept of intelligence is vague and unclear "most of us are sure that we can distinguish 'intelligent behavior' when we face one. However, it is unlikely that someone can give intelligence a definition specific enough to evaluate a supposedly intelligent computer program and at the same time reflect the viability and complexity of the human mind"; "intelligence is a very complicated field of knowledge that is impossible to describe with the help of one theory" (Morkhat, 2018, p. 65).

It is believed that the term "artificial intelligence" was first coined by computer scientist John McCarthy at the Dartmouth Seminar in 1956, but before that, there were speculations whether machines can think (Smith, 2006).

In 1945, in his work "As We May Think," Vannevar Bush offered a system that enhances the human ability to think (Ibid.). Five years later, in the article "Computing Machinery and Intelligence," Alan Turing first raised the question of the possibility of creating a full-fledged artificial imitation of human intelligence. According to Turing, if a machine can behave as intelligently as a human being can, then it is just as intelligent as a human being (Alexander, 2017).

Stuart Russel and Peter Norvig identify 4 main approaches to the definition of artificial intelligence:

- an approach based on human thinking, that is, it is assumed that artificial intelligence should be able to implement mental activity similar to a human one, for example, it is capable to make decisions, solve problems, and learn;
- an approach based on human behavior, that is, it is assumed that artificial intelligence should be able to perform the actions, the performance of which requires intelligence from a person;
- an approach based on rational thinking;
- an approach based on rational behavior (Morkhat, 2018, p. 78).

To understand the details of artificial intelligence, it is necessary to consider the concepts of artificial intelligence proposed by the doctrine.

Human intelligence usually follows a sequence known as "perception—cognition—action," that is, in the first stage, people perceive something in the world around them, think about what to do, and then, once they have considered the options, decide to act (Kostoeva, 2019, p. 50–51). Artificial intelligence is programmed to do something similar since the computer perceives the world, then it processes the obtained information with the help of algorithms of optimization and verification, and the choice of actions is made in the same way as in humans. But it should be noted that even though there are lots of similarities between human intelligence and artificial intelligence, there are significant differences.

Each autonomous system working in a dynamic environment should create a model of the world and constantly update it, that is, the world should be perceived (or felt with the help of cameras, micros, and/or touch sensors) and then reconstructed in such a way that the computer "brain" has an effective and updated model of the world in which it operates before it can make decisions. The accuracy of the world model and the timeliness of its updating are the key conditions for an effective autonomous system.

In modern science, there are two approaches to artificial intelligence: strong (general) artificial intelligence and weak (narrow) artificial intelligence (Gutenev, 2012, p. 78).

A strong version of artificial intelligence suggests that computers can acquire the ability to reflexive thinking and self-awareness, even if their thought process is different from the human one.

The term "strong artificial intelligence" was first used by American philosopher John Searle. In this case, artificial intelligence is considered not just as a model of the mind, but it is this mind, thus, it is assumed that there is no fundamental difference between natural intelligence (human) and artificial intelligence (machine) (Ableev, 2015, p. 60).

A weak version of artificial intelligence rejects any possibility of thinking for computers, emphasizing its limitation to one established (prescribed, imputed) task. Programs with weak artificial intelligence can have exceptional computing capabilities, but they are limited with a particular field, for example, as IBM Deep Blue beat Gary Kasparov at chess in 1997. We encounter more recent versions of weak artificial intelligence daily: virtual assistants as Siri and Alexa that are trained to recognize the voice and perform certain tasks set by the user, search systems, algorithms of social network platforms, web cookies that identify users on the Internet, etc.

Recently, a third approach to artificial intelligence has appeared—superintelligence (Duberry, 2019). This category refers to artificial intelligence, which surpasses the human brain in all tasks, including scientific creativity, general wisdom, and social skills. The appearance of this category is predicted because strong artificial intelligence learns and develops its skills exponentially, so it will reach the level of superintelligence.

It seems that thinking about the real risks of digitalization directs us precisely to the potential of such a "superintelligence."

We suggest, taking into account the above-mentioned doctrinal and technological features and the specifics of artificial intelligence, in particular of its "special type," to classify all problems and risks of digital transformation into three main groups, representing three basic scenario models for the development of robotic technologies and systems of artificial intelligence:

(1) "machine phylum"–the beginning of the machine era, robotic singularity or robotic event, after which systems of artificial intelligence and robotic technologies will gain full autonomy and will be able to launch the process of their self-reproduction and self-improvement, which will provoke a confrontation between humans and robots;
(2) 4th industrial revolution that implies a qualitative leap and change in the socio-economic structure of society, where systems of artificial intelligence and robotic technologies will become new perfect tools and will expand human capabilities, radically change our life, forms, and methods of interaction with social and natural objects (Matytsin & Rusakova, 2021);
(3) convergence—describes the processes of fusion, merging of human and artificial intelligence systems, the emergence of new human-robotic subjects, the "fusion" and subsequent integration of natural and artificial intelligence, the formation of a new stage of evolution (Inshakova et al., 2017);

Each of the mentioned categories needs a doctrinal understanding at the level of legal science. Law as a social regulator currently has the widest regulatory potential, which is expressed both in the actual opportunity to establish (formalize) the doctrinal and legal framework for digitalization and the opportunity of advancing the deontological coding of the development of innovative technologies (AI and RT systems). We think that the instrumentalist approach, which dominates in modern conceptual and legal versions, misses the fact that the right-wing reality itself is also subject to digital transformations, and it waits for a suitable moment when there will be a certain type of social relations, for the settlement of which the legal and technical arsenal will be used.

4.3.3 *Digital Transformation of Socio-economic and Political-Legal Relations*

The introduction of intelligent systems for the detection and prediction of violations of the law is generally positive, but it increases the chance of human rights violations and discrimination against small groups (Dremliuga, 2020, p. 12). And digitalization of social relations itself on the one hand seriously expanding the range of opportunities, in particular in the field of access to information, on the other hand, may serve as a tool for serious restriction of rights of, for example, older members of society who do not have the necessary digital skills.

Currently, for the stable development of stable socio-economic relations, it is vital to elaborate ethical standards and requirements that are adequate to the conditions and requirements of the modern digital era that will regulate the processes of software development and the introduction of autonomous systems in the life of society. After all, the emerging technologization of the individual's everyday life transforms not only his behavioral attitudes but also significantly modifies the very structure of social relations, including the forms of resolving social contradictions (socio-political conflicts) (Goncharov et al., 2019).

For instance, instead of traditional subject-object forms of external pressure (army, security system, economy, political institutions, natural resources) used in socio-political conflicts, today the direction of the accentuated negative informational impact is no longer objects of the physical world but the direct consciousness of the individual (values, beliefs, culture, behavioral attitudes, life strategies). I.e., earlier the objects of influence were institutions (army, territory, economy), then in the conditions of reformatting the direction of external pressure, specific technologies for activating the destabilizing potential that transfer the conflict from the latent sphere to the public space come to the fore (Postalovsky, 2019, p. 47).

First of all, to develop the above-mentioned list of requirements, it is necessary to analyze potential risks of digitalization for the subsequent systematic objectification of the pros and cons of the possible introduction of certain technological models.

Indeed, when looking at the above-mentioned differentiation of risks, it becomes obvious that each of the potential threats is managed and directed by a person, that is, it can be stopped by the same person. Recently, it has been massively overlooked that end-to-end digital technologies are just a product and an object of human activity and that it is people who create, program the shell, and use these *objects*, they lay valuable, rational, moral, emotional, and psychological components in the original code of the latter (for example, based on which machine learning, data array analysis, and the formation of a meaningful solution are carried out). Thus, the potential threats coming from artificial intelligence units are nothing more than a future mistake of the creator/user?

We suppose that, to neutralize the risks mentioned today, first of all, it is necessary to formalize ethical standards, ensuring both their suitability for regulation of specific relations and innovative processes, as well as the integrity of the latter with the current value-regulatory systems of the society (at the national and international legal levels). Moreover, there is a need for comprehensive work to predict and model the impact of ethical standards on the development of RT and AI, individual autonomous systems, and robotic technologies. Currently, projects of such ethical coding for the development of RT and AI have just started to emerge. For example, the version of the ethical standard "Ethically Aligned Design" for the creation of robots and artificial intelligence of the Institute of Electrical and Electronics Engineers (IEEE), which justifies that autonomous devices and intelligent systems should function based on a system of human value-regulatory and ethical regulators, in compliance with the universal standard of human rights and freedoms.

Some leading states have proposed the formation of a Universal Declaration of Robotics and ethical standards for the development of software based on systems of artificial intelligence. From the point of view of the research team, this generalized experience can become the basis for the development of a national ethical standard for the creation of robots and artificial intelligence, reflecting both global trends and international standards as well as national and cultural patterns of digitalization of the Russian society, aims, tasks, and specifics of the implementation of the program "Digital economy in Russia."

The latter is also since for the stable development of the Russian state and society, their reproduction in the future, it is necessary to harmonize various regulatory and legal systems that regulate the life of society and the functioning of robotic technologies. This is a key problem that should be raised during the development of any projects of ethical and normative mediation of relations. All various social regulators of all levels, including the regulatory one, should function in coordination and a consistent manner.

4.4 Conclusion

1. To neutralize main existing risks and threats of digitalization, first of all, it is necessary to conceptualize the key concepts and relations, to form the doctrinal and legal foundations and priority directions of the state's legal policy in the field of the development of end-to-end digital technologies, to distinguish and form the appropriate legal modes for the functioning of autonomous devices based on "weak artificial intelligence" (an autonomous device that performs certain tasks, set and controlled by the software and/or the operator) and "strong artificial intelligence" (an autonomous device that independently perceives the external environment, makes decisions, selects or corrects the interaction model, operating mode, and so on.);

2. At the legislative level, it is necessary to first form the foundations and specific socio-political programs related to stimulating the development of robotic and digital technologies, software, artificial intelligence, as well as their potential introduction to social processes to improve the life of people, to preserve and produce basic socio-cultural values;
3. It is necessary to elaborate deontological and ethical standards based on basic national and cultural values and moral standards, metric certificates, etc., regarding the development of end-to-end digital technologies, which should be followed by developers, manufacturers, and users of these innovative technologies;
4. The current informational and digital legislation is contradictory and incomplete, and it needs to be systematized and brought into a coherent, consistent state. One of the options for such systematization could be public and expert discussion, as well as the adoption of a strategic doctrinal act in the field of information development of society;
5. We suppose that it is necessary to elaborate a special state program for the preservation and reproduction in the society of metapolitical (traditions, customs, symbols, images, rituals, etc.) and metajuridic (mental, psychological, spiritual, moral, etc.) foundations for stability and constant development of political and legal organization of the society. We think that in the twenty-first century, the main competition will unfold between various socio-cultural images, information content, virtual images (which will be constructed based on the national-cultural material developed by society in the process of its development), and other symbolic resources of politics. In its turn, market competition (between goods, services, resources) will recede into the background and it will be replaced with the competition between projects socio-cultural plans that provide a semantic and ideological paradigm for the modern development of digital and social, intersubjective, and virtual reality. At the same time, the key resources that can ensure the harmonization of digital development trends and real socio-political relations, the integration of socio-cultural identities and virtualized forms of group interaction and online communities will be the traditional sign and symbolic systems and the dominant socio-cultural organizations.

Acknowledgements This work was financially supported by the Grant of the President of the Russian Federation No. HSH-2668-2020.6 "National-Cultural and Digital Trends in the Socio-Economic, Political and Legal Development of the Russian Federation in the 21st Century".

References

Ableev, S. R. (2015). Modeling of consciousness and artificial intelligence: Limits of possibilities. *Bulletin of the Academy of Economic Security of the Ministry of Internal Affairs of Russia, 3*, 58–64.

Alexander, F. M. (2017, June 1). The legal status of artificially intelligent robots: Personhood, taxation, and control. *Electronic Journal of the NRC* 1–68. https://papers.ssrn.com/abstract=2985466 Accessed: 16.09.2019.

Bennett, J. (2018). *Pulsating matter: The political ecology of things* (p. 220). Gile Press.

Brockman, J. (2017). What we think about machines that think: The world's leading scientists on artificial intelligence [Electronic resource] (p. 552). Alpina-Non-fiction. https://www.litmir.me/br/?b=592732&p=1 Accessed: 10.11.2020.

Bruno, L. (2014). *Reassembling the social: An introduction to actor-network theory* (p. 384). House of the Higher School of Economics.

Bruno, L. (2018). *The politics of nature. How to establish democracy in the Sciences* (p. 336). Ad Marginem.

Delanda, M. (2012). *The new philosophy of society: Assemblage theory and social complexity* (p. 170). Gile Press.

Deleuze, J., & Guattari, F. (2010). *Capitalism and schizophrenia.* Book 2. *Thousand plateaus* (p. 672). U-Factoriya, Astrel.

Digital agenda and digital initiatives during COVID-19 (review of the practices of the European Union, the Organization for Economic Cooperation and Development, as well as of other countries). (2020) (p. 19). National Research University Higher School of Economics.

Dremliuga, R. I. (2020). *How development of Artificial intelligence technology will cause changes in crime and criminal law. "AI for Everyone: Benefitting from and building trust in the technology"* (p. 85). AI Access.

Duberry, J. (2019). *Global environmental Governance in the Information Age: Civil society organizations and digital media.* [Electronic resource]. Routledge. https://www.researchgate.net/publication/330967399_Global_Environmental_Governance_in_the_Information_Age_Civil_Society_Organizations_and_Digital_Media Accessed: 11.09.2019.

Goncharov, A., Dolinskaya, V., Inshakova, A., Kalinina, A. (2019). Corporate conflicts in the system of public relations, law, and legislation of the Russian Federation. In E. G. Popkova (Ed.), *"Conflict-free" socio-economic systems* (pp. 75–88). Emerald Publishing Limited.

Goodbye, Covid. (2020). In K. Gaase, V. Danilov, I. Dudenkova, D. Kralechkin, & P. Safronov (Eds.) (p. 432). Publishing House of the Gaidar Institute.

Greenfield, A. (2018). *Radical technologies: The device of everyday life* (p. 424). Publishing House "Delo" RANEPA.

Gutenev, MYu. (2012). The problem of artificial intelligence in the philosophy of the XX century. *Bulletin of the Chelyabinsk State Academy of Culture and Arts, 4*, 77–80.

Inshakova, A., Goncharov, A., Smirenskaya, E., & Dolinskaya, V. (2017). Modern communication technologies in notification of notarial actions in Russia. *Journal of Advanced Research in Law and Economics, VIII, 7*(29), 2144–2151.

Inshakova, A. O., Frolova, E. E., Rusakova, E. P., Kovalev, S. I. (2020). The model of distribution of human and machine labor at intellectual products in Industry 4.0. *Journal of Intellectual Capital, 21*(4), 601–622.

Kaku, M. (2018). *The future of the mind* (4th ed., p. 646). Alpina non-fiction.

Kelly, K. (2017). *Inevitably. 12 technological trends that determine our future* (p. 347). Mann, Ivanov, and Ferber.

Kostoeva, A. A. (2019). Sigmund Freud on the nature of the unconscious in the human psyche. *Colloquium-Journal, 3*, 50–51.

Matytsin, D. E., & Rusakova, E. P. (2021). Strategy of quality management in industry 4.0 and formation of cognitive economy based on industrial and manufacturing engineering in the Russian

Federation and countries of the EU. *International Journal for Quality Research, 15*(4). https://doi.org/10.24874/IJQR15.04-03

Morkhat, P. M. (2018). The legal personality of artificial intelligence in the field of intellectual property law: Civil-legal problems (p. 405) Ph.D. in Law thesis Moscow, RGIIS.

Ozhegov, S. I. (2014). *Explanatory dictionary of the Russian language* (p. 315). Mir i obrazovanie [World and education].

Postalovsky, A. V. (2019). Sociology of destabilization: Theoretical and methodological aspects. Bulletin of the Mogilev State University named after A. A. Kuleshov. *Economics, Sociology, Law, 2*(54), 47–55.

Smith, C. (2006). *The history of artificial intelligence* [Electronic resource]. https://courses.cs.washington.edu/courses/csep590/06au/projects/history-ai.pdf Accessed: 28.10.2020.

Stuart, J. (2019). *Russell human compatible. Artificial intelligence and the problem of control* (p. 450). Alpina Digital.

Tarakanov, V. V., Inshakova, A. O., & Dolinskaya, V. V. (2019). Information society, digital economy, and law. Ubiquitous computing and the Internet of Things: Prerequisites for the development of ICT. *Studies in Computational Intelligence, 826*, 3–15.

Thacker, E. (2020). *Three texts about the infection* (p. 74). Gile Press.

The digital Agenda and initiatives in the field of digital technologies in the context of COVID-19 (review of the practice of the European Union, the Organization for Economic Cooperation and Development, and other countries) (p. 19). Higher School of Ecoomics.

The End of the Familiar World (2021). *The Knife Magazine's guide to new ethics, new relationships, and new justice* (p. 380). Alpina Non-fiction.

Chapter 5
Prospects for the Use of Cross-Cutting Digital Technologies in the Framework of Tax and Currency Control Over Cross-Border Financial Transactions

Ekaterina A. Tsepova

Abstract Information technologies provide the state with ample opportunities in countering financial offenses committed with the use of the cross-border movement of funds. The article examines the process of creating a global system of control over the payment of taxes on income received by residents of one state from sources in other states and the prospects for its further improvement. Particular attention is paid to the issues of using digital technologies in the field of financial control in the Russian Federation. The author comes to the conclusion that cross-cutting digital technologies used to control cross-border capital flows both at the global and domestic levels should be considered broader than one of the instruments of the state coercion system. Their use should contribute to the revision of the mechanisms of tax administration and currency exchange regulation aimed at improving the legal status of bona fide taxpayers and subjects of currency control.

Keywords Digitalization · Automatic exchange of financial account information · CRS · Common reporting standard · Financial control · Tax administration

JEL Codes K24 · K33 · K34 · O16

5.1 Introduction

For the tax authorities of most states, the problem of mass evasion of individuals and organizations from paying taxes by performing cross-border financial transactions, eroding the tax base and withdrawing funds to jurisdictions with a lower tax burden, including offshore zones, is becoming increasingly important (Eden & Kudrle, 2005; Mara, 2015). Therefore, the most important condition determining the further development of world legislation in the field of taxation was the protection of the integrity of the tax system of each state (Morten & Zeume, 2018; Pistone, 2014).

E. A. Tsepova (✉)
Law Firm "PharmConsulting", 3, bldg. 2 Ilimskaya St., 127575 Moscow, Russia

Vladivostok State University of Economics and Service, 41 Gogol St., 690014 Vladivostok, Russia

A. O. Inshakova and E. E. Frolova (eds.), *Smart Technologies for the Digitisation of Industry: Entrepreneurial Environment*, Smart Innovation, Systems and Technologies 254, https://doi.org/10.1007/978-981-16-4621-8_5

A turning point in the fight against tax havens and tax evasion was the financial and economic crisis of 2009, when, during a summit of leaders of the G20 states, it was announced, "the era of banking secrecy is over." One of the key decisions taken during the summit was to create the Global Forum on Transparency and Exchange of Information for Tax Purposes to strengthen the potential of cooperation in international tax matters (OECD, 2016a).

Countries that favor fair tax competition have agreed to create an OECD-based automatic exchange of tax information. On July 23, 2012, the OECD presented a report "Automatic Exchange of Information what it is, how it works, benefits, what remains to be done" (OECD, 2012). It summarizes the key aspects of an effective automatic information exchange model, including the development of a united standard for the collection, analysis, and transmission of information, the establishment of a legal basis for information exchange, and the preparation of appropriate technical means. The development of Standard for Automatic Exchange of Financial Account Information in Tax Matters Common Reporting Standard (hereinafter—Common Reporting Standard or CRS) was completed in February 2014 after its approval by the finance ministers and heads of central banks of G20 countries (OECD, 2017). CRS describes in detail the composition and process of collecting data that is intended to be transmitted as part of the automatic exchange of financial information.

In 2016, a major international scandal broke out around the large Panamanian law firm Mossack Fonseca. The leak of confidential documents received an informal name "Panama Papers." This story was associated with the hiding of assets of more than 140 politicians from around the world in 21 tax havens. This story became a kind of trigger for states to enter a new level of communication in the field of tax control (Obermaier & Obermayer, 2017).

Already, in 2017, about half of the countries participating in the exchange following the CRS, for the first time exchanged information on the financial accounts of residents of one country opened in financial institutions of other countries. The first data exchange with the participation of Russia took place in 2018 and since then has performed annually. According to the deputy head of the Federal Tax Service, Russia was one of the 25 states that co-financed the emergence of a unified data transmission system (Gaidar Forum, 2021). As a result, control over the payment of taxes by residents of one state, who own various assets in other states, has passed into the power of digital technologies.

5.2 Materials

The legal framework for the study was formed by international agreements concluded based on the Organization for Economic Cooperation and Development (OECD): the Multilateral Convention on Mutual Administrative Assistance in Tax Matters, Multilateral Competent Authority Agreement on Automatic Exchange of Financial Account Information, Foreign Account Tax Compliance Act, Multilateral Competent Authority Agreement on the Exchange of Country-by-Country Reports.

For the Russian Federation, these are the Tax Codes of the Russian Federation, Decree of the Government of the Russian Federation of June 16, 2018, No. 693 "On the implementation of the international automatic exchange of financial information with the competent authorities of foreign states (territories)," as well as departmental regulatory legal acts of state authorities, primarily Federal Tax Service.

The theoretical basis for studying the application of digital technologies in the field of public financial control and tax administration was provided by the OECD materials, such as Common Reporting Standard, Action Plan on Base Erosion and Profit Shifting, report "Automatic Exchange of Information what it is, how it works, benefits, what remains to be done," Tax Transparency 2016 Report on Progress, as well as works of foreign authors Pistone (2014), Jimenez et al. (2013), Hamilton and Stekelberg (2017), Johannesen et al. (2018); Li et al. (2020), Inshakova et al. (2018), Goncharov et al. (2016), and Matytsin (2021).

Thanks to research by authors such as Mishustin (2014), Khabrieva (2018), Levashenko and Koval (2018), the article discusses the development of cross-cutting digital technologies in the field of control over the transboundary movement of the capital of taxpayers in the Russian Federation.

What gaps exist in the mechanism of automatic exchange of information on the assets of taxpayers in the jurisdictions of foreign states are studied based on works Stratiev (2018), Kalinina et al. (2019), Morris (2017), Noked (2018a, 2018b), Drake et al. (2019), as well as materials published by the OECD.

Prospects for improving cross-cutting digital technologies in the field of financial control are considered, thanks to research Levashenko and Koval (2018), Saint-Amans (2016), Andres-Aucejo (2018), and Pross (2015).

5.3 Methods

The methodological basis of the study was the analysis methods, thanks to which it was possible to consider the process of the formation at the global level of a new system of financial control over the movement of the capital of residents of one state, allocated in the jurisdiction of the other states, based on the use of end-to-end digital technologies, and to outline the prospects for the development of this system.

5.4 Results

Exploring the legal framework that allowed the digitalization of tax and currency exchange control over cross-border financial transactions to take place, the following should be noted:

1. The legal framework for international tax transparency was created after the conclusion in Strasbourg on January 25, 1988, the Multilateral Convention on

Mutual Administrative Assistance in Tax Matters (hereinafter—MCMAATM) (OECD, 1988). The Convention provides for the possibility of various ways of exchanging tax information between states that have ratified it, including exchange of information on request, proactive exchange of information, and automatic exchange of information. Currently, 141 states and territories are parties to the convention (OECD, 2021a).

2. Until recently, the exchange of tax information between the MCMAATM member states was carried out mainly cross-cutting the exchange of information upon request based on bilateral intergovernmental agreements and was selective. As a result, the access of the tax authorities of the MCMAATM member states to information on the property and financial transactions of their tax residents in the other jurisdictions was significantly limited. The volume of information transferred under bilateral agreements was extremely low, and the terms of the agreements themselves often included significant restrictions on the information transferred. For example, they did not touch on many financial market instruments that allow hiding information about the real owners of companies, bearer shares (Chanel, 2013).

In addition, the banking community insisted on maintaining a banking secrecy regime beneficial to their clients (Inshakova et al., 2018). As a result, the tax authorities could not obtain sufficient information to make sure that taxpayers paid the proper amount of tax to the budget. For this reason, the effectiveness of state financial control in this area was very low.

To make the detection, prevention, and suppression of tax offenses effective, it was necessary to change the nature of state financial control from selective to continuous. The work on solving this problem on a global scale, which is carried out based on the OECD, can be divided into two key areas:

1. International cooperation in the tax area to create conditions for the automatic exchange of information, development, and implementation of an appropriate legal framework, in particular, allowing access to banking secrecy both at the domestic and international levels.
2. Development of digital technology solutions that allow the systematic collection, analysis, and transfer of tax information in the framework of international exchange, while ensuring the confidentiality of the transmitted data.

A huge role in the realization of the first direction has played the development of the Multilateral Competent Authority Agreement on Automatic Exchange of Financial Account Information dated October 29, 2014 (hereinafter—CRS MCAA) and some other international agreements (Tsepova, 2019). Currently, 110 states, including Russia, are members of CRS MCAA (OECD, 2021b).

Thus, almost half of the world's jurisdictions have joined the agreement; the total nominal value of their gross domestic product is 75% of the global value (International Monetary Fund, 2019).

The implementation of CRS MCAA, without exaggeration, became a breakthrough in the field of tax control over cross-border transactions of taxpayers since it creates a legal basis that allows the tax authorities of the member states to receive data on the accounts of their tax residents in foreign financial market organizations freely and regularly. At the same time, the amount of data is sufficient to monitor compliance with tax legislation (Levashenko & Koval, 2018).

The purpose of the state's participation in CRS MCAA, first of all, is to control the fulfillment of tax obligations by residents, however, countries that apply currency exchange restrictions can use the information received for currency exchange control purposes.

All CRS MCAA member states collect, analyze, and transfer to each other information on financial accounts of taxpayers following the Common Reporting Standard (hereinafter—CRS) (OECD, 2014, 2017). It can be argued that this agreement is an example of the successful creation of a unified field of financial control in international taxation.

Another agreement governing the international exchange of tax information is the Multilateral Competent Authority Agreement on the Exchange of Country-by-Country Reports (hereinafter—CBC MCAA) (OECD, 2016b), concluded as part of the implementation of the thirteenth paragraph developed by the OECD Base Erosion and Profit Shifting (OECD, 2013). Currently, this agreement has been signed by 89 states (OECD, 2021c). Accession to the agreement will allow the tax authorities of the CBC MCAA member state to receive on an annual basis and in automatic mode the information on the distribution of income and taxes between the members of international groups of companies, which is necessary to assess the risks associated with the erosion of the tax base and the withdrawal of profits in jurisdictions with a lower tax burden.

The successful implementation of the second direction was due to several circumstances:

1. Most of the data intended for international exchange was collected and processed by financial institutions in digital format. This meant that it was only necessary to summarize the data and ensure that the competent authorities of the relevant state have access to this information.
2. Modern technologies for storing and processing information allow maintaining huge databases and archives at relatively low costs, as well as ensuring the quality and high speed of data processing.
3. The development of Internet technologies has made possible high-speed access to large amounts of data from anywhere in the world.
4. It became possible to ensure reliable data protection from unauthorized access using various encryption technologies.

Fifteen years ago, the creation of a single digital field of financial control would have been unreasonably difficult and costly for all parties involved in the collection and processing of information. Banks exchanged information with clients mainly using paper media, and electronic media were much less compact than they are now. Server rooms, even in small banks, could occupy several rooms, and the duration of the process of electronic search for certain data in the database did not allow processing a large number of requests. The current level of development of information technology was a prerequisite for the creation of a global financial control system.

5.4.1 The Use of Cross-Cutting Technologies in the Field of Control Over Cross-Border Financial Transactions as a State Response to the Challenges of Digital Reality

Digital technologies in the area of capital management are becoming more affordable and more popular among individuals.

It is now possible to open an account in a foreign bank, buy and sell securities and even make business abroad without leaving your home (Matytsin, 2021). The development of financial relations between the state and its tax residents, who owned various assets in other countries, is of great scientific and practical interest all over the world. The mobility of people and capital, increased in the context of globalization, increases the risk of tax evasion through the blur of the tax base and cash withdrawal in a jurisdiction with a lower level of taxation (Frolova et al., 2020).

The state responds in different ways to the challenges of digital reality, trying to manage the ongoing transformations and regulate new, as well as changing social relations (Khabrieva, 2018; Mishustin, 2014). To counter the negative consequences of capital mobility for the tax system, it is important to develop effective control mechanisms for cross-border transactions of taxpayers.

CRS is a guarantor of openness and availability of information on financial accounts around the world (in compliance with the necessary data protection measures). We are talking not only about bank accounts but also about investment accounts, custody accounts, trust management, investment life insurance, etc. CRS also requires disclosure of the beneficiaries of trusts and foundations. To comply with this rule, financial institutions have to do a lot of work within the due diligence procedure for each client, but this is the only way to ensure the correct declaration of accounts.

Thus, we can conclude that the task of CRS is to teach organizations and individuals to fully transparently declare their assets and income abroad, pay taxes in accordance with international law and the legislation of the country of residence (Owens, 2015; Saint-Amans, 2016).

5.4.2 Development of Cross-Cutting Digital Technologies in the Russian Federation

The legal basis for the use of cross-cutting digital technologies in the framework of tax and currency control in Russia is the Federal Law No. 340-FZ of November

27, 2017 "On amendments to part one of the Tax Code of the Russian Federation in connection with the implementation of the international automatic exchange of information and documentation for international groups of companies" (Government of the Russian Federation, 2017). The law provides for the creation of a mechanism for the collection of information by the tax authorities of the Russian Federation necessary for exchange with the competent authorities of foreign countries within the framework of the fulfillment of the conditions of CRS MCAA and CbC MCAA.

In addition, a number of regulatory legal acts were developed and adopted that directly regulate the procedure and terms for processing and sending information intended for transmission by financial market organizations.

The key of them is the Decree of the Government of the Russian Federation of June 16, 2018, No. 693 "On the implementation of international automatic exchange of financial information with the competent authorities of foreign states (territories)," which entered into force on July 19, 2018. This resolution is a text of the CRS standard translated into Russian in a somewhat truncated version, but with the inclusion of the key conditions.

Deputy Chairman of the Government of the Russian Federation Alexei Logvinovich Overchuk (until January 21, 2020—Deputy Head of the Federal Tax Service of the Russian Federation) described the current state capabilities in the field of control over cross-border transactions of taxpayers: "This is a completely new level of tax transparency. If earlier it was necessary to make a request to the competent authority of a foreign jurisdiction, now the information about the foreign account of a particular taxpayer itself comes and is available by the push of a button" (RBC, 2019).

Today, the number of accounts of Russian residents in the foreign financial market has exceeded 700 thousand, and their total balance is estimated at 13 trillion rubles. In total, during the automatic exchange of tax information, the participating states exchange data on 84 million accounts, which hold a total of about 10 trillion euros (Gaidar Forum, 2021).

Russia has been using automated risk analysis tools for tax control for a long time (Mishustin, 2014). The use of information technologies has a positive effect on the development of a risk-oriented approach, including in the field of control over cross-border transactions of Russian residents. It is based on an assessment of the severity of damage to public interests as a result of tax evasion or other financial offenses and the likelihood of such damage by certain persons (Soloviev, 2017).

Analysis of the data obtained in the course of the implementation of international agreements on the exchange of tax information and comparing them with the data at the disposal of the tax authorities (provided by the residents themselves) makes it possible to create a high-quality system for selecting subjects for in-depth tax or currency control. Such measures will make it possible to expediently use the limited human resources of the tax authorities and obtain the maximum possible results (Savin, 2012).

Moreover, the use of new digital platforms for receiving and processing data will allow not only to increase the level of tax discipline but at the same time to dramatically reduce the tax and administrative burden, which will contribute to strengthening trust between the state and society (Mishustin, 2014).

5.4.3 What Are the Gaps in the Mechanism of Automatic Exchange of Information on the Assets of Taxpayers in Foreign Jurisdictions?

The list of countries participating in the global exchange of tax information is constantly growing. But about 40 jurisdictions have not yet joined it.

Currently, among developed countries, only the United States of America has not joined the CRS MCAA and CBC MCAA agreements. International exchange of financial information with the USA is carried out in accordance with the Internal Revenue Code based on amendments, collectively referred to as the Foreign Account Tax Compliance Act (hereinafter—FATCA) (United States Government, 2010), which were introduced much earlier than CRS MCAA and CBC MCAA.

Initially, the exchange of tax information following FATCA was supposed to form the basis of the model of an intergovernmental agreement on the improvement of international tax legislation, developed by a number of European countries (France, Germany, Italy, Spain, and the United Kingdom) and the United States of America ("Model 1 IGA"), but on April 19, 2013, the finance ministers and central bank governors of the G20 countries decided to develop a new standard for automatic exchange of information for tax purposes. This decision was driven by the need to develop a universal multilateral information exchange standard that combines the advantages of using FATCA, but at the same time does not contain the characteristic features of US tax legislation. For example, such as recognizing a person as a tax resident on the basis of citizenship and penalties for foreign financial institutions for refusing to apply FATCA.

Compared to the implementation of the exchange of tax information based on intergovernmental agreements in accordance with FATCA, the application of the CRS standard is more efficient and requires less cost from financial institutions. Nevertheless, the exchange of tax information based on intergovernmental agreements in accordance with FATCA is an independent system for the exchange of tax information, which has a significant number of similar principles to the exchange according to the CRS standard and is compatible with its application (Noked, 2018b).

It should be noted that the perimeter of automatic exchange of tax information should increase. Shortly, some states of the Eurasian Economic Union (EAEU) and the CIS, including Kazakhstan and Moldova, will begin to exchange data with Russia (Commonwealth of the Independent States, 2018).

It should be noted that CRS is constantly being improved. Particular attention is paid to addressing gaps, allowing evading taxation, including the use of digital assets and cryptocurrency (OECD, 2020; Stratiev, 2018).

One of the first threats, identified by experts, to the effectiveness of the financial control mechanism based on the use of automatic information exchange in accordance with the CRS was the abuse by individuals of state programs that provide citizenship or residency in exchange for investment.

At the time the CRS MCAA came into effect, such programs were operating in the UAE, Malta, Cyprus, Seychelles, etc. As a rule, program participants have a certain time frame for their stay in the country, but there is no strict control over their observance. As a result, a person who does not reside in the country can manipulate the status of a tax resident (Morris, 2017; Noked, 2018a).

These gaps reduce the effectiveness of measures to prevent crime in the financial sector since there is an inverse relationship between tax risks and the propensity of taxpayers to commit offenses (Drake et al., 2019).

The OECD's response to this threat followed quickly enough, recommendations were issued requiring financial market organizations to check the period of a citizen's real presence in a country where such programs are in place (OECD, 2018a, 2018b). In addition, under pressure from the international community, some countries stopped granting citizenship or residency for investment or significantly tightened the requirements for applicants to participate.

5.4.4 Prospects for Improving Cross-Cutting Digital Technologies in the Field of Financial Control

The formation of international tax transparency gives the member states of agreements on the exchange of tax information broad opportunities in the fight against tax evasion through cross-border financial transactions. But the use of these opportunities in practice depends on the perfection of the legal framework and the effectiveness of the administration of information obtained through the international exchange (Pross, 2015).

As one of the promising directions for the development of financial control over cross-border transactions of taxpayers, it is necessary to highlight the reduction of the time required for tax authorities to analyze the information received through the international exchange, and the prompt detection of offenses that cause the greatest damage to state interests (Levashenko & Koval, 2018).

Since the beginning of the international automatic exchange of information for tax purposes, the tax authorities annually receive a large amount of data on the financial transactions of their tax residents, carried out in other states—participants of the exchange.

The possibility of applying liability measures to taxpayers who have committed offenses depends on how promptly the tax authorities analyze the information received the impossibility of identifying the most significant offenses among the multitude of offenses will lead to the loss by the state of part of the budget revenues from tax collection and the application of liability measures for committing offenses.

In addition, the importance of the problem of determining the residence of individuals is increasing, since this is a key criterion in the implementation of the norms of international agreements that ensure the automatic exchange of tax information (Goncharov et al., 2016). Currently, in world practice, there is no uniform approach to establishing the status of a tax resident individual, and the criteria used are heterogeneous and difficult to assess in the context of increasing people's mobility (Zelinsky, 2017).

This situation not only creates unjustified tax risks for persons with international economic interests but also reduces the efficiency of financial institutions in identifying the residence of account holders in order to comply with CRS MCAA.

In such conditions, it is obvious that there is a need to harmonize individual norms of the tax law of different states, focused on the principles and approaches of international multilateral agreements in the field of taxation (Andres-Aucejo, 2018; Saint-Amans, 2016). Studies on the harmonization of tax systems in the EU countries can be taken as a basis (Safonova et al., 2016; Sinn, 1990).

Effective financial regulation of relations with individuals allocating capital in several jurisdictions, in the context of international tax transparency, is impossible without changes in domestic tax and currency exchange legislation. Improving the regulatory framework should primarily focus on optimizing information flows within the framework of financial control, eliminating the excessive procedural burden on individuals and regulatory authorities, as well as harmonizing approaches to determining the tax residence of individuals at the global level.

5.5 Conclusion

To summarize, it can be argued that digitalization in the field of control over cross-border capital movements has an impact on the development of financial legislation both at the international and domestic levels since it radically changes the system of control over the payment of taxes on income received by residents of one states from sources in other states. In the Russian Federation, digitalization also contributes to the liberalization of currency exchange restrictions (Frolova & Tsepova, 2020).

In the future, for the successful functioning and development of the international tax exchange system, the states that joined the CRS MCAA will need to continue work in the following areas: development and implementation of measures to improve the quality of transmitted data, standardization of technical requirements for hardware and software used for collection and transmission information in the course of international exchange, improvement of approaches to ensuring the confidentiality of transmitted data.

As a result of the implementation of international agreements such as the CRS MCAA, the volume of data processed by the tax authorities has increased and, in these conditions, the ordering of information flows becomes especially important.

In addition, due to the pandemic caused by the spread of the new coronavirus infection (2019-nCoV), many government agencies around the world have temporarily suspended the reception of visitors and the reduced presence of employees in the workplace, focusing on key functions.

In such a situation, it is important to not only provide taxpayers with the opportunity to perform their duties remotely using digital technologies but also to evaluate each report, notification, or message from the point of view of expediency. The data received by the tax authorities should be sufficient to carry out risk-oriented control measures, at the same time, uninformative and duplicating information and reports should be excluded from the information flow.

The introduction of new digital tools endows the state with unprecedentedly broad opportunities in the fight against financial violations. But in such conditions, it is very important to maintain the balance of interests of the parties of the financial relationship. The achievements of digitalization must be used not only to improve the efficiency of the tax collection but also to reduce the administrative burden, as well as to improve the situation of conscientious taxpayers.

Digitization should be viewed more broadly than a tool for replacing paper-based reports with electronic reporting. Due to the information on financial accounts being regularly and automatically submitted to the tax authorities under the CRS MCAA, many of the previously existing procedural obligations of subjects have lost their relevance and require deregulation.

Improvement of the current regulation in this area is necessary to: firstly, reduce the procedural burden on individuals and organizations; and secondly, to make monitoring compliance with financial legislation and enforcement activities of the tax authorities effectively, as well as reduce the cost of development of the state financial control system.

Acknowledgements This work was financially supported by Grant of the President of the Russian Federation No. NSH-2668-2020.6 "National-Cultural and Digital Trends in the Socio-Economic, Political and Legal Development of the Russian Federation in the 21st Century".

References

Andres-Aucejo, E. (2018). Towards an international code for administrative cooperation in tax matter and international tax governance. *Revista Derecho Del Estado, 40*, 45–85.

Chanel, G. (2013). *Transparency and global tax: Clearing the way*. Organisation for Economic Cooperation and Development. The OECD Observer, (295), 24

Commonwealth of Independent States. (2018). A protocol on the exchange of information electronically between the CIS member states for the implementation of the tax administration on November 2, 2018. Retrieved from: http://www.cis.minsk.by/reestr/ru/index.html#reestr/view/text?doc=5963. Accessed: 15.02.2021

Drake, K. D., Lusch, S. J., & Stekelberg, J. (2019). Does tax risk affect investor valuation of tax avoidance? *Journal of Accounting, Auditing & Finance, 34*(1), 151–176.

Eden, L., & Kudrle, R. T. (2005). Tax havens: Renegade states in the international tax regime? *Law & Policy, 27*(1), 100–127.

Frolova, E. E., & Tsepova, E. A. (2020). 30 years of currency liberalization in Russia: Prospects for the abolition of restrictions on the movement of capital of individuals. *State and Law, 4*, 108–124.

Frolova, E. E., Ermakova, E. P., & Protopopova, O. V. (2020). Consumer protection of digital financial services in Russia and abroad. *Advances in Intelligent Systems and Computing, 1100*, 76–87.

Gaidar Forum. (2021). Expert discussion foreign assets of Russians: Disclosure will show. Retrieved from https://gaidarforum.ru/ru/program/946/. Accessed: 06.02.2021.

Goncharov, A. I., Inshakova, A. O., Kazachenok, O. P., & Dikarev, I. S. (2016). Unfair "housing regulation of major construction" in the Russian Federation. *International Journal of Environmental & Science Education, 11*(7), 9741–9752.

Government of the Russian Federation. Federal law "On amendments to part one of the Tax Code of the Russian Federation in connection with the implementation of the international automatic exchange of information and documentation for international groups of companies" dated of November 27, 2017, No. 340.

Hamilton, R., & Stekelberg, J. (2017). The effect of high-quality information technology on corporate tax avoidance and tax risk. *Journal of Information Systems, 31*(2), 83–106.

Inshakova, A. O., Goncharov, A. I., Kazachenok, O. P. (2018). Legal regulation of the syndicated lending: Banking innovations in contemporary Russia. In E. Popkova & V. Ostrovskaya (Eds.), *Perspectives on the use of new information and communication technology (ICT) in the modern economy* (pp. 368–377). Springer. https://doi.org/10.1007/978-3-319-90835-9_43

International Monetary Fund. (2019). World economic outlook database, October 2019 Edition. https://www.imf.org/external/pubs/ft/weo/2019/02/weodata/index.aspx. Accessed: 16.02.2021

Jimenez, G., Mac, Sionnaigh, N., & Kamenov, A. (2013). Information Technology for Tax Administrations. USAID Bureau for Economic Growth, Education and Environment, Office of Economic Policy, database, October 2019 Edition. http://pdf.usaid.gov/pdf_docs/PNAEA485.Pdf. Accessed: 02.02.2021

Johannesen, N., Langetieg, P., Reck, D., Risch, M., & Slemrod, J. (2018). *Taxing hidden wealth: The consequences of US enforcement initiatives on evasive foreign accounts*. National Bureau of Economic Research, w24366. https://www.nber.org/papers/w24366. Accessed: 03.02.2021

Kalinina, A. E., Inshakova, A. O., & Goncharov, A. I. (2019). Polysubject jurisdictional blockchain: Electronic registration of facts to reduce economic conflicts. *Studies in Computational Intelligence, 826*, 205–213.

Khabrieva, T. Y. (2018). The right to the challenges of digital reality. *Journal of Russian Law, 9*(261), 5–16.

Levashenko, A. D., & Koval, A. A. (2018). How international automatic exchange of tax information works. *Economic Development of Russia, 25*(3), 71–73.

Li, J., Wang, X., & Wu, Y. (2020). Can government improve tax compliance by adopting advanced information technology? Evidence from the Golden Tax Project III in China. *Economic Modelling, 93*, 384–397.

Mara, E. R. (2015). Determinants of tax havens. *Procedia Economics and Finance, 32*, 1638–1646.

Matytsin, D. E. (2021). Internet-investing as a remote algorithm of the retail investment financing. In *Modern global economic system: Evolutional development versus revolutionary leap*. Springer Book Series Lecture Notes in Networks and Systems. Springer Nature.

Mishustin, M. V. (2014). Improving tax administration tools to ensure stable incomes of the state budget. *Economics. Taxes. Right, 4*, 4–8.

Morris, M. (2017). The 26 OECD common reporting standard loopholes. http://www.the-best-of-both-worlds.com/support-files/oecd-crs-loopholes-report.pdf. Accessed: 16.02.2021

Morten, B., & Zeume, S. (2018). Corporate tax havens and transparency. *The Review of Financial Studies, 31*(4), 1221–1264.

Noked, N. (2018a). Tax evasion and incomplete tax transparency. *Laws, 7*(3), 31. https://doi.org/10.3390/laws7030031. Accessed: 17.02.2021

Noked N. (2018b). FATCA, CRS, and the wrong choice of who to regulate. *Florida Tax Review, 22* (forthcoming). https://papers.ssrn.com/sol3/papers.cfm?abstract_id=3143663 Accessed: 11.02.2021

Obermaier, F., & Obermayer, B. (2017). *The Panama Papers: Breaking the story of how the rich and powerful hide their money*. Oneworld Publications.

OECD. (1988). Convention on mutual administrative assistance in tax matters. http://www.oecd.org/ctp/exchange-of-tax-information/convention-on-mutual-administrative-assistance-in-tax-matters.htm. Accessed: 05.12.2020

OECD. (2012). Automatic exchange of information what it is, how it works, benefits, what remains to be done. www.oecd.org/ctp/exchange-of-tax-information/automatic-exchange-of-information-report.pdf Accessed: 05.12.2020

OECD. (2013). Action plan on base erosion and profit shifting. https://doi.org/10.1787/9789264202719-en. Accessed: 15.02.2021

OECD. (2014). Multilateral competent authority agreement on automatic exchange of financial account information. https://www.oecd.org/tax/automatic-exchange/international-framework-for-the-crs/multilateral-competent-authority-agreement.pdf. Accessed: 03.02.2021

OECD. (2016a). Tax transparency 2016 report on progress. http://www.oecd.org/tax/transparency/GF-annual-report-2016.pdf Accessed: 05.12.2020.

OECD. (2016b). Multilateral competent authority agreement on the exchange of country-by-country reports. https://www.oecd.org/tax/automatic-exchange/about-automatic-exchange/cbc-mcaa.pdf. Accessed: 15.02.2021

OECD. (2017). Standard for automatic exchange of financial account information in tax matters, (2nd ed.). https://www.oecd-ilibrary.org/taxation/standard-for-automatic-exchange-of-financial-account-information-in-tax-matters-second-edition_9789264267992-en;jsessionid=IW0_TZBRjv88sDJiI20E2i1C.ip-10-240-5-179 Accessed: 05.12.2020

OECD. (2018a). Consultation document: Preventing abuse of residence by investment schemes to circumvent CRS. http://www.oecd.org/tax/exchange-of-tax-information/consultation-document-preventing-abuse-of-residence-by-investment-schemes.pdf Accessed: 05.02.2021

OECD. (2018b). Model mandatory disclosure rules for CRS avoidance arrangements and Opaque offshore structures. http://www.oecd.org/tax/exchange-of-tax-information/model-mandatory-disclosure-rules-for-crs-avoidance-arrangements-and-opaque-offshore-structures.pdf. Accessed: 16.01.2021

OECD. (2020). Taxing virtual currencies: An overview of tax treatments and emerging tax policy issues, OECD, and Paris. www.oecd.org/tax/tax-policy/taxing-virtual-currencies-an-overview-of-tax-treatments-and-emerging-tax-policyissues.htm. Accessed: 18.02.2021

OECD. (2021a). Convention on mutual administrative assistance in tax matters, chart of participating jurisdictions. http://www.oecd.org/tax/exchange-of-tax-information/Status_of_convention.pdf Accessed: 05.12.2020

OECD. (2021b). List of CRS MCAA signatories. http://www.oecd.org/tax/automatic-exchange/international-framework-for-the-crs/MCAA-Signatories.pdf

OECD. (2021c). List of CbC MCAA signatories. https://www.oecd.org/tax/automatic-exchange/about-automatic-exchange/CbC-MCAA-Signatories.pdf. Accessed: 01.02.2021

Owens, J. (2015). Tax transparency and BEPS. *Journal of Tax Administration, 1*(2), 1–10.

Pistone, P. (2014). Coordinating the actions of regional and global players during the shift from bilateralism to multilateralism in international tax law. *World Tax Journal, 6*(1), 3–9.

Pross A. (2015, March 25). How tax transparency went global—The new automatic exchange standard: From concept to reality. *International Tax Review*. https://www.internationaltaxreview.com/Article/3439573/How-tax-transparency-went-global-the-new-automatic-exchange-standard-from-concept-to-reality.html?ArticleId=3439573 Accessed: 05.12.2020

RBC. (2019). Tax authorities have disclosed foreign accounts and assets of Russians in 58 countries. https://www.rbc.ru/economics/05/03/2019/5c78f4b39a794706e35338f8 Accessed: 03.02.2021

Safonova, M. F., Reznichenko, D. S., Melnichuk, M. V., Karaev, A. K., & Litvinova, S. F. (2016). Taxes harmonization features in the European Union countries. *International Journal of Economics and Financial Issues, 6*(S8), 154–159.

Saint-Amans, P. (2016). Global tax and transparency: We have the tools, now we must make them work. OECD Observer. https://doi.org/10.1787/15615529. Accessed: 03.02.2021

Savin, D. A. (2012). Foreign experience of organizing tax control services. *Service in Russia and Abroad, 7*, 138–151.

Sinn, H. W. (1990). Tax harmonization and tax competition in Europe. *European Economic Review, 34*, 489–504.

Soloviev, A. I. (2017). Risk-oriented approach in the system of state control and supervision in the tax sphere. *Economics. Taxes. Right, 6*, 139–146.

Stratiev, O. (2018). Cryptocurrency and blockchain: How to regulate something we do not understand. *Banking & Finance Law Review, 33*(2), 173–212.

Tsepova, E. A. (2019). The legal nature of international tax transparency and its role in the development of financial relations with the owners of property abroad. *Eurasian Law Journal, 1*(128), 262–266.

United States Government. (2010). Public Law "Hiring Incentives to Restore Employment Act" (March 18, 2010 No. 111-147). Washington, USA.

Zelinsky, E.A. (2017). Defining residence for income tax purposes: Domicile as gap-filler, citizenship as proxy and gap-filler, 38 Mich. *Journal of International Law, 271*. http://repository.law.umich.edu/mjil/vol38/iss2/5. Accessed: 15.02.2021

Chapter 6
Medical Cluster as a Tool of Innovative Economy: Legal Aspects

Natalia S. Volkova and Natalia V. Putilo

Abstract The cluster as a special approach to the organization of innovative long-term business projects has successfully established itself in the economies of foreign countries. Currently, medical clusters (medical hubs, bio hubs) are used to develop breakthrough technologies in medical, genomic, and pharmaceutical activities and to attract investments in these areas. Being unified research, technological, and production infrastructure, the medical cluster allows providing a full cycle of creation and application of advanced technologies and the latest medical and pharmaceutical products. At the same time, due to the creation of special economic and legal regimes on the territory of the cluster, it is an attractive investment object and an effective tool for sustainable development of the territory. This work examines the features of the development of medical clusters on the territory of the Russian Federation. The analysis of legal obstacles to the development of innovative medical activities in the cluster is carried out; the legal possibilities of further intensification of the use of the cluster approach in medical practice are being studied. The article reveals the essential and specific features of medical clusters. It has been proven that the differences between the cluster and other institutional elements of health care are its innovative potential, innovative infrastructure, and special goal orientation. The last one implies that the ultimate goal of the functioning of the medical cluster is to improve the quality of medical care provided to the public. The study describes various forms of clusters operating on the territory of the Russian Federation; revealed their characteristic features, differences, and development prospects. It shows that in Russian realities the conceptual approaches to understanding the cluster are modified: the medical cluster has acquired the features of an artificially formed territory with a specific mode of activity. It is concluded that the prerequisites for the effective development of medical clusters have been created in the Russian Federation. The cumulative effect of their activities should contribute to the formation of innovative health care.

N. S. Volkova (✉) · N. V. Putilo
Deputy Head of the Department of Social Legislation Institute of Legislation and Comparative Law Under the Government of the Russian Federation, 34 Bolshaya Cheremushkinskaya St., 117218 Moscow, Russia
e-mail: n_volkova@inbox.ru

A. O. Inshakova and E. E. Frolova (eds.), *Smart Technologies for the Digitisation of Industry: Entrepreneurial Environment*, Smart Innovation, Systems and Technologies 254, https://doi.org/10.1007/978-981-16-4621-8_6

Keywords Medical cluster · Innovation activity · Medical activity · Investments · Cluster infrastructure · Innovative healthcare

JEL Codes I15 · I18 · K32 · O35

6.1 Introduction

Currently, the Russian state is facing the tasks of innovative development of the medical and pharmaceutical industry, improving medical care, and increasing the quality of medical services, one of the ways to achieve this is to create special regimes for objects of medical activity, including medical clusters. A number of strategic and policy documents of an economic orientation that have a direct impact on the formation of new economic and other relations (Tikhomirov, 2019), foresees the creation of special territories with specific status—clusters. Such territories are formed for the breakthrough development of certain spheres of life of the Russian state. As the analysis of legislative regulation shows, the number of areas in need of new organizational approaches in order to ensure the mobility of their adaptation to the modern needs of the economy and scientific and technological development is constantly expanding (Inshakova et al., 2020).

In recent years, in addition to the industrial sector, the "cluster movement" includes medicine, pharmaceuticals, IT, aircraft construction, tourism, etc. The Russian state has placed particular emphasis on the healthcare sector; the adoption in 2015 of a special law regulating the operation medical cluster in Moscow is evidence of this. In many ways, this interest is due to the importance that the state attaches to the innovative development of healthcare (Dudin et al., 2017) as well as the functional potential of the cluster. At the same time, in the conditions of Russian reality, the cluster approach in healthcare has acquired specific features that distinguish Russian clusters from classical foreign models.

The cluster as an institutional unit of high-tech business has been used long and very successfully abroad. It is a combination of geographically close interconnected organizations of diverse, but complementary functional orientation ("geographic concentrations of interconnected companies and institutions in a particular field") (Porter, 2000). As a rule, clusters unite companies of the same industry with a common infrastructure, suppliers, and distribution networks.

Geographic localization, the interconnectedness of the activities of the participants provide the cluster with certain competitive advantages due to the increased productivity of its member organizations due to the creation of a single infrastructure chain and facilitated communication between cluster members. These features allow clusters to successfully develop innovative industries and make them an attractive investment. At the same time, the effective activity of clusters becomes a factor in the territorial development of the localization region.

Despite the widespread use of this form of business organization, special legislation on the functioning of clusters and containing their legal definition is rare. At the same time, such legal regulation can be universal (for example, the Republic of Korea) or be personalized, establishing the legal regime of a specific cluster (for example, the Republic of Kazakhstan).

There is a special Industrial Cluster Development and Factory Establishment Act, 2003 in the Republic of Korea. It regulates, among other things, the creation and operation of an industrial cluster, which is understood as a cluster that creates synergy through mutual links by concentrating companies, research institutes, universities, and corporate-support facilities in a specific area. At the same time, the law provides not only the legal aspects of the organization and activities of the relevant industrial structures (factories, plants, clusters, etc.) but also measures to support them.

In the Republic of Kazakhstan, the functioning of the innovation cluster is provided for by the 2014 Law "On the innovation cluster" Park of innovative technologies." The law defines a cluster as an association of participants to stimulate industrial and innovative activities by interacting and sharing existing opportunities, sharing knowledge and experience, conducting research, effective technology transfer, establishing sustainable partnerships, and disseminating information. This act establishes the goals of the function of the innovation cluster, its legal status, and management features. At the same time, other innovative clusters can also function in Kazakhstan, formed in accordance with the Entrepreneurial Code of 2015.

In the Russian Federation, the range of acts regulating various aspects of the functioning of clusters in various sectors of the economy is very wide. The legislation operates with the concepts of "industrial cluster," "agrarian cluster," "oil and gas chemical cluster," "tourist cluster," etc. However, on the legal level, as an act of supreme power, only the status of industrial clusters (Federal Law of The Russian Federation of December 31, 2014, No. 488-FZ "On industrial policy in the Russian Federation") and the medical cluster (Federal Law of The Russian Federation of June 29, 2015, No. 160-FZ "About the international medical cluster and modification of separate legal acts of the Russian Federation") are determined. Thus, the Russian Federation combines universal and personal approaches to regulating the status of clusters.

In the European Union, the advantages of the cluster strategy for economic development are well studied and recognized. As highlighted by Regulation (Eu) No. 1287/2013 Of The European Parliament And Of The Council Establishing A Program For The Competitiveness Of Enterprises And Small And Medium-Sized Enterprises (Cosme) (2014–2020) And Repealing Decision No. 1639/2006/Ec, the clustering of small and medium-sized enterprises can be a key means of strengthening their capacity to innovate and to begin operating in overseas markets. At the same time, in European law, clusters are limited geographical areas with a relatively large number of firms and employees within a small number of related industrial sectors) (Regional Clusters in Europe, 2002).

The aforementioned advantages of the cluster approach are actively used in the formation of "innovative healthcare" as a special branch of the economy, where, in addition to achieving an economic effect, an important goal is to ensure a balance of multidirectional interests: the interest of the patient, the interest of a medical organization and the public interest of the state. At the same time, despite the obvious antagonism of these interests, one cannot but recognize the relationship between health and economic growth. On the one hand, health is an important factor in economic development; and on the other, economic growth has a significant positive impact on the health of the population, affecting the overall level of well-being (Romaniuk et al., 2016). These factors predetermine the importance of medical clusters, which, due to their essential characteristics, can become an effective tool for the innovative development of both healthcare and the economy as a whole.

6.2 Materials

A comprehensive analysis of the legal regulation of the activities of medical clusters on the territory of the Russian Federation is based on the study of the regulatory legal acts of the Russian Federation, including strategic and programmatic type. The established practice of legalizing medical clusters in the Russian Federation was investigated due to the provision of such acts.

The doctrinal positions that formed the theoretical basis of the research are based on materials from works related to the general understanding of the cluster as an institutional formation (Porter, 2000), economic, organizational, and legal problems of cluster development of healthcare (Dudin et al., 2017; Lebedinskaya & Yakovets, 2016; Melnikov, 2017; Romaniuk et al., 2016; Zhavoronkov, 2013, etc.); affecting the theoretical foundations of investment activities (Gutnikov, 2019, Inshakova et al., 2019–2021), as well as dedicated to the study of special legal regimes in the Russian Federation (Mokhov, 2017; Mokhov et al., 2018; Nozdrachev, 2017; Zyryanov & Kalmykova, 2020).

6.3 Methods

A review of the literature has shown that despite the many works devoted to various aspects of the activities of medical clusters, there are many unresolved issues. This is primarily due to the fact that most of the research is carried out from an economic, organizational, and managerial point of view. Legal scholars rarely turn to the study of the essential characteristics of the cluster. In this regard, the issues of legalization of the institutional foundations of the cluster, the peculiarities of the legal regulation of the activities of medical clusters, the sufficiency of legal measures to ensure the effective functioning of the cluster have not been properly studied.

Within the framework of the study, the methods of systematic analysis became basic (made it possible to compare information from the historical perspective and in relation to various legal systems), methods of structural and functional analysis (were used to different positions and their grouping according to various classification criteria), the hermeneutic method (helped to identify concepts and approaches present in the studied texts, including the texts of legal acts).

The informational and analytical basis of the study is made up of information materials on the official websites of the government, ministries, and departments.

6.4 Results

Medical clusters (medical hubs) are used to develop breakthrough medical and pharmaceutical technologies and attract investment in this area in the United States of America, Australia, Germany, France, Switzerland, Italy, and other developed countries (Klein et al., 2015). At the same time, in some European countries, medical clusters are rather an exception than a traditional tool of the medical business, which is often explained by the underestimation of the eventual capabilities of the cluster by potential participants (Romaniuk et al., 2016).

In the Russian Federation, for a long time, the cluster approach was not considered as fundamental for the development of the economy, although in the USSR production associations (organizations uniting industrial enterprises and scientific institutions), they were a very common organizational form of cooperation between science and production, including in healthcare. For example, in the 1987 document "The main directions for the development of public health protection and the restructuring of health care in the USSR in the twelfth five-year plan and the period up to 2000" as one of the directions for the development of health care, the expansion of the number of organizations uniting scientific centers, laboratories and industrial enterprises for intensifying implementation of scientific research results into practice.

In modern Russian legislation, the term "cluster" itself, implying the cooperation of organizations—suppliers of equipment, components, specialized production and maintenance services, research and educational organizations, has become widespread since 2006, when the Government of the Russian Federation approved the Program of Socio-Economic Development of the Russian Federation for 2006–2008. One of the goals of the program was the development of industrial clusters.

A legal basis for the development of the pharmaceutical cluster began to be created with the adoption of the Strategy for the Development of the Pharmaceutical Industry of the Russian Federation for the period up to 2020 by the Ministry of Industry and Trade of the Russian Federation in 2009. At the same time, the lack of a systematic approach in the legislation in terms of pharmaceutical clusters led to their formation as self-regulatory systems operating based on agreements between economic entities of the sphere and (or) business entities and authorities (Mokhov, 2017) focused mainly on increasing competitiveness the pharmaceutical market in a particular region.

At the present, certain conditions have been created for the formation of a cluster approach in healthcare. The essences are in the concentration on a small area of interconnected innovative industries, research and development bases and pharmaceutical organizations, involving the creation of a single infrastructure, additional jobs, and the achievement of economic benefits through coordinated activities and large-scale cooperation between scientific organizations, clinics, and industries.

There are different types of cluster-type associations in the Russian Federation: pharmaceutical clusters, scientific and educational medical clusters, an international medical cluster, world-class genomic centers, etc. Such institutional diversity entails a wide discussion in the literature about what a medical cluster is.

In the Federal Law "About the international medical cluster and modification of separate legal acts of the Russian Federation," a medical cluster is defined as a kind of aggregate that includes the territory of the cluster, its members, and mechanisms of interaction. This rather abstract legal definition has started a discussion in the scientific literature about what the medical cluster is. Despite the generality of approaches to understanding the key parameters (geographic proximity, interconnection and complementarity, industry affiliation), a universal definition of the concept of "medical cluster" has not been developed in the literature. At the same time, we are not talking only about the medical cluster in the context of the aforementioned law—the definitions formulated in the literature are of a general nature.

As a rule, all scientific definitions are based on an organizational principle: a medical cluster is defined as a "group of interrelated organizations" (Zakharova & Kovaleva, 2013), as "an association of competitive organizations" (Zhavoronkov, 2013), as an "organizational environment" (Tolstopyatenko & Zinkovskaya, 2013), as an "innovative model for organizing medical care" (Meshcheryakova, 2017) and so on. There is also a broader approach to understanding the cluster as a "functional economic system" (Balashov et al., 2018). Meanwhile, the basic features of the medical cluster are obvious which makes it possible to isolate it among other institutional elements of the health care system.

First of all, it is necessary to highlight the essential characteristics of such cluster that reflect its nature: (1) functional unity and interrelation of the activities carried out by the subjects of the cluster; (2) territorial proximity of the subjects of the cluster; (3) synergy (interaction) of the subjects of the cluster, which generates the effectiveness of the cluster. All these points are fully inherent in the medical cluster.

However, there are also specific characteristics, compliance with which makes it possible to position one or another institutional element of the economy as a medical cluster. First of all, we are talking about the innovative potential of the cluster—the creation and implementation of innovative projects are prerequisites for the functioning of the cluster. The result of the cluster's activity can be an innovative drug product, innovative medical technology (Inshakova et al., 2021), an innovative system of continuous professional education of medical and pharmaceutical workers, etc. This also implies the presence of innovative infrastructure, i.e., infrastructure, equipment, laboratories, production facilities, software, and information support, etc., that meet the modern needs of science and technology.

Another important feature of the medical cluster is in its goal orientation. The final goal of the functioning of the medical cluster is to improve the quality of medical care provided to the population. This means that the effect of synergy is expected not only from the commercial benefit of the cluster subject but also on the creation of a public good in the form of preserving (maintaining) the health of consumers of services, products, technologies created by the subjects of the cluster.

As a result, the cumulative effect of the clusters should contribute to the formation of innovative health care that meets both the interests of subjects of medical activity, consumers of their services and products, and the state that has a social function. This means that the medical cluster has every reason to be recognized as an instrument of the innovative economy.

The presence of a wide range of collaborations, bearing the name "cluster" and, being one, provides a basis for the typology of clusters in health care. Based on such a parameter as the specificity of activities, the following are distinguished: (1) clusters of the pharmaceutical industry, the purpose of which is to increase competitiveness in the pharmaceutical sector; (2) scientific and educational medical clusters targeting training personnel that meets modern health care needs; (3) medical clusters providing general medical services (Melnikov, 2017). Another classification is based on the nature of the services provided within the frame of the cluster: (1) clusters of medical services, (2) clusters of medical equipment and medicine supply, (3) clusters of biotechnology and innovations in medicine (Meshcheryakova, 2017). Meanwhile, the proposed typologies are not complete, since the state: first, constantly modifies the legislation that regulates the functioning of clusters, blurring clear boundaries between clusters of different types. Secondly, it stimulates the emergence of new forms of unification of sectoral and inter-sectoral structures that aim at the development and implementation of innovative products and technologies.

Initially, due to the Strategy for the Development of the Pharmaceutical Industry of the Russian Federation for the Period up to 2020, the pharmaceutical clusters in the industry (pharmaceutical clusters) were legalized. In the legal system, a pharmaceutical cluster is understood as a group of geographically localized interconnected innovative firms—medicine developers, manufacturing companies; equipment suppliers, components, specialized services; infrastructure facilities: research institutes, universities, technoparks, business incubators, and other organizations that fulfill each other and enhance the competitive advantages of individual companies and the cluster as a whole. Due to the territorial specifics of the Russian state, pharmaceutical clusters seek regionalization and largely depend on the innovative potential of a particular region, the willingness of local authorities to support cluster members. The requirements of the legislation on the availability of a formed infrastructure, highly qualified personnel and an educational environment are added to this. As fairly noted in the literature, it is critical to link with science, the creation of new products, and their subsequent commercialization, rather than the production of a finished product for a pharmaceutical cluster (Mokhov, 2017).

Scientific and educational medical clusters are formed as part of the implementation of the Strategy for the Development of Medical Science in the Russian Federation for the Period up to, 2025, approved by the Government of the Russian Federation in 2012. These clusters represent an association of educational and health care organizations to interact to solve strategic problems in the field of public health protection.

In the literature, scientific and educational medical clusters are considered as aggregate structures of the health care system, which create the basis for the integration of joint projects on actual problems of medicine (Vardomatskaya & Kuznetsova, 2019).

In such clusters, due to the combination of research potential, faster exchange of technologies and information creates an opportunity for continuous generation of innovations. The synergy effect that appears should contribute to the emergence and strengthening of new ties between cluster members, the exchange and sharing of resources and technologies, which should lead not only to the emergence of breakthrough results but also to reduce costs (Vardomatskaya & Kuznetsova, 2019). However, in its essence, such clusters are not business units, they are rather an "intellectual union" of the resources, leading educational and scientific organizations of medical focus to develop innovative educational programs for the training of highly qualified personnel. In this regard, their investment potential is not high but can play a significant role in raising the level of training of medical and pharmaceutical personnel focused on the use of the latest scientific technologies and developments in medicine and pharmaceuticals.

Medical clusters created as part of the implementation of the Federal Law "About the international medical cluster and modification of separate legal acts of the Russian Federation" have a different focus. It should be mentioned that currently, only one International Medical Cluster (IMC) is functioning in the Russian Federation—in Moscow, although there are fundamentals for the creation of similar projects in other territories. For example, the Strategy for the Development of Tourism in the North Caucasian Federal District until 2035 mentions the implementation of a project to create an innovative international medical cluster, including the construction of a high-tech federal medical center, specialized clinics, and rehabilitation centers in the territory of the Karachay-Cherkess Republic.

The objectives of the creation and functioning of the IMC are (a) the development of medical activities for the provision of medical care, (b) improving its quality; (c) promoting the development of drugs for medical use, medical technologies, and medical devices; (d) developing educational activities and conducting scientific research in the sphere of health protection; (e) development of international cooperation.

The priority task of modern health care, in general, is the achievement of these goals, and the advantages that legislation provides to cluster members lays the foundation for a joint search by cluster members (scientific, educational, medical, industrial organizations) for breakthrough innovative solutions in medicine and pharmaceuticals for the subsequent their consistent introduction into medical practice throughout the country (Mokhov et al., 2018).

Analysis of the legislation governing the activities of the IMC shows that, in general, a special (preferential) regime for the implementation of medical activities is being created on its territory, which consists: (1) rejection of certain instruments of state control; (2) admission to the implementation of activities by foreign organizations on the basis of permits from foreign states; (3) carrying out activities following an agreement with the management company; (4) admission to the use of drugs and medical devices that are not duly registered on the territory of the Russian Federation but have permits from foreign member states of the Organization for Economic Co-operation and Development, OECD; (5) exemption from taxation of medical services provided by IMC members, etc.

As rightly noted in the literature, the essence of a special regime is to provide subjects with the right of various preferences: additional rights, state guarantees, various favorable opportunities for them to exercise their rights and private interests (Nozdrachev, 2017). The main goal of introducing such a regime is to attract potential investors and participants who will be spared from bureaucratic costs and additional costs when obtaining permits and carrying out activities on the territory of the cluster.

Meanwhile, studies show that such special regimes, aimed at stimulating entrepreneurial activity and attracting investment, often do not bring immediate tangible results. This is due to the fact that, despite the presence of economic indulgences, the main method of the regulation still is state regulation by administrative and legal means (Zyryanov & Kalmykova, 2020), and this means that a certain number of barriers remain, reducing the entrepreneurial interest of potential investors.

One of such barriers is the imposition of several legal regimes in the territory where the medical cluster operates. The IMC is created on the territory of the Skolkovo innovation center, the status of which is regulated by Federal Law No. 244-FZ of September 28, 2010 "About the Skolkovo Innovation Center." This Law sets out the specifics of carrying out activities at Skolkovo. For example, medical activity on the territory of Skolkovo does not require a license, unlike IMC, where OECD permits are not enough to carry out certain types of medical activities (for example, related to blood transfusion) and a Russian license is required. In these conditions, transparency of permitting mechanisms, clarity, consistency of legislation (Gutnikov, 2019), lack of competition in legal regimes are the key to the investment attractiveness of the cluster.

It must be mentioned that the cluster approach in the development of medicine and pharmaceuticals in the Russian realities has acquired a certain specificity that distinguishes it from the well-established foreign approaches. In the functioning of medical clusters abroad, the main emphasis is paid to the geographical and infrastructural proximity of scientific centers and industries, the interconnected and fulfillment activities of which allow the implementation and application of innovative developments in medicine and pharmaceuticals. As a rule, clusters are naturally formed around leading research centers and universities, a kind of innovation territory.

In the Russian understanding, with the preserved essence of the medical cluster as a geographically limited concentration of functionally interconnected organizations, the IMC acquired the features of an artificially formed territory with a specific mode of activity.

At the same time, there is another very specific form of cooperation between science and industry in Russia, the so-called centers of the world standing. The basis for their creation is the national project "Science" and the Act of the Government of the Russian Federation of 2019 "On measures of state support for the creation and development of world-class scientific centers." Such centers are being established in three areas: mathematics, genomic research, and priority scientific research. In 2020, three world-class genomic centers were founded (within the framework of the Science project and the Federal Scientific and Technical Program for the Development of Genetic Technologies for 2019–2027) in four main areas: biosafety, medicine, agriculture, and industry. Genomic centers are consortia that unite the potential of research institutes, universities, and organizations of the real sector of the economy working in the field of genetic research.

An important feature of the activities of such genomic centers is that the efforts of state organizations and within the framework of predominantly state funding (in the form of grants) are cooperated, which means state control and a kind of ideological dictate in the form of the state's predetermination of priorities in activities. In the classical sense, the aforementioned genomic centers are not clusters; however, their sequential modification into more traditional forms of clusters cannot be ruled out (Matytsin, 2021). This is because the acts regulating the status of genomic centers do not exclude the attraction of non-state funding as well as the gradual expansion of the scope of activities of these centers. Currently, they are at the stage of formation, and as the scale of their activities increases, there is a possibility of their transformation into more global cluster structures.

Thus, today there are certain institutional and organizational problems of the functioning of medical clusters.

First of all, they are associated with the lack of consistency of legal regulation. In the context of a variety of forms of medical clusters, only the activities of the IMC have a transparent and understandable regulation. At the same time, its functioning is also not free from organizational and legal problems arising from the constant modification of the relevant law. Often, such changes are significant and can affect the stability of the activities of IMC members. Concerning other forms of clusters, legislation is usually schematic, reflecting mainly financial issues.

In addition, the limited tools and insufficient mechanisms for financing the development of the cluster at the initial stage are obvious. These problems are caused by the lack of a systematic approach to regulation to a large extend. In such conditions, the investment attractiveness of the cluster decreases, and its competitive advantages are leveled. In this regard, the state sets new tasks focused on the transition of the pharmaceutical and medical industry to an innovative development model.

6.5 Conclusion

The legislation of the Russian Federation regulating relations in connection with the creation and functioning of international medical clusters is currently in the stage of intensive development. The emergence of numerous forms of cluster organization of medical activity indicates that the state is actively looking for the most effective models for organizing innovative healthcare. Variations of medical clusters are diverse: from interdisciplinary cooperation to improve the quality of medical education to independently functioning territories with an innovative environment. At the same time, the state and its institutions play an important role: they must ensure the creation of the necessary infrastructure, contribute to building relationships between the subjects of the cluster, and determine measures of state support.

Numerous legislative initiatives aimed at changing the current legislation on the medical cluster indicate the intensification of the use of the cluster approach in medical practice. This is mostly because the cluster as a tool for the development and improvement of medical and pharmaceutical activities, testing of innovative technologies is an attractive business project, including for foreign participants. In addition, the creation of clusters is an effective mechanism for the development of territories, since the interest of the business community in the region where the cluster is created and operates is stimulated due to the emergence of additional legal, institutional, and economic opportunities.

The cluster allows to concentrate technological and production capacities, research and development, and advanced human resources within one territory, which contributes to the development of medical technologies that meet modern scientific achievements, the emergence of new medical services and goods, the creation of the newest medicines, and the facilitated mode of implementation of developments contributes to the economic attractiveness of clusters as an investment object. Thus, the medical cluster is a promising tool for an innovative economy and innovative healthcare.

References

Decree of the Government of the Russian Federation of April 30, 2019, No. 538 "On measures of state support for the creation and development of world-class scientific centers" http://publication.pravo.gov.ru/Document/View/0001201905060038 Accessed: 29.12.2020.

Dudin, M. N., Sertakova, O. V., Frolova, E. E., Artemieva, J. A., & Galkina, M. V. (2017). Development of methodological approaches to assessing the quality of healthcare services. *Quality—Access to Success, 18*(158), 71–78.

Entrepreneurial Code of the Republic of Kazakhstan 2015. (2021). https://online.zakon.kz/document/?doc_id=38259854. Accessed: 02.02.2021.

Gutnikov, O. V. (2019). Trends of development of the investment legislation In The Russian Federation. Investments in National and International Law: Balance of Public and Private Interests: The Collection of Materials for the XIV Annual Scientific Readings in Memoriam of Professor S. N. Bratus, L. V. Andreeva, L. V. Andrichenko, V. R. Bairashev, etc.; The Institute of Legislation and Comparative Law under the Government of the Russian Federation. Moscow (pp. 165–179) (in Russian)

Industrial Cluster Development and Factory Establishment Act, Korea (2003). https://elaw.klri.re.kr/kor_service/lawViewTitle.do?hseq=13125 Accessed: 02.02.2021.

Inshakova, A. O., Goncharov, A. I., Ershova, I. V. (2020). Digital blockchain registration of legally significant stages of complex good's export-import supplies by business entities of the EAEU and BRICS Jurisdictions. In A. Inshakov, E. Inshakova (Eds.), *Competitive Russia: Foresight model of economic and legal development in the digital age. CRFMELD 2019*. Lecture Notes in Networks and Systems (vol. 110, pp. 328–336). Springer. https://doi.org/10.1007/978-3-030-45913-0_37

Inshakova, A. O., Baltutite, I. V., Epifanov, A. E., & Abesalashvili, M. Z. (2021). Biotechnology as a type of converged technologies in Industry 4.0 and a source of increased danger in civil

law of the Russian Federation. Modern global economic system: Evolutional development versus revolutionary leap. Lecture Notes in Networks and Systems (pp. 1558–1567).
Klein, Th., Banga, B., & Martelli, A. (2015). The seven most important Medtech clusters in Europe, 2015. https://www.mddionline.com/seven-most-important-medtech-clusters-europe-0. Access: 29.12.2020.
Federal Law of The Russian Federation of 28 of September 2010, No. 244-FZ "On Innovative Center, "Skolkovo" http://pravo.gov.ru/proxy/ips/?docbody=&nd=102141548 Accessed: 29.12.2020.
Federal Law of The Russian Federation of December 31, 2014, No. 488-FZ "On industrial policy in Russian Federation" http://pravo.gov.ru/proxy/ips/?docbody=&nd=102365303 Accessed: 29.12.2020.
Federal Law of The Russian Federation of June 29, 2015, No. 160-FZ "About the international medical cluster and modification of separate legal acts of the Russian Federation" https://cis-legislation.com/document.fwx?rgn=76881 Accessed: 29.12.2020.
Law of the Republic of Kazakhstan 2014 "On the innovation cluster" Park of innovative technologies" https://online.zakon.kz/document/?doc_id=31565313 Accessed: 02.02.2021.
Lebedinskaya, Yu. S., & Yakovets, O. N. (2016). Medical cluster: Concept and special features. The azimuth of Scientific Research. *Economics and Administration, 2*, 170–173. (in Russian).
Matytsin, D. E. (2021). Internet-investing as a remote algorithm of retail investment financing. Modern global economic system: Evolutional development versus revolutionary leap. Lecture Notes in Networks and Systems. Springer Nature.
Melnikov, A. Y. (2017). Financial Mechanism of the Healthcare Field's Cluster Development: Research of Foreign and Domestic Experience. *Bulletin of Chelyabinsk State University, 10*, 88–93. (in Russian).
Meshcheryakova, Zh. V. (2017). Medical cluster in the health system of the region as an innovative model of integration of subjects of public-private partnerships. *Economy and Business: Theory and Practice, 3*, 43–45. (in Russian).
Mokhov, A. A. (2017). Pharmaceutical clusters: Theory, legislation, practice. *The Lawyer, 10*, 41–46. (in Russian).
Mokhov, A. A., Balashov, A. E., Shevchenko, O. A. et al. (2018). *Innovation clusters: Doctrine, legislation*. In A. A. Mokhov (Ed.), KONTRAKT [Contract], 100. (in Russian)
Nozdrachev, A. F. (2017). Modern contents of the theory "Administrative and Legal Regime." *Journal of Russian Law, 2*, 98–99. (in Russian).
Porter, M. E. (2000). *Locations, clusters, and company strategy*. Oxford University Press.
Regional clusters in Europe. (2002). Observatory of European SMEs 3/2002. https://ec.europa.eu/regional_policy/archive/innovation/pdf/library/regional_clusters.pdf
Regulation (Eu) No. 1287/2013 Of The European Parliament And Of The Council Establishing A Programme For The Competitiveness Of Enterprises And Small And Medium-Sized Enterprises (Cosme) (2014 - 2020) And Repealing Decision No. 1639/2006/Ec https://eur-lex.europa.eu/legal-content/EN/TXT/?uri=CELEX%3A32013R1287. Accessed: 29.12.2020.
Romaniuk, P., Holecki, T., & Woźniak-Holecka, J. (2016, June 21). Using new instruments of clustering policy in the health care system. The Case of Poland. *Frontiers in Pharmacology*. https://doi.org/10.3389/fphar.2016.00177 Access 29.12.2020.
Strategy for the Development of Medical Science in the Russian Federation for the period up to 2025 (order of the Government of the Russian Federation dated December 28, 2012, No. 2580-r) http://government.ru/docs/3/ Accessed: 29.12.2020.
The main directions of the development of public health protection and the restructuring of health care in the USSR in the twelfth five-year plan and for the period up to 2000 (Resolution of the Council of Ministers of the USSR dated November 19, 1987, No. 1318).
The program of socio-economic development of the Russian Federation for 2006–2008 (order of the Government of the Russian Federation of January 19, 2006, No. 38-r) http://old.economy.gov.ru/minec/activity/sections/strategicplanning/doc20060118938 Accessed: 29.12.2020.
The strategy for the development of the pharmaceutical industry of the Russian Federation for the period up to 2020 (order of the Ministry of Industry and Trade of the Russian Federation

of 23.10.2009 No. 965) https://minpromtorg.gov.ru/common/upload/files/docs/MinProm_02.06.14.pdf. Accessed: 29.12.2020.

Tikhomirov, Yu. A. (2019). Control vectors in the focus of law. *Public Administration Issues, 1*, 136–159. (in Russian).

Titova, NYu., & Titova, Yu. V. (2017). Medical cluster: Concept and features. *The Azimuth of Scientific Research: Economics and Administration, 4*, 227–229. (in Russian).

Tolstopyatenko, M. A., & Zinkovskaya, N. V. (2013). Pharma-medical cluster: A model and formation problems. Collection of Scientific Works of Russian Universities "Problems of Economy, Finance, and Production Management", 33, 94–102 (in Russian).

Tourism development strategy in the North Caucasus Federal District until 2035. (2020). (Order of the Government of the Russian Federation dated 07.03.2019 No. 369-r) http://government.ru/docs/36037/. Accessed: 29.12.2020.

Vardomatskaya, L. P., & Kuznetsova, V. P. (2019). Scientific and educational innovative medical cluster as an aggregate structure of health care system. *State and Municipal Management. Scholar Notes, 2*, 105–111. https://doi.org/10.22394/2079-1690-2019-1-2-105-111

Zakharova, E. N., & Kovaleva, I. P. (2013). Formation of the medical cluster as the direction of integration interaction of subjects in regional medical area. *The Bulletin of Adyghe State University: Internet Scientific Journal, 4*, 216–222. (in Russian).

Zhavoronkov, E. P. (2013). Cluster strategy in development of medical organizations. *Medicine and Education in Siberia, 1*, 1. (in Russian).

Zyryanov, S. M., & Kalmykova, A. V. (2020). Public regulated models for business. *International Journal of Economics and Business Administration, VIII*(SI 1), 84–91.

Chapter 7
International Cooperation in the Field of Economic and Legal Regulation of End-To-End Technologies

Maryia V. Miashchanava

Abstract The scientific research purpose of this article is to analyze various manifestations of international cooperation aimed at bridging the global challenge of end-to-end digitalization. It is concluded that the integration factor is one of the key factors in ensuring the sustainable growth of national economies in the context of digital transformation. Various mechanisms of international cooperation in the field of bridging the digital divide and equitable distribution of the benefits of using end-to-end technologies among all countries were studied. Barriers to the integration of developing countries into the digital economy, as well as ways to eliminate these barriers, were determined. It is justified that the economic crisis caused by COVID-19 in 2020 leveled the efforts of the international community to bridge the digital divide between countries. A way out of this situation is seen in operating a policy of multilateralism, taking into account the interests of states with digital lagging, as well as in reforming the system of WTO agreements, which should take into account the need to expand the space for development policy. The documents adopted within the framework of the Eurasian Economic Union and aimed at the development of digital technologies were considered. It is concluded that the implementation of the "Own Center of Power" scenario fully meets the purposes of preserving the digital sovereignty of the Union member states, which will allow them to adequately representing their common interests within the framework of the global digital agenda.

Keywords End-to-end technologies · Synergistic effect · Multilateralism · International technology transfer · Digital platform · Digital infrastructure · Digital divide

JEL Codes F15 · K15 · K24 · O14 · O33

M. V. Miashchanava (✉)
Department of Civil Law, Belarusian State University, 4 Nezavisimosti Avenue, 220030 Minsk, Belarus
e-mail: civillaw@bsu.by

A. O. Inshakova and E. E. Frolova (eds.), *Smart Technologies for the Digitisation of Industry: Entrepreneurial Environment*, Smart Innovation, Systems and Technologies 254, https://doi.org/10.1007/978-981-16-4621-8_7

7.1 Introduction

The modern world economy is at the stage of sweeping structural transformations, which predetermines the need for multifaceted international cooperation in the field of end-to-end technologies, which are the main driving force behind such transformations. According to the Deputy Chairman of the Board of the Eurasian Development Bank (hereinafter referred to as "EDB") Tigran Sarkisyan: "It is necessary to develop such digital solutions and products that will be in demand in neighboring countries and will allow building a normal infrastructure for mutual trade, for cooperation projects. It is simply impossible to close within the framework of one state. It is also impossible to develop standards in the field of digitalization for only one country. The standards must be international, understandable for everyone to ensure interoperability" (BELTA, 2021)

The most accepted way to consolidate agreements, achieved at the interstate level, is the adoption of international treaties. However, as you know, international treaties go through a rather difficult and long way from the moment the initiative for the development of the corresponding project emerges, then the opening for signing and, finally, the expressing consent by each state, signed a treaty, to be bound by this international document. In addition, domestic procedures that implement the provisions of treaties into domestic law cannot be disregarded. Even if the state accepts the direct application of the provisions of international treaties, "the national system of law must, through its general implementing rule, and in some cases also through special reference, authorize the specified direct use of the provisions of international treaties" (Miashchanava, 2020a). As a result, the international treaty, despite its effectiveness, is simply not able to keep up with the digitalization processes that are evolving at super speeds. Other, more flexible, mechanisms of international cooperation come to the fore, including the adoption of non-regulatory and program acts within the framework of international organizations and supranational associations. In this regard, in the framework of this study, practically no attention was paid to the regulatory impact of international treaties on the field of end-to-end digitalization.

Currently, many well-established concepts and categories are revised, and many challenges require the decisive intervention of the international community. Thus, researchers of digital problems draw attention to the need to create a "global system of international acts, regulating most of the existing issues of human interaction in cyberspace and including a clear mechanism for rapid response to the emergence of new relations in this area [...] the concept of sovereignty in international law will include not the only geographic territory, but also a scope of potential actions and opportunities, that a particular state has the right to realize, as a subject in cyberspace" (Mamychev et al., 2019).

On June 11, 2020, the Report of the Secretary-General "Roadmap for Digital Cooperation" (2020), was officially released to the public. It assesses the situation in the field of digitalization and global digital cooperation and notes the corresponding gaps and proposes measures to consolidate international cooperation. In 2019, the

High-Level Group on Digital Cooperation put forward a number of recommendations, the main focus of which was to overcome the negative consequences of digitalization processes, such as cyber threats, digital inequality, violation of human rights on the Internet, including infringements on personal data. "Roadmap for Digital Cooperation" aims to outline ways to solve these problems. It contains eight directions for action to create a safer and more equitable digital world in the context of the Sustainable Development Goals. The designated directions include:

(1) ensuring Internet access for every adult by 2030;
(2) the creation of digital public goods, including open-source software, open data, standards, and content;
(3) ensuring universal digital technology coverage, closing various digital divides;
(4) capacity development in the field of digital technologies through the development of skills and abilities of the population;
(5) protection of human rights in the digital age through the development of appropriate regulatory frameworks and legislation;
(6) supporting global cooperation in the field of artificial intelligence;
(7) ensuring trust and security in the digital environment;
(8) building a more effective architecture of global digital cooperation, including overcoming the consequences of the coronavirus pandemic and contribute to the achievement of sustainable development.

To implement each of the indicated directions, both the efforts of individual states and measures to be taken by the international community are required. In this regard, the High-Level Group proposed the creation of the place of minister of UN Secretary-General for technology in 2021, drawing attention to the need to ensure the development of mechanisms for global digital cooperation and interaction between the UN system and the technology industry.

Each of the named eight directions of international cooperation is of parity importance, and while within the framework of this study, it seems reasonable to pay attention to the following positions.

1. Bridging the digital divide and equitable distribution of the benefits of end-to-end technologies among all countries.
2. The synergistic effect of regional integration in the field of digitalization on the example of cooperation between the EAEU member states.

7.2 Materials

In order to determine the priority directions of this study, the topic of state sovereignty in cyberspace highlighted in the publication by Mamychev et al. (2019) was touched.

Eight directions of action for creating a safer and more equitable digital world in the context of the Sustainable Development Goals, defined in the report of the UN Secretary-General entitled "Roadmap for digital cooperation," 2020, served as the basis for highlighting the most vital tasks of international cooperation in the field of economic and legal regulation of end-to-end technologies.

The problem of bridging the digital divide and equitable distribution of the benefits of using end-to-end technologies among all countries was considered based on the analysis of the annual UNCTAD "Trade and Development Report" for the period of 2018–2020 (UNCTAD TDR, 2018, 2019, 2020). The 2018 report paid special attention to bridging the digital divide, the 2019 report demonstrates some positive changes in this direction, but there were no improvements in 2020 due to the difficult epidemiological situation, which led to serious economic consequences.

The problem of digital inequality in relation to developing countries was considered in economic research by Smirnov (2019), who highlighted the main prerequisites for the emergence of such inequality, and also proposed ways to bridge it.

International technology transfer (hereinafter referred to as "ITT") is considered an effective tool for digital convergence. The problems of the modern use of this tool can be traced in the studies of Kowalski et al. (2017).

Cooperation of states in the field of international trade, in terms of tariff, non-tariff regulation, application of trade preferences, is mainly accumulated within the framework of the agreements of the World Trade Organization (hereinafter referred to as "WTO"). The possibility of using preferential trade agreements to bridge the digital divide is considered through the lens of the works of Rodrik (2018).

The study touches on the politics of multilateralism as a likely way of overcoming the consequences of the economic crisis caused by COVID-19. The idea of operating such a policy, based on the concerted actions of a number of states to achieve common goals, appeared in the early 90s of the twentieth century and is presented in the works of Ruggie (1992). The current view on the development of the digital economy and the information society is analyzed in the focus of the works of Frolova et al. (2018), Inshakova et al. (2020), Matytsin and Rusakova (2021), and Tarakanov et al. (2019).

The analysis of cooperation between the member states of the Eurasian Economic Union (hereinafter referred to as "EAEU") in the field of economic and legal regulation of end-to-end technologies was carried out on the basis of regulatory and program acts adopted within the framework of this integration association. The normative basis for EAEU actions is the Treaty on the Eurasian Economic Union of 05.29.2014 (revised at 10.01.2019) (hereinafter referred to as "EAEU Treaty"). In addition, a number of decisions of the Supreme Eurasian Economic Council were considered, the Main Directions for the implementation of the digital agenda of the Eurasian Economic Union until 2025 were analyzed, and examples of the practical implementation of these Main Directions were given.

7.3 Methods

The study is based on the general theory of systems, which helps to consider the interaction of the processes of economic and legal regulation of end-to-end technologies and international cooperation aimed at overcoming the negative developments of digitalization. Among the universal methods were also used: the method of dialectical materialism, the phenomenological, axiological, and hermeneutical methods. In addition, in the work, the method of comparative studies and a specific sociological method necessary for assessing risks in digitalization processes in relation to individual states and the international community as a whole are used.

7.4 Results

International cooperation in the field of economic and legal regulation of end-to-end technologies is multi-aspect and requires detailed analysis. The beginning of the third decade of the twenty-first century was marked by many cardinal changes and shocks for the world community: Health protection and overcoming the consequences of the economic crisis caused by COVID-19 became a priority; the values of globalization were replaced by the values of multilateralism. In this work, we focus on such a manifestation of international cooperation in the field of digitalization as bridging the digital divide between different countries, taking into account the special situation of developing countries. We will also pay attention to cooperation within the framework of regional integration associations, using the example of EAEU, aimed at the steady and even development of digital technologies in the member states of the association. International cooperation in the field of economic and legal regulation of end-to-end technologies is not limited to these directions, however, it seems that the digital lagging and the subsequent loss of "digital sovereignty" are one of the most significant political and economic threats of our time, and overcoming this threat is possible precisely through exploiting advantages of macro-regional and regional international cooperation.

7.4.1 Bridging the Digital Divide and Equitable Distribution of the Benefits of Using End-to-End Technologies Among All Countries

The world community at present is making significant efforts to bridge all forms of inequality between countries and, above all, economic inequality, which inevitably affects the political processes in the world. However, over the past five years, there has been a serious escalation of digital inequality, which has an impact on the position of developing countries and requires immediate intervention from the international community and a policy of equitable distribution of the benefits of using end-to-end technologies among all countries. Depending on the level of development of a country, its unpreparedness for end-to-end digitalization processes can take several forms—from a lack of skills and infrastructure to non-existent or fragmented adaptation policy—and can have many adverse effects, including further lagging in the technological level, stopped economic development, or even marginalization from the global economy (UNCTAD TDR, 2018).

The following is accepted as barriers to integrating developing countries into the digital economy: "the lack of development of digital infrastructure in developing countries, namely, broadband infrastructure, information and communication technologies (ICT), software, data" (Smirnov, 2019). To solve each of these problems, large-scale public investments are required, as well as an increase in the general level of digital literacy of the population and "professional digital literacy."

An important component in closing the digital divide is a policy of "digital sovereignty" combined with ITT. Some countries (for example, Vietnam and Indonesia) adopted a "localization" policy, which provides for the requirement to store data on servers, located within the borders of the state to which this data belongs. This policy stimulates foreign investment in digital infrastructure and its development.

For developing countries, ITT is of particular importance, since end-to-end digitalization must directly touch the production field in order to close the divide. The import of technology is traditionally accompanied by a whole set of promotional tools, for example, licensing, the foundation of joint ventures, and conclusion of investment agreements. At one point, such efforts were very effective, but in the context of digitalization, technologies, and data became more and more is the subject of commercial secret (Kowalski et al., 2017). Commercial secret protection provisions are incorporated into trade agreements and investment promotion and mutual protection agreements, which limits ITT. At the same time, a positive factor affecting ITT is the so-called policy of technological neutrality, which includes a set of measures, aimed at prohibiting preferences for local technologies. Such a policy may also result from the WTO requirements although the question of whether "technology neutrality" is a mandatory requirement of the General Agreement on Trade in Services remains open (Inshakova et al., 2020).

When carrying out international trade, one can't help but take into account the mechanisms of preferential trade agreements (hereinafter referred to as "PTA"); however, in terms of bridging the digital inequality, they are unlikely to play a significant role, since they mainly level obstacles for exporters (Rodrik, 2018).

The 2018 annual report of the United Nations Commission on Trade and Development (UNCTAD) focuses on ways to bridge the digital divide for developing countries. At the same time, the following most important directions are traditionally highlighted:

- Building a digital infrastructure.

In this regard, UNCTAD pays attention not only to the need to create a "physical" infrastructure, but also to ensure access to it for the general population through the support of the financial and banking sectors, as well as to the inadmissibility of internal imbalances that may emerge, for example, between urban and village residents.

- Development of a national regulatory/data protection policy.

Currently, such a policy does not exist in most developing countries and, de facto, the data belongs to whoever collects and stores it. These entities are mainly foreign digital supergiants, who then acquire full, exclusive, and unlimited rights to use the data obtained. National data policy, according to UNCTAD, should be developed to solve four main problems:

- who can accumulate data;
- what methods can be used to collect data;
- who can use the collected data;
- under what conditions the data can be used.

It is also necessary to consider the problem of "data sovereignty" associated with the answer to the question—what data can leave the country and not be regulated by national legislation.

- Regulating digital platforms and developing national marketing platforms.

Regulating digital platforms is important for developing countries as these platforms allow collecting benefits of e-commerce. At this stage, super platforms dominate, which accumulate a large amount of data, thereby flourishing and facilitating the access of foreign companies-owners of the super platform to domestic markets (Tarakanov et al., 2020). Stricter regulation of limitation of competition; prohibition of the activities of large firms responsible for market monopolization; creation of regulated digital platforms as public enterprises with direct public provision of digitized services; strict monitoring and administration at the international level—these are some of the options of regulating the activities of super platforms in the interests of developing countries.

- Taxation of super companies.

It is proposed to tax such companies at the place of their main activities, and not at the location of the administrative center/headquarters. This measure is intended to redistribute tax revenues and increase public revenues in developing countries.

- Development of digital industrial policy.

For developing countries, initially, it is necessary to ensure control of data, as well as limit the monopolization of the market by super platforms. The next step can be the formation of a digital industrial policy, which should be aimed at expanding the use of digital technologies and digital services in production, as well as developing digital competencies in all sectors (Matytsin & Rusakova, 2021).

- Using digital startups.

Digital startups are gaining increasing recognition, including in developing countries. At the same time, UNCTAD recommends a more comprehensive digital start-up policy aimed at overcoming the trend of innovation "flowing out" from the developer country. Recommended practices include direct investment by governments in corporate stock to support digital innovation, increased use of advanced technologies, and promotion of "reverse innovations." The phenomenon of "reverse" ("reverse") innovations emerged back in the 90s but gained the greatest recognition precisely during the period of digitalization: Scientific and research activities to create an innovative product are carried out in a developing country, with a focus on the relevant market. At the same time, as practice shows, such innovative products can subsequently interest consumers in developed countries. "Reverse" innovations can also be a component of the internationalization strategy of companies in large developing countries: "At the initial stage, they satisfy the demand in the domestic market, and then enter the 'customized' market segment in a developed country" (Smirnov, 2019).

- Development of digital competencies.

Digital competency development activities should be comprehensive and goal oriented. It should cover all levels of education—from secondary (basic) to university and also include various options for training and advanced training for adults.

Thus, back in 2018, the main directions for the activities of developing countries to bridge the digital divide (UNCTAD TDR, 2018), to operate a coordinated policy aimed at development, were determined. In 2019, there were some positive changes noted in the UNCTAD annual report (UNCTAD TDR, 2019), but 2020 was a significant disappointment and eliminated the fruits of many put efforts to close the digital divide. The economic crisis, caused by Covid-19, had the greatest impact on developing countries, where solving economic and public health service problems had to be combined with servicing large debt and austerity regime (UNCTAD TDR, 2020). At the same time, the international community sees a way out of this situation in the reappearance of multilateralism.

A multilateralism is a special form of institutional interaction between different states, designed to solve common most significant problems in an interdependent world. According to a specialist in international affairs John Ruggie, "multilateralism is an institutional form, that coordinates relations between three or more states based on 'common' principles of behavior [...] without taking into account the particular interests of the parties or strategic needs, that may exist in each particular case" (Ruggie, 1992). This form of interaction implies the cession by states of a number of their powers to international organizations for the implementation of coordinated actions, aimed at fulfilling the assigned tasks.

Recovering from the recession, caused by COVID-19, requires a new vision of an interdependent world. At the same time, both overcoming the consequences of the pandemic and closing the digital divide can only be carried out with the expansion of the powers and influence of the public sector.

"The reappearance of multilateralism is not a matter of gradual changes of existing rules and practices, it implies a system-wide transformation that links immediate aid efforts to mitigate the severe damage to life and means of living caused by the pandemic and related countermeasures" (UNCTAD TDR, 2020). At the same time, a sustainable recovery program should be carried out, which includes the more promising directions laid down in the Agenda for the period up to 2030. In terms of regulatory forms of international cooperation, this will mean:

(1) rejection of a destructive form of integration in favor of a less expensive one, as it was in the 1930s;
(2) promoting a collective approach that can help establish new rules, principles, and policies in line with the Agenda for the period up to 2030;
(3) building a more democratic (and less hegemonic) model of global control.

In September 2019 German Minister of Foreign Affairs, Heiko Maas put forward an initiative to create an "alliance for multilateralism," which was supported by representatives of more than 50 states. Multilateralism is based on the mechanism of multilateral agreements that allow each state to enjoy benefits in relations with all partners. This model does not imply a limited membership, as is the case with the G7 or G20, but it provides for the reform of international institutions and the consolidation of international agreements (DW.COM, 2021). It seems that the approaches of multilateralism will allow solving the problems associated with the penetration of digital transformations into all spheres of life of modern society and the emergence of digital inequality to the greatest extent.

The system of international trade, regulated by the WTO, did not confirm its viability in terms of ensuring the interests of developing countries and closing the economic gap. In this regard, a reform of the named system is planned, which should take into account the need to expand the space for development policy. "Within the framework of confidence restoration, new issues such as digital rules should not take the form of a multilateral agreement until developing countries are aware of their development aspects and thus increase their digital competitiveness" (UNCTAD TDR, 2020).

7.4.2 Synergistic Effect of Regional Integration in the Field of Digitalization, on the Example of Cooperation Between EAEU Member States

First of all, it should be noted that any regional integration entity, including the EAEU, has certain goals of its activities, for the achievement of which the named entity is entrusted with the necessary competence. The supranational nature of the EAEU is most evident in the activities of the Eurasian Economic Commission (hereinafter referred to as "EEC"). At the same time, "taking into account the scope of the EEC's competence, it can be concluded that supranational within the framework of EAEU is functional and does not cover all fields of activity of the member states" (Nazarova, 2017). According to Article 4 of the EAEU Treaty, the main goals of the Union include the craving for forming a single market of goods, services, capital, and labor resources within the framework of the Union. This goal is systemically important for all EAEU activities; therefore, the efforts of the association aimed at promoting end-to-end digitalization, obviously, also lie in the plane of the achievement of this goal.

Following clause 3 of Article 23 of the EAEU Treaty, to ensure effective interaction and coordination of state information resources and information systems, the member states shall operate a coordinated policy in the field of informatization and information technology. These activities of states are also regulated by the Protocol on Information and Communication Technologies and Information Interaction within the framework of the Eurasian Economic Union (Appendix No. 3 to the EAEU Treaty). In addition, end-to-end digitalization processes have a much wider sphere of penetration than the interaction of state information resources and information systems. The EAEU needs to synchronize the digital transformations taking place in the member states, achieve a synergistic effect, and enhance the roles of the member states and the Union as a whole in shaping the global digital agenda. Within the framework of the EAEU, the escalation of the digital divide must not be allowed, which could lay down any achievement of integration processes at the stake: political, economic, social, or cultural.

In October 2015, by the Decision of the Supreme Eurasian Economic Council No. 28, the Main Directions of Economic Development of the Eurasian Economic Union were approved, which reflected the scenario "Own Center of Power," the implementation of which is impossible without joint actions of states in the field of end-to-end digitalization. In this regard, on October 11, 2017, by Decision of the Supreme Eurasian Economic Council No. 12, the Main Directions for the implementation of the digital agenda of the Eurasian Economic Union until 2025 (hereinafter referred to as the "Main Directions") were approved. This document is the EAEU's basic program act in the field of digital cooperation among member states at the moment.

Achieving the goals of the digital agenda should lead, among other things, to accelerating the processes of free movement of goods, services, capital, and labor resources within the framework of the Union, as well as to leveling the readiness levels of the member states for the development of the digital economy (Frolova et al., 2018).

One of the ways to achieve goals within the framework of the implementation of the digital agenda is the improvement of the Union's law, taking into account the trends of global digital transformation. At the same time, the scope of powers of the EAEU bodies to adopt legal acts will directly depend on which field of competence the issue under consideration belongs to a single policy, a systematic policy, or a coordinated policy. For example, the field of informatization and information technology belongs to the systematic policy of the EAEU, where harmonization of law of the member states is possible, but not its unification. "Harmonization, as a method of approximation of national legislation, involves the implementation of appropriate measures by the member states within the framework of their domestic regulation [...] while a coordinated policy does not imply either unification or harmonization of national legislation" (Miashchanava, 2020b).

Within the framework of the Main Directions, vectors for the development of the digital economy are determined, which include: digital transformation of economic sectors and cross-sector transformation, digital transformation of markets of goods, services, capital, and labor, digital transformation of integration management processes, development of digital infrastructure and ensuring the security of digital processes. For each of the selected vectors, a program of the most significant actions of the member states and the Union was determined that should eventually create a synergistic effect.

It is important to realize that international cooperation itself must be digitally transformed (Tarakanov et al., 2019). Thus, the Main Directions provide for the digital transformation of the integration management processes in the Union, including through the creation of a digital platform of the Union.

At the moment, processes, aimed at implementing the digital agenda, are being actively implemented. The EEC Board approved the draft decision on the implementation of the project "Digital technical regulation of the Eurasian Economic Union." "The proposed initiative involves the transformation of unified mandatory requirements and procedures for admitting products to the market in a machine-understandable format, which will allow the requirements of technical regulations to be squared with the provisions of documents aimed at their implementation—standards, practices, and others—and, thus, to ensure free access to concentrated, accessible, complete and reliable information for all market participants",—EEC Collegium Member, EEC Minister for Technical Regulation Viktor Nazarenko emphasized (EEC News, 2021).

Thus, regional digital cooperation, covering countries with an allowably insignificant digital divide, can be recognized as an effective mechanism that provides a controlled process of end-to-end digitalization of various spheres of public life.

7.5 Conclusion

1. The international treaty, due to the long and complex procedure of adoption and national implementation, cannot effectively respond to the ultra-dynamic processes of digitalization, therefore, other, more flexible mechanisms of international cooperation come to the fore, including the adoption of recommendatory and program acts within the framework of international organizations and supranational associations.
2. Bridging the digital inequality and equitable distribution of the benefits of using end-to-end technologies among all countries is one of the priority directions of international cooperation in the field of digitalization. Digital inequality has a significant impact on the economic situation of developing countries; moreover, there is an objective impossibility of overcoming the digital development lagging of these states without the intervention of the world community. An important component in closing the digital divide is a policy of "digital sovereignty" combined with international technology transfer. ITT at present is preferable to ensure through a policy of "technology neutrality," as well as the use of reverse innovation.

3. The economic crisis caused by COVID-19 had the greatest impact on developing countries, including devaluing many efforts to bridge the digital divide. At the same time, the international community sees a way out of this situation in the reappearance of multilateralism, as a special form of institutional interaction between different states, designed to solve common, most significant problems in an interdependent world.
4. There is an urgent need to reform the rules regulating international trade, primarily within the framework of the WTO. It seems that the approval at the level of multilateral agreements of provisions in the field of end-to-end digitalization of the global economic and legal space cannot be carried out until a set of measures, aimed at eliminating the digital lagging in developing countries, is prepared.
5. The EAEU's prospects in the field of creating the digital space of the Union directly depend on the concerted efforts of the EAEU bodies and the member states, since most of the significant areas of economic and legal regulation of end-to-end digitalization are not included in the Union's single policy.
6. The implementation of the "Own Center of Power" scenario fully meets the purposes of preserving the digital sovereignty of the EAEU member states, which will allow representing their common interests within the framework of the global digital agenda.
7. International cooperation itself must be digitized, ensuring the fastest possible response to ongoing technological changes.

Acknowledgements The study was carried out with the financial support of the Belarussian Foundation for Basic Research in the framework of research project No. Г20Р-080.

References

BELTA. *EDB: digitalizing the economy needs international cooperation and common standards.* Accessed January 23, 2021. https://www.belta.by/economics/view/eabr-dlja-tsifrovizatsii-ekonomiki-neobhodimo-mezhdunarodnoe-sotrudnichestvo-i-obschie-standarty-418624-2020/

DW.COM (2021, January 23) 50 states established an "alliance for multilateralism". https://www.dw.com/ru/50-государств-учредили-альянс-за-мультилатерализм/a-50599763

EEC News (2021, January 29). *The EAEU will digitize technical regulation.* http://eec.eaeunion.org/news/eaes-otsifruet-tehregulirovanie/ Accessed: 30.01.2021

Frolova, E. E., Zankovsky, S. S., Dudin, M. N., Zinkovsky, S. B., & Kirsanov, A. N. (2018). Studying concepts of the breakthrough economic reforms in selected developed and developing countries and regions of the world: economic and legal aspect. *Journal of Advanced Research in Law and Economics, 9*(4(34)), 1236–1242.

Inshakova, A.O., Goncharov, A.I., & Salikov, D.A. (2020). Electronic-digital smart contracts: modernization of legal tools for foreign economic activity. In: *Digital economy: Complexity and variety versus rationality.* Lecture Notes in Networks and Systems. Springer Science.

Kowalski, P., Rabaioli, D., & Vallejo, S. (2017). International technology transfer measures in an interconnected world: Lessons and policy implications. In *OECD Trade Policy Papers, 206.* OECD Publishing.

Mamychev, A. Y., Zolochevskaya, E. Y., Miroshnichenko, O. I., Yevtushenko, S. A., & Kerimov, O. Y. (2019). National sovereignty: Socio-political transformation in context of modern cyber realities. *Journal of Politics and Law, 12*(3), 11–17.

Matytsin, D. E., & Rusakova, E. P. (2021). Strategy of quality management in Industry 4.0 and formation of cognitive economy based on industrial and manufacturing engineering in the Russian Federation and countries of the EU. *International Journal for Quality Research, 15*(4). https://doi.org/10.24874/IJQR15.04-03

Miashchanava, M. V. (2020a). Place of international treaties in the system of sources of international private law. *Journal of the Belarusian State University Law, 3*, 68–76. (In Rus.).

Miashchanova, M. V. (2020b). System of law of the Eurasian Economic Union and the prospects for the development of private international law. *Eurasian Law Journal, 5*(144), 18–22. (In Rus.).

Nazarova, O. V. (2017). Participation of the Republic of Belarus in the Eurasian Economic Union: Constitutional law evalution. *Journal of the Belarusian State University Law, 2*, 24–27. (In Rus.).

Report of the Secretary-General "Roadmap for Digital Cooperation". (2020). Accessed January 23, 2021. https://www.un.org/en/content/digital-cooperation-roadmap/assets/pdf/Roadmap_for_Digital_Cooperation_EN.pdf

Rodrik, D. (2018). What do trade agreements really do? *Journal of Economic Perspectives, 32*(2), 73–90.

Ruggie, J. (1992). Multilateralism: The anatomy of an institution. *International Organization, 46*(3), 561–598.

Smirnov, E. (2019). International economic cooperation for overcoming of digital inequality. *Trade Policy, 2*(18), 72–80. (In Rus.).

Tarakanov, V. V., Inshakova, A. O., & Dolinskaya, V. V. (2019). Information society, digital economy, and law. *Studies in Computational Intelligence, 826*, 3–15.

Tarakanov, V.V., Inshakova, A.O., & Goncharov, A.I. (2020). Interactive demo on-line maintenance foreign trade deals of business entities of BRICS jurisdictions. In A. Inshakova, E. Inshakova (Eds.) *Competitive Russia: Foresight model of economic and legal development in the digital age. CRFMELD 2019*. Lecture notes in networks and systems (Vol. 110, pp. 319–327). Springer. https://doi.org/10.1007/978-3-030-45913-0_36

UNCTAD TDR. (2018). *Trade and development report, 2018*. Accessed January 23, 2021. https://unctad.org/webflyer/trade-and-development-report-2018

UNCTAD TDR. (2019). *Trade and development report, 2019*. Accessed January 23, 2021. https://unctad.org/webflyer/trade-and-development-report-2019

UNCTAD TDR. (2020). *Trade and development report, 2020*. Accessed January 23, 2021. https://unctad.org/system/files/official-document/tdr2020_en.pdf

Chapter 8
New Challenges of International Legal Regime and Use of Artificial Intelligence

Natalia N. Emelianova and Andrey A. Dementev

Abstract Artificial intelligence is increasingly entering our life in all its spheres. AI-based technologies are used everywhere: in medicine, entertainment, and government services. A lot of multinational corporations build part of their inside processes on the neural networks. Increasingly, one can come across arguments that artificial intelligence will gradually (and already can) replace human workers. On the other hand, international organizations, in particular the United Nations, represented by the Secretary General, talk about the need for a balanced and rational, but still development and creation of such technologies, since they can be used for the benefit of humanity. Artificial intelligence technologies can be used, among other things, to achieve the Sustainable Development Goals. In this article, the authors analyze such an application of technologies built using artificial intelligence, review the current state of the international legal regulation of such technologies (their creation and application), and assess the existing risks and threats that accompany these technologies.

Keywords Artificial intelligence · Risk · EU · Smart technologies · Digitalization · Legal regimes

JEL Codes K10 · K24 · K33 · K38

8.1 Introduction

"Success in creating effective AI could be the biggest event in the history of our civilization.

Or the worst. We just don't know.

So we cannot know if we will be infinitely helped by AI,

N. N. Emelianova (✉) · A. A. Dementev
RUDN University, Moscow, Russia
e-mail: emelianova-nn@rudn.ru

A. A. Dementev
e-mail: dementyev-aa@rudn.ru

A. O. Inshakova and E. E. Frolova (eds.), *Smart Technologies for the Digitisation of Industry: Entrepreneurial Environment*, Smart Innovation, Systems and Technologies 254, https://doi.org/10.1007/978-981-16-4621-8_8

or ignored by it and side-lined,
or conceivably destroyed by it,"
—Steven Hawking.

There are many artificial intelligence concepts (AI) in different areas (engineering, IT, law) and representatives of different schools of doctrinal studies. So, for example, Averkin A.N. and Gaase-Rapoport M.G. in their explanatory dictionary defined AI as "the property of intelligent systems to perform creative functions that are traditionally considered the prerogative of man."

Artificial intelligence, in most general concept, is the simulation of human intelligence by machines. In simple wording, artificial intelligence is a system or machine that can mimic human behavior to perform tasks and gradually learn from the information it gathers (Matytsin & Rusakova, 2021). With artificial intelligence, machines perform functions such as learning, planning, reasoning, and problem-solving.

Data analytics, including big data, has become the main driver for AI development and is the next step in business analytics and classical machine learning (ML), making it possible to obtain specific, meaningful units from an array of raw data.

However, deep learning (DL) principles being the extensions of sight and language leading to the human perception of information are likely to become a much stronger driver for AI in the long-term due to a wide range of possible application ways.

8.2 Methodology

The study's methodological basis involves a combination of general scientific (dialectical, historical, inductive, deductive, analytical, synthetic) and specific methods (formal-legal, comparative-legal, interpretative, statistical, procedural, and dynamic).

The specificity and features of systems using AI arise in connection with the possibility of individual systems with a high degree of autonomy to independently form a decision-making algorithm simultaneously with the development of such a decision, which distinguishes them from most of the currently existing systems that have a fixed, previously described algorithm of functioning. At the same time, this property makes it possible to use AI systems successfully in areas where previously only a man could operate.

At the same time, given the impressive possibilities of performing complex calculations based on large-dimensional data, the AI system cannot obtain results corresponding to the abilities of a person especially when it comes to issues where a person acts based on life experience, upbringing, socially accepted norms, rules, and traditions. The computational process (computational logic) of systems and a poor-quality set of initial data can cause unfair, biased AI system decisions.

Thus, in other words, artificial intelligence is a collective term for different algorithms, models, and technologies. Products commonly referred to as "AI systems" may share common characteristics such as the ability to learn or recognize themselves, but they can take many forms and operate based on different architectures and algorithms.

8.3 Results

The vast set of modern technologies, techniques, and applications that fall under the concept of artificial intelligence represents in two groups (Deryugina et al., 2019).

The first group of technologies consists of knowledge-based systems that are "committed to creating behavior through deduction from a set of axioms."

These include "expert systems" that use formal logic and coded rules to participate in mindlike operations. Such systems sometimes referred to the "closed rule algorithms," including everything from commercial tax reporting software to the first generation of decision support algorithms for healthcare diagnostics.

These systems are useful for making specific situations and making optimal decisions based on certain rules in a specific area. However, they cannot learn or automatically use the information they have accumulated over time to improve decision-making (unless they are combined with some of the methods described below).

The second group of technologies uses statistical learning to improve their decision-making ability continually. This new wave of technologies, which includes widely discussed methods known as "machine learning" and "deep learning," has been made possible thanks to the exponential growth of computer's power, a significant reduction of the cost of digital storage, and, as a result, acceleration of the data collection efforts.

Systems in this category include:

- self-driving cars,
- facial recognition systems,
- natural language processing techniques automating the translation and modification of content, and
- even algorithms that tell a user what to watch next in video streaming services.

While these systems are impressive in their aggregate capabilities, they are probabilistic and, therefore, may not be reliable at the individual level.

For example, deep learning computer vision systems can classify an image almost as accurate as human; however, they will make mistakes from time to time that no human would make, such as taking a photograph of a turtle for a gun. They are also prone to mislead "examples of competition" which drive algorithm to incorrect response with a high degree of confidence.

Interesting, in our opinion, is the categorization of AI (Accenture Future of AI Economic Growth, 2018) proposed by professor Hintze of Swedish University of Dalarna (Hintze, 2020). This categorization is simple and, however, cover all possible application scenarios and development of AI.

Type 1: Reactive machines—these machines can react to situations. A famous example would be the Deep Blue, an IBM chess program that beat chess legend Garry Kasparov. The principle of operation of such machines is to analyze all possible alternatives and choose the best option. However, such machines do not have enough memory. These machines definitely cannot use past experiences for the future. Also, machines of this type are made for very narrow tasks and cannot be used for other purposes.

Type 2: Limited memory—the artificial intelligence system is capable of using past experiences to solve problems in the future. Self-driving cars equipped with decision-making systems are a good example. The vehicle performs actions similar to changing lanes. It is noteworthy that these actions are based on observations. However, there is no permanent storage of these observations.

Type 3: Theory of Mind—this refers to understanding others. First of all, it means realizing that others have their own beliefs, intentions, desires, and opinions. However, this type of AI does not yet exist.

Type 4: Self-awareness is the highest and most complex level of artificial intelligence. Such systems have self-esteem. They also have awareness, consciousness, and emotions. Yet this technology does not still exist, and this technology is sure to be a revolution.

Why do we talk about risks related to AI?

Artificial intelligence technology itself is neither a risk nor a threat. It can be risky when applied due to the usually self-directed, self-learning nature of many AI products. However, what is this risk and how to describe its severity and consequences?

When we talk about the risks associated with artificial intelligence technologies, another phenomenon arises: the world does not know standstill. Many of the significant risks associated with AI technology could appear or disappear by looking 5 or 10 years into the future. The ethical standards that we are currently trying to establish about AI technology may no longer matter in the (near) future due to changes in the perception of AI and/or a different structure of the legislation. Besides, society itself will change in the long term, and therefore, wholly different risks may arise.

What can humanity do at the present stage of development of AI technologies to predict and prevent the risks and dangers associated with AI development? In our opinion, these risks directly relate to AI development, when AI reaches such a level of human self-awareness that will allow it to make independent decisions. And then will these decisions take into account the interests of man and all humankind?

However, at the present stage of AI development, AI "market" players and other stakeholders still presented the greatest danger. It is crucial who and for what purposes will create AI tools, what goals will be set and achieved.

Many countries are already aware of the need to develop a coherent state policy for the development of AI, where all parties' interests would be equally taken into account (while not taking into account the interests of AI). One of the fundamental steps in this direction will be the need to investigate the potential risks and threats associated with AI development.

In January 2020, the Netherlands Organization for Applied Scientific Research (TNO) prepared and posted on the RIVM website (RIVM, 2019) the Report "Artificial Intelligence in the Context of National Security." The TNO—a respected Dutch Organization for Applied Scientific Research deals with risk analysis in critical infrastructure violations, cyber threats, and technological developments.

The report notes that we (human society) cannot accurately assess the potential risks and threats associated with artificial intelligence technologies. The report stressed that to understand how the state's national security can potentially be affected; we should better realize the concept of AI and the context of using AI technology.

The report is an alleged list of various risks, and threats associated with the use of technologies developed based on the AI, as already available at the disposal of humankind and the future, but at the same time notes that have not yet been studied, whether they can damage national safety in all cases. Potential threats and risks included in the report set below. This description is abstract, so we tried to supplement it with examples where, in our opinion, it attracts the most attention to the problem.

Thus, the report contains the following list.

- *The emergence and use of programs and algorithms that can determine a person's intention to commit a crime.* In this case, people are arrested based on a prediction that they will commit a crime, in connection with which it is impossible to determine whether a person would have committed a crime if he had not been arrested. As follows from the OECD Artificial Intelligence and Society (2019) report, the world has already created and operates AI-based systems designed to predict the commission of crimes.

There are two main forecasting methods. The first one, *crime scene prediction* when historical crime data are used to predict the likelihood crime. These can include liquor stores, bars, and parks where crimes have been committed in the past. Law enforcement agencies can try to prevent future crimes by sending an officer to patrol these areas on a specific day of the week or time of day. The second one is personal prediction, when law enforcement agencies use crime statistics to help predict which individuals or groups are most likely to be involved as victims or criminals. AI predictive policing initiatives are being tested in cities worldwide, including Manchester, Durham, Bogota, London, Madrid, Copenhagen, and Singapore. In the United Kingdom, Greater Manchester Police developed a crime forecasting system in 2012. Since 2013, the Kent Police has been using a system called PredPol. These two systems assess the likelihood of a crime being committed at specific locations over a specified period of time. They use an algorithm initially developed to predict earthquakes. In Colombia, the Data-Pop Alliance uses crime and traffic data to predict crime spots in Bogota.

- *The use of autonomous armed drones to attack vital infrastructure.*
- *Deliberately overloading hardware (used, for example, in critical government processes) by hacking operating systems using AI.*

- *Spreading disinformation on the Internet through automated processes.*
- *"Weapon of math destruction": the idea that opaque algorithms make completely autonomous decisions, for example, regarding a person's creditworthiness.*
- *"Caste system":* the application of artificial intelligence, operating with big data, can put people into particular "cell," for example, in the field of health insurance premiums, training, and mortgages. This can lead to creating of an undeniable feedback system in which it is no longer possible to change the profile of a person created about him or her based on big data.

The last two assumptions can be illustrated by the following example from the OECD Artificial Intelligence and Society report. Credit Rating Company Alipay uses consumer data to determine a credit score. These include purchase history, type of phone used, games, and social media friends (Inshakova et al., 2018). In addition to the traditional credit rating for making loans, Chinese social credit rating can influence decisions such as the deposit level when renting an apartment or even the likelihood of matching profiles when dating online. A person who plays video games for hours every day may receive, for example, a lower social credit score than a person who buys diapers, who is considered as a responsible parent.

- *Loss of control*: the creation of technology with the ability to self-learning, that people are no longer able to control, which can lead to autonomous weapons to fight with people, or with each other. This may not necessarily be military-related. For example, the report mentioned example of flash-crash, when automated processes in the financial sector to respond to each other, and billions of moneys disappeared for a few seconds. The military equivalent of this would mean that autonomous weapons react to each other and thus provoke communication disruptions or physical damage.
- *"Selling FEAR"*: delay in informing the public about the impact of AI and use technology provoking social threat images that do not exist or are not relevant. Criminals can use this situation for criminal purposes.
- *Over-reliance on AI*: it may be too fast the first growth of confidence to the system with AI, but people do not realize that their operation rights still needs much work.
- *Malicious software is controlled by the AI, for the destruction, such as critical infrastructure.*
- *Dependence on overseas market for (patented) key technologies.*
- *Legislation Lagging*: opportunities for drafting and enforcing legislation lag behind the rapid development of artificial intelligence technologies. It should be noted here that the recommendation on artificial intelligence (AI)—the first intergovernmental standard on AI—was adopted by the OECD Ministerial Council on May 22, 2019, at the suggestion of the Committee on Digital Economy Policy (CDEP). The recommendation aims to foster innovation and trust in AI by encouraging responsible governance of AI while ensuring respect for human rights and democratic values.
- *"Backdoors"*: AI systems purchased from a vendor in the free market, the information about which is not transparent, which provides strategic dependence and

the ability for the outsiders to use personal data, for example, discussion around 5G and Huawei.

- *Hostile AI*: the manipulation of an AI system or data that the system must transmit to achieve adverse results.

As we can see, the above list gives only the most general idea of the risks and threats that humanity will face when using AI.

At the same time, it is evident that the most rapid development of AI will go in industries where it will bring clear and quick commercial or practical benefits, then in industries where the use of AI requires financial investments without precise guarantees of profitability or even break-even.

We cannot stop progress, whereas AI is one of its essential parts. However, now it is worth to pay attention to the most critical sectors for the humankind development, while the use of AI is still at its initial stage of development.

Before proceeding to the analysis of AI's use in specific spheres of human life and the risks that arise from such use, it is also worth pointing out the existing international legal regulation of the developing and use of systems using AI algorithms.

First of all, we should focus on the United Nations as the main universal regulator. In 2018 was prepared the special report of the United Nations Activities on Artificial Intelligence, which addressed, among other things, the impact of new technologies and artificial intelligence on human rights. This document also contains several recommendations: creation of professional standards for engineers developing and implementing technologies using AI; the creation of a system for the consideration of appeals and complaints, and the data on such complaints and the responses to them should be regularly published (United Nations Activities on Artificial Intelligence, 2018).

The Commission on Economic and Social Development and the UN Conference on Trade and Development also contributed to the development of regulation of the developing and application of AI technologies. They primarily deal the digital economy's problems, which also includes issues related to AI technologies.

Under the auspices of UNCTAD, there are various forums created to join representatives of the private sector and the states and academics participate. The forums aim to minimize the harmful effects of the digitalization of the economy, AI technologies and bridge the gap on these issues between developed and developing countries (UNCTAD Digital Economy Report, 2019).

We cannot overlook the issues of AI by UNESCO, which encourages everyone to use the so-called R-O-A-M principles developed by it. The organization also insists on developing ethical principles in working with AI, including social and digital equality. Also, the World Commission on the Ethics of Scientific Knowledge and Technology operates within UNESCO's framework, which deals with the development of ethical principles. Moreover, in 2019, the general conference adopted a resolution on the development of ethical standards in the field of AI. Currently, a group of experts under UNESCO's auspices is finalizing its work on the preparation of the relevant documents (UNESCO, 2019).

Within the framework of the International Telecommunication Union, several expert groups and conferences developed to study possible directions for standardizing technical aspects in the field of AI: a group on machine learning for future networks, including 5G; Joint ITU-World Health Organization (WHO) group "Artificial Intelligence for Health"; group on unmanned vehicles: The effectiveness of artificial intelligence and other new technologies from the point of view of protection of the surrounding environment".

Questions about trust in AI systems and defining complementary principles for responsible governance were posed by the OECD (AI Guidelines). These principles were designated as follows:

- AI must benefit people and the planet by driving inclusive growth, sustainable development, and well-being;
- AI systems should be designed to respect the rule of law, human rights, democratic values, and diversity, with adequate safeguards, for example, where necessary, human intervention in the system should be provided to ensure fair and honest society (human-centered approach and justice);
- it is necessary to ensure transparency and responsible disclosure of information about AI systems so that people understand the results of using AI and can control it (transparency and explainability);
- artificial intelligence systems must function reliably, guaranteed, and safely throughout the entire life cycle; potential risks must be constantly monitored, and appropriate control measures must be taken (stability/robustness/safety and reliability);
- organizations and individuals developing, implementing, or operating AI systems should be held accountable for their proper functioning in accordance with the above principles (responsibility).

To implement these principles, the OECD makes five key recommendations to governments:

- promote public and private investment in research and development to stimulate innovation in robust AI.
- to promote the development of accessible artificial intelligence ecosystems with digital infrastructure, modern technologies, and mechanisms to exchange data and knowledge.
- provide a policy environment that will pave the way for the deployment and implementation of robust AI systems.
- facilitate human acquisition of AI skills and support workers to make a just transition to AI.
- engage in cross-border and cross-sectoral collaboration to make progress in the responsible management of trusted AI.

The International Organization for Standardization and the International Electrotechnical Commission are developing several AI standards, which can be divided into the following main areas and types:

- fundamental standards (terminology and concepts).
- big data standards for AI systems.
- trust in AI systems.
- practical use cases and applications.
- computational methods and characteristics of AI systems.

WIPO is also actively involved in developing of standards and common solutions, considering the specifics of the organization itself—the patent landscape. In 2019, the organization prepared a corresponding report on AI.

Several trends in the patent landscape publication help understand the comparative scope of interest in intellectual property management in AI.

The World Trade Organization, despite internal problems, is actively developing universal rules for the movement of AI technologies.

Moreover, for the WTO, this issue is fundamental and, at the same time, challenging. Since the package of WTO agreements itself was adopted when no one spoke about digitalization (especially on such a scale) from the point of view of transforming economies. Accordingly, all of the essential documents of the WTO cannot help states modify international trade to the development of e-commerce and the digital economy.

In the 2018 WTO annual report, WTO experts noted that AI, along with blockchain, IoT and 3D technologies, carries the potential for a significant transformation of international trade and e-commerce (Inshakova et al., 2020a, b). It is noted that these technologies can have a significant impact on computer services and the movement of intellectual property.

In January 2019, 76 countries, including the US, EU, China, and Russia, participated in developing of the e-commerce agreement. Some of the discussed provisions will causally relate to the development of the global market for AI technologies: market access in digital commerce, information and data transfer, consumer trust, and protection.

Thus, we see extensive work to standardize and regulate the creation and application of AI technologies globally. However, of course, this work requires unswerving adherence to the principles of the priority of human rights, especially in those areas where it concerns health and education.

Let us move on to specific examples.

8.3.1 Using AI in Healthcare

The WHO Regional Office for Europe has prepared the report "Health (2020): a European policy framework and strategy for the twenty-first century" (WHO, 2020). This publication is a summary of two documents approved in 2012 by the WHO European Regional Committee under the strategy of Health (2020).

In many countries, the report noted that the share of the state budget on health care is now more than ever high, while the cost of services is growing faster than GDP (Matytsin & Plaksunova, 2020).

However, at least some of them show no correlation between health expenditure and the final useful result in health. Costs are increasing primarily due to increased prices from suppliers; we are talking about the new AI methods in treatment and technology, making people expect protection from health risks and easier access to high-quality governmental health services.

An important measure to support the development of health care is the use of cutting-edge solutions from information technology.

These technologies can provide access to the full scope of clinical information by which health care providers and consumers of services can make correct and timely decisions without delay and duplication of services and unnecessary use of public and private spending.

The report notes that now we observe trends leading in the opposite direction. Information technology solutions in primary care and hospital level are various and provide unsatisfactory compatibility, leading to poor coordination between the two levels. The other significant challenge of medical service digitalization is paying close attention to personal data protection and safety.

Based on this, it is worth considering whether the widespread introduction of AI in the healthcare sector will lead to such a degree of commercialization when the strategic goal of the industry is to generate profit under the guise of introducing high AI technologies, rather than curing patients and preventing diseases.

In December 2020, The Economist published an article "The dawn of the digital medicine" dedicated to the problems of digitalization of healthcare (The Economist, 2020).

The article referring to The Global Institute of McKinsey, the brain center of the same consulting firm, says that when it comes to digitizing healthcare lags not only banking but also on travel, retail, automotive and even products packaging.

About 70% of American hospitals still fax and mail medical records. CEO of a large hospital in Madrid said that during the first wave of coronavirus, it became clear that the exchange of electronic records in Spain's regions is almost non-existent.

McKinsey estimates that global digital health revenues—from telemedicine, online pharmacies, wearables, and more—will grow from $350 billion in 2019 to $600 billion in 2024. A significant portion of the $3.6 trillion US healthcare market will undergo digital transformation. The same thing happens in China, Europe, and most other places where doctors practice their profession.

Patients are also passionate about new technologies, have high expectations, and are willing to pay for their use.

Entrepreneurial firms, from health app startups and hospitals to insurance companies, pharmacies, and tech giants like Amazon, Apple, and Google, are struggling to provide such services.

Amazon wants Alexa, its digital assistant, to be able (with the person's permission) to analyze his cough and tell him if he is it flu or COVID. At Apple, there are hours and about 50,000 applications for the healthcare of the iPhone. Google's parent company, Alphabet, has a biosciences division, Verily.

All of this could lead to helping medicine evolve from "data-driven clinical science to clinically supported data science," says Pamela Spence of EY consulting.

Is healthcare making the IT giants and small startups interested in medicine like a big technology? Will it lead to the increase of the cost of the creation of technologies that use AI in the health system to improve the quality of services and increase the efficiency of the treatment?

Medicine is a regulatory minefield with influential players, primarily the state, for which the business models of large technologies, especially those supported by advertising, are not suitable. The 2020 pandemic also showed that complex equipment and expensive services not always do improve health. For the next generation of digital technology to thrive, it must "improve health, not increase costs," says Verily's Vivian Leigh. Her firm is shifting from service fees to risk-based contracts, which are paid when results improve (for example, if people with diabetes have blood sugar control or more people get eye exams) (The Economist, 2020).

Also, the use of AI in the health sector raises one of the main issues when using AI technology—protecting personal data.

The collection, storage, and use of healthcare information generate a high degree of responsibility and the need to take adequate protection measures.

As noted in a 2019 OECD report by the Artificial Intelligence and Society, issues related to the use of data extracted from implantable medical devices such as pacemakers can be used as evidence in court.

As these types of medical devices become sophisticated, they increase security risks such as malicious hijacking and malicious use. This report helps build a shared understanding of AI in the present and near future by mapping the AI technical, economic cases of use and policy landscape and identifying primary public policy considerations. It is also intended to help coordination and consistency with discussions in other national and international fora.

Another example is the use of biological samples (like tissues) for ML (machine learning), which raises complex questions of consent and ownership.

As a result of these concerns, the report notes, many OECD countries report legal barriers to the use of personal health data. These obstacles include the prohibition of data communication and the obstacle to developing databases built on electronic health records.

8.3.2 Education and AI

Today, much attention is also paid to the use of AI in education (Global News Wire, 2020).

According to a marketing research report published by P&S Intelligence, the global market of AI in education sector reached $1.1 billion in 2019 and is expected to reach $25.7 billion in 2030, while the forecast period (2020–2030) will register an average annual growth rate of 32.9% (Arend Hintze, n.d.).

Undoubtedly, modern technology can do a lot to help teachers make the teaching process more exciting and comprehensive. AI can support the teacher by automating some of the tasks and thus reducing the workload. For example, the AI can significantly automate and/or ease in the near future, exposure estimates for the work, for example, to identify the essay's strengths and weaknesses, then a teacher in the first place can spare the lot it time evaluating of these moments.

AI can be used in personalized teaching: teachers have limited time and attention, and for these reasons they cannot teach each student individually. AI does not have this limitation, allowing AI to align education with learners' desires better. As a result, if necessary, the teacher can focus his attention on the "problem students," and the capable student advances through the curriculum at his own pace and level.

AI can correct possible (subconscious) teacher bias and make education fairer about ethnicity or gender: an AI that does not have information about a student's ethnicity or gender cannot include these variables in assessing knowledge.

However, when the AI becomes so smart that it can replace the teacher, it remains fantastic. Developing an AI equal to the level of human intelligence will take at least several decades. However, this does not change the fact that the less intelligent AI that is already available today can support the teacher to devote more time to each student or work more efficiently.

Optimists argue that breakthroughs in deep learning are advancing so rapidly that AI's implementation comparable to human is only a matter of computing power. Pessimists argue that a fundamental breakthrough is needed first so that AI can "think like a human." Polling experts in the field of AI, conducted in 2013 by Nick Bostrom, showed that the optimistic forecast (probability 10%) for the AGI—2022, a realistic prognosis (probability 50%)—2040, and the pessimistic forecast (probability 90%) in 2075 (Müller & Bostrom, 2014).

However, educational goals can be compromised when placing too much emphasis on technology.

When education is fully automated, and students go through the educational process without human intervention, one may wonder if education in this way will become inhuman.

This risk is increasing when simulations are used to convey teaching materials. There is a chance that the knowledge and skills gained in the simulation will not be related to reality.

There is also a risk that humans' bias is transformed into data, which is then taken as facts by AI. Historically data often contain biases and stereotypes. There is a risk that AI will even reinforce these biases by creating a feedback loop.

Another real risk is lower prospects in the teaching profession (for current and future teachers). Undoubtedly, the teacher may need new skills to use AI correctly. Data-driven learning is currently not part of teacher training courses, while AI expects basic knowledge from the user in this area. If teachers cannot keep up with this development, the chances are that the future teacher will be a data scientist with less strong educational skills.

Besides, there is a risk that we will become dependent on artificial intelligence systems that we do not fully understand. In this case, the teacher cannot take over responsibility for the system, which cannot be understood.

Thus, AI requires several conditions to be met before it can be successfully applied, primarily the relevant technical infrastructure and stakeholder awareness. When AI comes to teaching process too quickly, there is a chance that it will fail to live up to expectations. Ultimately, this frustration can lead to an aversion to AI in education and possibly hinder other innovations.

Also, another serious concern is that the introduction of AI will become the prerogative of large IT companies, which will control educational policy even at the state level through technical means.

8.3.3 Impact on Employment and Tax Problems

However, it is not only health care and education that are areas of human life that require close attention when introducing AI technologies.

In the article Cities of the Future The Potential Impact of Artificial Intelligence, Eva Kassens-Noor and Arend Hintze note that AI can radically change the way people work since about half of today's professions are very vulnerable to technical automation (Research Gate).

Consulting company McKinsey studied 2000 species of 800 professions and activities and found that 47% of these tasks can be automated, by merely adapting existing technologies. McKinsey also found that 1/3 of all tasks out of 60% of all professions can already be automated. This study demonstrates the impact of AI on potential job displacement and job loss.

At the same time, the development of AI-based technologies can turn into a positive effect on humanity.

So, at the World Conference on Artificial Intelligence in Shanghai in 2019, Alibaba founder Jack Ma suggested that it would be enough for people to work three days a week for four hours, that artificial intelligence would give people more time to be human. It will create jobs that will make people happier. For example, the invention of electricity made it possible to "allocate more time for people so that you can dance in the evening or go to a karaoke bar" (CNBC, 2019).

How the process of replacing human labor with machines will develop depends on many factors.

As mentioned above, AI can surpass humans and replace them in various types of jobs. Accordingly, there are many concerns about government revenues, as AI/robots could trigger widespread unemployment, resulting in less tax revenue.

The tax on robots, proposed by Bill Gates in 2017, has sparked heated debate among tax professionals and academics on the taxation of income derived from the use of robots or robots' labor.

However, even without taking into account the difficulties associated with the definition of the very concept of "robot," it should be recognized that the collection of revenue through taxation of robots should be considered only as a last resort, and in any case, temporary, if there is no other alternative.

Any justification for such an economic measure of government revenue growth should outweigh fears that a tax on robots could lead to distortions, complexities, and slower growth, especially in a globalized economy with intense international tax competition. In any case, any tax legislation relating to robots must solve a difficult question to define its scope so that it is about corresponded to the target and provides a flexible cover designed for the future with sufficient legal certainty (SSRN, 2019).

8.4 Conclusion

We are still in the early stages of developing and using AI.

Most likely, there will be no significant changes soon, and we will see, first a wall, many new products, and new services based on simple technologies with AI.

However, if we move the time range longer, the outlook becomes more challenging to predict.

It is interesting to note that the concept of what counts as artificial intelligence changes over time. It is known as the "AI effect" or "strange paradox": previously advanced innovations become mundane and routine, losing the privilege of being AI, while new technologies with more impressive capabilities are instead labeled as AI (McCorduck, 2004).

What does it mean when we say that AI will change the world? Will we see new ecosystems? Will we see a new relationship between people in society, between societies and governments, between poor and rich, between developed and developing?

Which social groups or countries will be most vulnerable if the foreseeable risks associated with AI use to find their confirmation in the real future?

Unfortunately, considering the events of the last year, it must be admitted that humanity is still at a stage of development when long-term planning and even medium-term planning cannot consider all degrees of risk from the primary links of human life to the global level.

On the one hand, we can observe the international community's growing interest in the problems of AI. All the major international organizations, including the United Nations, are actively engaged in developing criteria, standards, codes of conduct about AI itself and those who develop such technologies and apply them. However, on the other hand, we face the existing problems due to the lack of a common understanding of what is AI and what functions we endow with such a definition. Besides, all documents mentioned above are of a purely recommendatory or even doctrinal and technical nature, and it is still too early to talk about creating a full-fledged universal regulatory mechanism.

It is only evident that there is still no clear idea of what AI will bring to humanity, just as it is obvious that AI can raise questions and problems related, particularly to human values, fairness, privacy, security, and accountability.

References

Accenture Future of AI economic Growth (2018). *Report. AI applications, OECD Library*. Accessed December 31, 2020. https://www.oecd-ilibrary.org/sites/eedfee77-en/1/2/3/index.html?itemId=/content/publication/eedfee77-en&_csp_=5c39a73676a331d76fa56f36ff0d4aca&itemIGO=oecd&itemContentType=book

Arend Hintze. *Cities of the future? The potential impact of artificial intelligence*. Researchgate official site. Accessed December 31, 2020. https://www.researchgate.net/publication/341376323_Cities_of_the_Future_The_Potential_Impact_of_Artificial_Intelligence

Arend Hintze. Researchgate. Accessed December 31, 2020. https://www.researchgate.net/profile/Arend_Hintze

Deryugina, T. V., Inshakova, A.O., Sevostyanov, M.V. (2019). Systemic issues of legal regulation for information aggregator operation. In Popkova, E. G. (Ed.) *Ubiquitous computing and the internet of things: Prerequisites for the development of ICT/Studies in computational intelligence* (Vol. 826, pp. 27–34). Springer Science + Business Media.

Englisch, J. (2019). *Digitalisation and the future of National tax systems: Taxing robots?* SSRN site. Accessed December 31, 2020. https://ssrn.com/abstract=32446701 Accessed 31.12.2020

Global News Wire. (2020). *AI in education market is poised to post $25.7 billion revenue by 2030: P&S Intelligence*. Global news wire site. Accessed December 31, 2020. https://www.globenewswire.com/news-release/2020/02/26/1990599/0/en/AI-in-Education-Market-is-Poised-to-Post-25-7-Billion-Revenue-by-2030-P-S-Intelligence.html

Hawking, S. *A.I. could be 'worst event in the history of our civilization'*. Accessed December 31, 2020. https://www.cnbc.com/2017/11/06/stephen-hawking-ai-could-be-worst-event-in-civilization.html

Inshakova, A.O., Goncharov, A.I., Kazachenok, O.P. (2018). Legal regulation of the syndicated lending: banking innovations in contemporary Russia. In Popkova, E., Ostrovskaya, V. (Eds.). Perspectives on the use of new information and communication technology (ICT) in the modern economy (pp. 368–377). Springer. https://doi.org/10.1007/978-3-319-90835-9_43

Inshakova, A.O., Frolova, E.E., Rusakova, E.P. and Kovalev, S.I. (2020a). The model of distribution of human and machine labor at intellectual production in Industry 4.0. *Journal of Intellectual Capital, 21*(4), 601 – 622.

Inshakova, A.O., Goncharov, A.I., Ershova, I.V. (2020b). Digital blockchain registration of legally significant stages of complex good's export-import supplies by business entities of the EAEU and BRICS Jurisdictions. In Inshakova, A., Inshakova, E. (Eds) *Competitive Russia: Foresight model of economic and legal development in the digital age. CRFMELD 2019*. Lecture notes in networks and systems (Vol. 110, pp. 328–336). Springer. https://doi.org/10.1007/978-3-030-45913-0_37

Kunstmatige intelligentie (AI) in het onderwijs, Rijksoverheid site. Accessed December 31, 2020. https://www.rijksoverheid.nl/documenten/rapporten/2019/02/28/kunstmatige-intelligentie-ai-in-het-onderwijs

Kunstmatige intelligentie in de context van national veiligheid. Accessed December 31, 2020. https://www.rivm.nl/sites/default/files/2020-10/AI%20in%20de%20context%20van%20nationale%20veiligheid_0.pdf

Matytsin, E., & Plaksunova, T. A. (2020). The content of the contract on the provision of paid medical services: theoretical and applied analysis. *Legal Concept Legal Paradigm, 4*, 89–99.

Matytsin, D. E., & Rusakova, E. P. (2021). Strategy of quality management in industry 4.0 and formation of cognitive economy based on industrial and manufacturing engineering in the Russian Federation and countries of the EU. *International Journal for Quality Research, 15*(4). https://doi.org/10.24874/IJQR15.04-03

McCorduck, P. (2004). *Machines who think: A personal inquiry into the history and prospects of artificial intelligence* (2nd edn.). A.K. Peters, Ltd.

Müller, V.C., & Bostrom, N. (2014). *Future progress in artificial intelligence: A survey of expert opinion*. Springer.

Taylor, L., Jack, M. A.I. could give us 12-hour work weeks. CNBC site. Accessed December 31, 2020. https://www.cnbc.com/2019/08/30/jack-ma-ai-could-give-us-12-hour-work-weeks.html
The Economist. (2020). The dawn of digital medicine. Accessed December 31, 2020. https://www.economist.com/business/2020/12/02/the-dawn-of-digital-medicine
UNESCO. (2019). ROAM-X indicators, official UNESCO site. Accessed December 31, 2020. https://en.unesco.org/internet-universality-indicators/roamx-indicators
United Nations Activities on Artificial Intelligence (AI). (2018). ITU official site. Accessed December 31, 2020. http://handle.itu.int/11.1002/pub/8120d5d5-en
UNCTAD Digital Economy Report. (2019). UNCTAD official site. Accessed December 31, 2020. https://unctad.org/system/files/official-document/der2019_overview_ru.pdf
Will de Freitas. The conversation official site. Accessed December 31, 2020. https://theconversation.com/understanding-the-four-types-of-ai-from-reactive-robots-to-self-aware-beings-67616
World Health Organization. (2020). *A European policy framework, and strategy for the 21st century*. WHO official site. Accessed December 31, 2020. https://www.euro.who.int/__data/assets/pdf_file/0017/215432/Health2020-Long-Rus.pdf
WTO Annual Report. (2019). Official WTO site. Accessed December 31, 2020. https://www.wto.org/english/res_e/booksp_e/00_wtr19_e.pdf

Part II
Digital Technologies for Regulating the Business Environment: Protecting the Rights of Participants in Economic Activity

Chapter 9
Legal Mechanisms for Protecting the Rights of Economic Entities in the Conditions of Development Penetrative Technologies

Natalia V. Antonova, Dmitry A. Pashentsev, and Yuliya N. Kashevarova

Abstract The modern era of economic development features active implementation of digital and penetrative technologies, which determine the new technological behavior and alter social relationships greatly. They include blockchain, Big Data, quantum technologies, neurotechnologies, AI, and robotics, augmented reality, 3d modeling, etc. The deflection and synthesis of penetrative technologies form the new digital economy, which provides a wide array of development opportunities, but also carries certain risks and threats for participants of economic activities. In the first instance, new relationships arise, which are not covered by effective legal acts, this leads to multiple legal gaps. New technologies, penetrating the sphere of law, tend to change its formal parameters. At the same time, legal actions are subject to massive automation, which raises questions of algorithmized and machine-readable law. Certain radical suggestions are made in relation to future replacement of law by technologies. The question concerning efficient application of algorithmized law in the field of economic relationships remains open today. In certain cases, we may speak about emerging new law subjects—robots. The application of such subjects representing "smart" cyber-physical systems begets uncertainty and unpredictability for all economic entities. At the same time, the development of digital technologies in some cases may allow economic entities to build their relationships beyond effective legal regulation (Inshakova et al., 2020). For instance, concluding smart contracts by

N. V. Antonova (✉)
Department of Social Legislation, Institute of Legislation and Comparative Law Under the Government of the Russian Federation, 34 Bolshaya Cheremushkinskaya St, 117218 Moscow, Russia

D. A. Pashentsev
Department of Theory Law and Interdisciplinary Studies of Legislation, Institute of Legislation and Comparative Law Under the Government of the Russian Federation, 34 Bolshaya Cheremushkinskaya St, 117218 Moscow, Russia

Y. N. Kashevarova
Department of Civil Law and Process, Institute of Legislation and Comparative Law Under the Government of the Russian Federation, 34 Bolshaya Cheremushkinskaya St, 117218 Moscow, Russia

A. O. Inshakova and E. E. Frolova (eds.), *Smart Technologies for the Digitisation of Industry: Entrepreneurial Environment*, Smart Innovation, Systems and Technologies 254, https://doi.org/10.1007/978-981-16-4621-8_9

the means of blockchain technologies allows securing obligations without addressing legal instruments. As a result, a whole new economic sector is building up, which isn't covered by relevant legal regulation. In fact, this is a new model of economic regulation, based upon new technological solutions and regulators of non-legal nature. Such alteration of the role of law in social relationships leaves the question of stable legal status of economic entities open and demands for a search of legal protection mechanisms, corresponding with the new digital agenda. Those processes require further fundamental analysis from the viewpoint of modern research methodology.

Keywords Digital technologies · Social relationships · Penetrative technologies · Legal regulation · Blockchain · Smart contracts · Digital economy · Participants in economic activity

JEL Codes K10 · K15 · K24 · L14 · L52 · O14 · O15 · O30 · O33 · P11 · Z18

9.1 Introduction

The history of human civilization shows clearly that one of the factors of social development is the certain technological behavior. New technologies are altering economical basis of the society and influence greatly social development and legal regulation as well. While being a regulator of social relationships, the law, on one hand, is changing alongside too and on another hand it calls for their stabilization. It is in the law that one should look for the mechanisms of restraint that allow taking into account the interests of all economic entities, providing them with an adequate level of legal protection in order to maintain the balance necessary for the development of a new industrial environment (Inshakova et al., 2020a, b, c; Pashentsev, 2019, 2020).

The transition from the fifth to the sixth technological behavior is, first of all, a synthesis of digital and so-called penetrative technologies that connect the real world with the digital world, creating the foundations for a single digital economy (Frolova et al., 2018; Pashentsev, 2020).

The list of penetrative technologies is defined by the Russian government and is legally binding. They include blockchain, Big Data, quantum technologies, neurotechnologies, AI, augmented reality, etc. The development of these technologies is provided by the government program "Digital Economy" (Pashentsev, 2019, 2020). Its implementation is oriented toward improving labor efficiency and forming new economic branches along with ensuring social economic breakthrough, intended to enhance country's population well-being.

The solution of these tasks is not possible without a significant modernization of the legal system, the development, and improvement of mechanisms for legal protection of all participants in economic activity.

The matters of upgrading legal regulation of using digital and penetrative technologies in economic sphere have been researched by such Russian authors as Zaloilo (2019), Khabrieva (2018), and Khabrieva and Chernogor (2018).

9.2 Materials

The grounds for the present research of legal base include the effective Russian legislation over the social economic relationships, and other relationships connected directly with modern digital technologies: the Civil Code of the Russian Federation, the Civil Procedure Code of the Russian Federation, the Federal Law "On Information, Information Technologies, and Protection of Information."

As the example of program documents of strategic and recommendation nature, devoted to the matters of law development as the necessary condition for economic resilience in frames of digitalization—the following documents were evaluated: The Doctrine of Information Security of the Russian Federation (2016), The Strategy of Information Society Development in the Russian Federation for 2017–2030 (2017), The State Program of the Russian Federation "Information Society (2011–2020)" (2014), etc.

The doctrinal positions of the theoretical basis for the current research have been analyzed owing to Russian and foreign scientists, who have devoted their research to digital technologies and their influence over the legal regulation of economic relationships.

The methodological grounds for researching the mentioned processes are disclosed in works of RAS academician Chestnov (2012) and Stepin (2003). The influence of digital economy upon the Russian law in general and its development vectors has been evaluated in the works of the following authors: Alimova et al. (2019), Inshakova and Goncharov (2019), Khabrieva (2018), Khabrieva and Chernogor (2018), Matytsin (2021), Pashentsev (2019, 2020), Pashentsev and Zaloilo (2018) and Tikhomirov et al. (2019).

9.3 Methods

The grounds of the research made include traditional legal scientific methods (formal legal, system structural) as well as certain new methodological approaches typical for post-classical science—i.e., constructivism and anthropocentrism. By the means of these methods, it was possible to analyze the dynamics of the legal normative system, which regulates the use of digital technologies in the economic social sphere as well as peculiarities of their implementation to the new technological way for participants in economic activity.

9.4 Results

The role of law in ensuring digital economy has two general aspects. The first one implies that the introduction of new technologies into economy will require relevant legal support of new relationship forms, which are not covered by law yet, or covered insufficiently (Eidenmueller, 2017; Hawken et al., 1999; Inshakova & Goncharov,

2019; Schwab & Davis, 2018). It means that in order to overcome legal gaps we might need certain new norms or even legal acts, which would consider the properties of subjects' interaction in the newly formed system of social values.

The second aspect presumes that new technologies, penetrating all possible social aspects, affect the law itself: change its shape, determine new requirements for legal engineering (Pashentsev, 2019, p. 115), provide technological opportunities for modernizing rulemaking, and enforcement processes.

Therefore, the creation of a new efficient model of legal regulation in the frames of the sixth-gen technological behavior would be impossible without taking into consideration the peculiarities of implementing modern technology into the legal sphere as well as without preserving the law as the basic regulator of social relationships or as the guarantee for resilience in the new digital environment (Legal Concept of Robotization, 2019, pp. 11–21).

One of the vectors of influence of digital technologies on the legal aspects of economic activity, which creates risks for its traditional participants, is connected with automation and algorithmization of the legal sphere (Pashentsev, 2020; Putilo et al., 2020; Schwab & Davis, 2018; Tapscott, 2016). The involvement of robots and AI into social and economic enforcement processes poses new questions concerning algorithms and algorithmization. The mentioned question was asked many years ago for the first time. American sociologist D. Bell, who was the first one to define and explain the concept of a post-industrial era, has seen among its basic principles "the replacement of intuitive thoughts with algorithms" (Bell, 1973, p. 32). Along with the development of digital technologies, the algorithmization of law turned from a mere speculation into reality. Robots now don't just perform mechanical industrial activities, but instead, they can do human work—they succeed in drafting documents and even serve for adjudication purposes, becoming sole subjects of law enforcement.

The relevant examples are already provided in Latvian and partly Chinese legislation. At the same time, such activities would be impossible to perform without relevant algorithms, which raises questions on the necessary creation and introduction of machine-readable legal texts (Khabrieva & Chernogor, 2018, p. 89). Yet such texts imply their "digitalization" as a machine is only capable of processing information expressed in binary code. Therefore, the transition from literal symbols to machine-readable or digital ones (or simply—to digits) would greatly facilitate robotic enforcement. However, there are no guarantees that such transformation won't hinder traditional enforcement made by human beings as general subjects of enforcement activities because of complex comprehension of legal acts, expressed in digital form.

The emergence of robots as new subjects of enforcement (law application) activities may significantly hinder human enforcement activities, which may exclude under certain conditions some resilient participants from economic relationships. At the same time, simple limitation of automation and robotization would definitely be not enough. The use of robots as enforcement subjects will beget a number of complications in economy anyway (Inshakova et al., 2020a, pp. 601–622).

Firstly, the enforcement processes in their general sense cannot be fully algorithmized, as the written law never has "all necessary information for solving a certain case" (Socio-cultural anthropology of law, 2015, p.263). It means that a human, who is interpreting legal norms in order to recover legal gaps and adjust an abstract norm to a certain situation, will either way remain a permanent companion of enforcement processes. Therefore, one of the main goals would be the development of legal mechanisms for due interaction between a human and a robot not only in industrial sectors but also in enforcement activities, which presumes combining algorithmized artificial thinking with creative and intuitive sources (Legal Concept of Robotization, 2019).

Secondly, there is a need in determining the legal status of an enforcement robot, which in contrast with a human is unable to carry legal liability for its actions. Moreover, defining its status raises fundamental issues as it implies providing robots with legal capacity. There already are certain initiatives aimed at recognizing robots as the subjects of civil law (Khabrieva, 2018, p. 12). Yet nowadays, such initiatives are subject to motivated criticism as robots have no psyche—and therefore, they cannot be held guilty from the psychological viewpoint. Without guilt, there is no delictual capacity, which is a necessary element of legal capacity. The question on possible legal liability of an enforcement robot, which has been vested with legal capacity, remains open, while the precise and unambiguous solution of the matters related to the liability of all subjects in civil-law relationships in conditions of applying penetrative technologies influences directly on the formation of mechanisms for protecting rights.

Among the penetrative technologies, which allow disputing upon the creation of flexible mechanisms and means of legal regulation in the field of economy we should note blockchain and smart contracts, concluded on its basis. This technology is reviewed by some experts as multileveled and multifunctional, as well as covering various application spheres (Swon, 2017, p. 19).

Being a dispersed database, existing as a consequent chain of blocks, the blockchain technology may greatly alter the functioning of the whole banking and insurance sectors. Moreover, under the influence of this technology, the political system in the society may change, including electoral systems and technologies of political decision making (Genkin and Mikheev, 2018, p. 149).

Advanced potential of blockchain technologies, including its alternate side chains, allow us to talk about it as of a new organizational paradigm of coordinating human activities (Swon, 2017, p. 19). On one hand, the technology provides for the automation of multiple actions, including the legal ones (i.e., concluding and affirming transactions, etc.). On another hand, this is the very technology to secure due confidentiality of all interaction processes, which would attract even more economic entities and give them impression of total control.

Thus, the revolutionary transformations in law are only a matter of time. They will happen exactly when the society is ready to reject traditional legal institutes (such as notaries, banks, registration chambers, etc.) or even—to refuse the central controlling entity—the state itself (Barraud, 2018). At the same time, it is quite possible that traditional institutes will learn to exist within penetrative technologies or will be merged with them in frames of partner relationships (e.g., the cooperation between cryptocurrency sector and partner banks, etc.).

It is a common belief that blockchain technologies may reduce the "risks of corruption, undue use of administrative resources and legal collisions" (Blockchain at the peak of HYIP: legal risks and opportunities, 2017, p.20). Blockchain also tends to alter the structure of relationships in the field of digital property. Smart contracts allow exchanging all possible sorts of goods and services with a due level of validity and confidentiality of contracts, which may beget a new type of economic relationships (Frolova et al., 2018; Schwab & Davis, 2018; Zaloilo, 2019). Eventually, blockchain claims itself as the main rival to law, since it creates alternate dispute resolution mechanisms and prevents their emergence in economic relationships. Decentralized turnover of property (including digital property) upon blockchain technologies allows in many cases to manage without law.

Smart contracts were explained by a cryptographer Nick Sabo in 1994 for the first time. Their goal is to satisfy common contractual terms (overpayment, security of obligations, transaction execution, etc.), to minimize random or intentional withdrawals, and to meet the need in reliable suppliers (Genkin & Mikheev, 2018, p. 209). Thus, the first steps, taken by the subjects of social relationships in this direction, show good potential of smart contracts from the viewpoint of taking a vital place in the system of law enforcement.

By the means of smart contracts, decentralized commercial relationships may be established. The electronic system itself guarantees contract validity and fairness without related risks of fake copies and doubling delivery—thus, the so-called Internet of values is born (Schwab & Davis, 2018; Tapscott, 2016). All this eventually means the formation of a new group of social relationships in economy, which exist beyond legal regulation.

For the society, which traditionally utilizes law as a universal regulator, the emergence of such relationships can be compared to a global revolution. In frames of the traditional system of values any relationships, which exist beyond legal regulation may greatly affect the resilience of economic relationships, create new threats, which could only be eliminated after the introduction of relevant legal protection mechanisms for economic entities. It is too early to speak about discarding the law as the basic regulator of economy, but we cannot deny that the model of legal regulation in the sphere of economy may in future alter significantly and demand adequate resilience guarantees.

Blockchain technologies may be used not only in concluding contracts between economic entities but also in the field of public contracting (procurement). It would be wise to introduce such technologies to the judiciary system. Generally, in the near future, penetrative technologies will be able to raise the quality and pace of taking procedural actions; they will also contribute to the more efficient operation of domestic judiciary system, which will positively affect resilience of economic relationships.

At the same time, legal protection mechanisms for the participants of economic activities in course of using blockchain technologies should emanate from two technological issues, which haven't been resolved yet. The first one is the vulnerability of blockchain technology, which may lead to members of the new ecosystem losing control over all conducted operations (Matytsin & Rusakova, 2021). The second one is the quasi-anonymity of blockchain. It is known that the chain of blocks itself can be tracked down mathematically by accessing identification data of any contracting party. Without certain security measures, it can be very easy to attach user pseudonyms to an IP address, from which the transaction was generated.

In order to spread regulatory legal impact upon the blockchain-based ecosystem for the sake of protecting rights of all participants, the central regulator should succeed in the following tasks:

- development of information-related documents with warnings and recommendations concerning existing risks;
- institutional regulation of rights and guarantees of certain participants of economic activities in the blockchain infrastructure;
- assessment of applying effective regulation in relation to blockchain infrastructure, including cryptocurrency ecosystem;
- expansion of traditional regulatory requirements for cryptocurrency and blockchain infrastructures (e.g., over the spread of consumer and competition protection legislation and concerning the spread of traditional law over smart contracts, etc.);
- prohibitory means of the state on behalf of its authorities, introducing any unjust limitations into the field of digital technologies and their application by participants of economic relationships;
- development of norms concerning the implementation of a central depository within blockchain infrastructure, while the relevant regulation would include both international and national legal acts (Genkin and Mikheev, 2018, p.139; The digitalization of law-making: the search for new solutions, 2019, p. 65).

Economic relationships should first of all include the principles of self-regulation and selection of trustworthy arbiters or agents by the subjects of economic activities. The mechanisms of legal protection in the field of economy in new conditions should also allow introducing arbitration algorithms and opportunities for applying traditional law to every smart contract (Pashentsev, 2020, p. 18).

We should note that "the Internet of values," born by new penetrative technologies (Schwab & Davis, 2018; Tapscott, 2016, Putilo et al., 2020), appears before the society in full scale from the viewpoint of validity of unique digital objects and elimination of such risks as duplication and unauthorized reproduction, which generally secures their transparency, verifiability, permanency, and indispensability in future. The sought-after system is based upon "trust technologies" which make cryptocurrency and blockchain infrastructures unique.

At the same time, foreign researchers (Schwab & Davis, 2018, p.113) note that the advantages of using penetrative technologies (particularly blockchain) wouldn't be fully beneficial for potential economic entities without reaching consensus over the following matters:

- the parameters of value (value calculation units);
- technical architecture (private or public origin, which serves as a basis for confirming transactions, forming and generating new markers of value);

- checking initial status of a chain;
- correlation and mutual provision of material objects and digital transactions, implying precise identification of material objects in the digital field, etc.

In other words, the protection of rights and legitimate interests of economic entities along with the implementation of penetrative technologies would be impossible without detailed calculation of all possible risks, connected with the "adjustment" to the radical technological shift and without a well-developed new "decentralized technology of trust" as well as the mechanisms of its legal regulation (Schwab & Davis, 2018, pp.106–114).

9.5 Conclusion

The research made shows that active implementation of penetrative technologies into economic relationships significantly affects law as the basic modern regulator of these relationships.

Such technologies change multiple basic law parameters, expand the circle of legal application subjects, may alter the role of the law itself in regulating economic social relationships.

Blockchain technologies, providing for smart-contracting allows economic subjects to manage without law in making transactions.

The ability to secure obligations fulfillment beyond the legal sphere begets economic relationships, able to exist and develop apart from the relevant legal regulation and brings economic risks, which are hard to predict and investigate.

In order to spread regulatory legal impact upon the blockchain-based ecosystem for the sake of protecting rights of all participants, the central regulator should succeed in the following tasks:

- development of information-related documents with warnings and recommendations concerning existing risks;
- institutional regulation of rights and guarantees of certain participants of economic activities in the blockchain infrastructure;
- assessment of applying effective regulation in relation to blockchain infrastructure, including cryptocurrency ecosystem;
- expansion of traditional regulatory requirements for cryptocurrency and blockchain infrastructures (e.g., over the spread of consumer and competition protection legislation and concerning the spread of traditional law over smart contracts, etc.);
- prohibitory means of the state on behalf of its authorities, introducing any unjust limitations into the field of digital technologies and their application by participants of economic relationships;
- development of norms concerning the implementation of a central depository within blockchain infrastructure, while the relevant regulation would include both international and national legal acts.

All this allows us saying that the process of economic digitalization and the relevant development of penetrative technologies entails not only new opportunities for economic growth but also brings up certain risks for economical resilience due to the possible change of regulatory role of law in social relationships.

Acknowledgements The part of the reported study written by D. A. Pashentsev was funded by RFBR according to the research project № 18-29-16219.

References

Barraud, B. (2018). Les blockchains et le droit. Revue Lamy Droit de l'immatériel, Lamy (imprimé). Wolters Kluwer édition életronique.

Bell, D. (1973). *The coming of post-industrial society*. A Venture of Social Forcasting.

Chestnov, I. L. (2012). *Postclassical theory of law* (p. 650). Alef-press.

Civil Procedure Code of the Russian Federation. (2002, November). No. 138-FZ, No. 46. Art. 4532. Accessed February 2, 2021.

Civil Code of the Russian Federation (Part one). (1994, November). No. 51-FZ, No. 32. Art. 3301. Accessed February 2, 2021.

Civil Code of the Russian Federation (Part two). (1996, January). No. 14-FZ, No. 5, Art. 410. Accessed February 2, 2021.

Civil Code of the Russian Federation (Part four). (2006, December). No. 230-FZ, No. 52 (1 part). Art. 5496. Accessed February 4, 2021.

Eidenmueller, H. (2017). The rise of robots and the law of humans. *Oxford Legal Studies Research Paper, 27*, 15.

Federal Law No. 149-FZ of July 27, 2006 (as amended on December 02, 2019). On information, information technologies and information protection. No. 31 (1 part). Art. 3448. Accessed December 20, 2020.

Frolova, E. E., Polyakova, T. A., Dudin, M. N., Rusakova, E. P., & Kucherenko, P. A. (2018). Information security of Russia in the digital economy: The economic and legal aspects. *Journal of Advanced Research in Law and Economics, 9*(1), 89–95.

Genkin, A., & Mikheev, A. (2020). *Blockchain* (p. 600). LAP Lambert Academic Publishing.

Hawken, H., Lovins, A.B., & Lovins, L. H. (1999). Natural capitalism: Creating the next industrial revolution (p. 396). Little Brown and Company.

Inshakova, A. O., Goncharov, A. I. (2019). Innovation as an integral condition for the development of modern foreign trade turnover of energy resources and its legal regulation. In: O. V. Inshakov, A. O. Inshakova, & E. G. Popkova, (Eds.) *Energy sector: A systemic analysis of economy, foreign trade, and legal regulations* (pp. 67–78). Springer Science + Business Media.

Inshakova, A. O., Frolova, E. E., Galkina, M. V., & Rusakova, E. P. (2020a). Development of social market economy under the influence of noneconomic factors: modeling and regulation. *International Journal of Sociology and Social Policy*, 1–11.

Inshakova, A. O., Frolova, E. E., Rusakova, E. P., & Kovalev, S. I. (2020b). The model of distribution of human and machine labor at intellectual production in industry 4.0. *Journal of Intellectual Capital, 21*(4), 601–622.

Inshakova, A. O., Goncharov, A. I., & Salikov, D. A. (2020c). Electronic-digital smart contracts: modernization of legal tools for foreign economic activity. In *Digital economy: Complexity and variety vs. rationality*. Lecture notes in networks and systems. Springer Science.

Isaev, N. I., & Chestnov, I. L. (2015). *Socio-cultural anthropology of law: Collective monograph* (p. 840). ALEF-PRESS. (In Rus.).

Ivanov, A. Yu., Bashkatov, M. L., Galkova, E. V., et al. (2017). *Blockchain at the peak of HYIP: Legal risks and opportunities* (p. 237). Higher School of Economics Publishing House.

Khabrieva, T. Y. (2018). Right before the challenges of digital reality. *Journal of Russian Law, 9*, 5–16. (In Rus.).

Khabrieva, T. Y., & Chernogor, N. N. (2018). Law in the conditions of digital reality. *Journal of Russian Law, 1*, 85–102. (In Rus.).

Matytsin, D. E., & Rusakova, E. P. (2021). Strategy of quality management in industry 4.0 and formation of cognitive economy based on industrial and manufacturing engineering in the Russian Federation and countries of the EU. *International Journal for Quality Research, 15*(4). https://doi.org/10.24874/IJQR15.04-03

Pashentsev, D. A., Zaloilo, M. V., Ivanyuk, O. A., & Golovina, A. A. (2019). *The digitalization of law-making: the search for new solutions: monograph* (D.A. Pashentsev (Ed)). The Institute of Legislation and Comparative Law under the Government of the Russian Federation: INFRA-M, 234. (In Rus.).

Pashentsev, D. A., & Zaloilo, M. V. (2018). Impact of modern digital technologies on the content and nature of lawmaking: Theoretical and legal aspect. *Legal Science and Practice: Bulletin of the Nizhny Novgorod Academy of the Ministry of Internal Affairs of Russia, 4*, 231–236. (In Rus.).

Pashentsev, D.A. (2020). Features of law enforcement in the context of digitalization of public relations. *Bulletin of St. Petersburg State University. Right, 11*(1), 35–48. (In Rus.).

Putilo, N. V., Volkova, N. S., & Antonova, N. V. (2020). Robotization in the area of labor and employment: On the verge of the fourth industrial revolution. In E. Popkova, B. Sergi, (Eds.) *Artificial Intelligence: Anthropogenic nature vs. social origin. ISC Conference, Volgograd 2020.* Advances in intelligent systems and computing (1100). Springer.

Schwab, K., & Davis, N. (2018). *Shaping the fourth industrial revolution.* World Economic Forum.

Stepin, V. S. (2003). *Theoretical knowledge* (p. 744). Progress-Tradition. (In Rus.).

Swan, M. (2017). *Blockchain: A new economy outline* (p. 240). Olymp-Business.

Tapscott, D., & Tapscott, A. (2016). *Blockchain revolution: How the technology behind Bitcoin is changing money, business.* First published in Great Britain by Portfolio Penguin.

The Doctrine of information Security of the Russian Federation (2016). Approved by the Decree of the President of the Russian Federation of 05.12.2016 No. 646, No. 50. Art. 7074. Accessed February 2, 2021.

The State Program of the Russian Federation "Information Society (2011–2020)". (2014). Approved by the Decree of the Government of the Russian Federation of 15.04.2014 No. 313) No. 18 (part II). Art. 2159. Accessed January 18, 2021.

The Strategy of the Development of the Information Society in the Russian Federation for 2017–2030. (2017). Approved by Decree of the President of the Russian Federation No. 203 of 09.05.2017, No. 20. Art. 2901. Accessed February 2, 2021.

Tikhomirov, Yu. A., Nanba S. B. (2019). *Legal concept of robotization: Monograph* (p. 240). Prospekt. (In Rus.).

Zaloilo, M. V. (2019). *Theory of law before the challenges of digitalization (on the example of smart contracts). Legal regulation of contractual relations arising in connection with the development of digital technologies (smart contracts): Collection of materials of the round table* (December 19, 2018) (pp. 168–173). Yurist Publishing House. (In Rus.).

Chapter 10
Introduction of Digital Methods of Protection of Rights as a Legal Guarantee of Business Activity in the Modern World (on the Example of China)

Ekaterina P. Rusakova and Evgenia E. Frolova

Abstract Digital technologies are now integrated into all spheres of society's life. Business activities are increasingly carried out on special platforms, the adoption of the law "On Electronic Commerce," which came into force on January 1, 2019, in China, in addition to regulating public relations arising in the process of conducting electronic commercial activities, also settled the issue related to the methods of dispute resolution. Modern electronic technologies make it possible to simplify the process of concluding transactions, so dispute resolution mechanisms should also comply with these trends. China was the first to introduce digital tools in all dispute resolution mechanisms, both in the judicial system and in alternative dispute resolution methods. The research purpose of this article is to study various methods of dispute resolution, which are used in the process of modern electronic technologies. It is proved that their application: (1) increases the economic attractiveness of China for business activities; (2) dispute resolution methods acquire such qualities as efficiency, transparency, and accessibility; (3) the accumulation of various types of activity on electronic trading platforms makes them more attractive not only for Chinese citizens, but also for foreign persons. The current situation around the world related to coronavirus infection has had the least negative impact on China's economic situation, and in many areas, on the contrary, it has gained leadership, thanks to the

E. P. Rusakova (✉) · E. E. Frolova
Department of Civil Law and Procedure and International Private Law, Peoples' Friendship University of Russia (RUDN University), 6 Miklukho-Maklaya Str., 117198 Moscow, Russia
e-mail: rusakova_ep@pfur.ru

E. E. Frolova
e-mail: frolova_ee@rudn.ru

E. E. Frolova
Leading Researcher, Vladivostok State University of Economics and Service, 41 Gogol St., 690014 Vladivostok, Russia

Department of Theory and History of State and Law, Far Eastern Federal University, 10 Ajax Bay, Russky Island, 690922 Vladivostok, Russia

A. O. Inshakova and E. E. Frolova (eds.), *Smart Technologies for the Digitisation of Industry: Entrepreneurial Environment*, Smart Innovation, Systems and Technologies 254, https://doi.org/10.1007/978-981-16-4621-8_10

correct state policy, which has systematically and for many years implemented the country's digital agenda.

Keywords Civil procedure · Platform · Court · Artificial intelligence · Mediation · Smart court · Internet court · Arbitration · Digital justice · "One Belt, One Road"

JEL Codes K40 · D18 · K38 · L50

10.1 Introduction

The changes taking place in the world have had an impact on all spheres of society's life (Artemyeva et al., 2019). New modern technologies are actively integrated into business activities, which have changed it qualitatively. Often, the entire process from the moment of conclusion of the transaction to its completion is carried out in the electronic form. Many countries create special electronic platforms that focus on all the necessary resources for fast transactions (Matytsin, 2021). In addition, for the convenience of entrepreneurs, a service is provided to resolve disputes arising from contracts concluded on this platform. For the most effective economic cooperation, it is necessary to create fast and flexible ways to resolve disputes. It is possible to achieve these goals by integrating modern information and telecommunications technologies into this process (Dudin et al., 2019).

In the world practice of dispute resolution, China holds the lead, and this is due to the active growth of the economy and the systematic policy of implementing the "One Belt, One Road" strategy, which implies achieving leadership in all spheres of public life.

A common trend in the dispute resolution process is the transition to electronic form. Many methods of dispute resolution no longer involve the personal presence of the parties to the dispute in the process. Moreover, a number of countries are considering the possibility of replacing a person in this process with a computer program, introducing so-called artificial intelligence technologies. The most flexible procedure from the point of view of law is mediation when a neutral third party offers solutions to the current situation to the disputing parties, and the form of this procedure can be any. In addition, for the most simple cases, this role can be performed by a special computer program.

10.2 Materials

The regulatory framework of the study is based on clarifications and other acts of the Supreme People's Court of the People's Republic of China, which regulate various ways of resolving disputes in the process of resolving which various modern digital technologies are used: "Opinion Concerning the Establishment of the Belt And Road

International Commercial Dispute Resolution Mechanism and Institutions 2018," "Provisions of the Supreme People's Court on Several Issues concerning the Establishment of International Commercial Courts" 2018, "Notice of the General Office of the Supreme People's Court on Issuing the Working Rules of the International Commercial Expert Committee of the Supreme People's Court (for Trial Implementation) 2018," "Notice of the General Office of the Supreme Peoples' Court on Inclusion of the First Group of International Commercial Arbitration and Mediation Institutions in the "One-Stop" Diversified International Commercial Dispute Resolution Mechanism" No. 212 of 2018, "Notice of the General Office of the Supreme People's Court on Issuing the Procedural Rules for the China International Commercial Court of the Supreme People's Court (for Trial Implementation)" No. 13 of 2018, "Opinions of the Supreme People's Court Regarding Further Providing Judicial Services and Guarantees by the People's Courts for the Belt and Road Initiative" No. 29 of 2019, "Notice by the Supreme People's Court of Issuing the Guiding Opinions on the Services and Guarantees of the People's Courts for Further Expanding Opening-up" No. 37 of 2020, "Opinions of the Supreme People's Court on Providing Support and Guarantee for Shenzhen to Build Itself into a Pilot Demonstration Zone for Socialism with Chinese Characteristics" No. 29 of 2020.

The doctrinal positions, that formed the theoretical basis of the study, were studied thanks to scientific works devoted to the phenomena of digital (electronic, virtual, network) development of legal proceedings, which have become a reality and a necessity, based on generally recognized values that are mediated by law: Frolova (2019), Rusakova and Inshakova (2021), Rusakova et al. (2019), and Tarakanov et al. (2019). Objective factors and prerequisites for the formation of the need for creating digital legal proceedings are studied in the works: Artemyeva et al. (2019), Frolova et al. (2018), Hunter (2019), Mamychev and Sklyarova (2020), and Matytsin (2021).

10.3 Methods

The study used the method of comparative analysis, which was used to analyze the process of integrating digital technologies into various methods of dispute resolution, as well as methods of cause-effect and system analysis, which allowed us to identify the main conceptual features of this process. On the basis of dialectical methods of cognition, the process of digitalization of civil proceedings, arbitration, mediation, and other alternative methods of dispute resolution in the People's Republic of China was studied, and directions for improving this process were formulated. Based on the results of the analysis, it can be concluded that the introduction of modern technical means in the dispute resolution process makes it more efficient, simpler, and less costly. The existing advantages make China more attractive for the country of its commercial activity.

10.4 Results

At the 19th Party Congress, President Xi Jinping stressed the importance of ensuring the interests of both Chinese and foreign parties in the project implementation process by creating a fair business environment, providing effective ways to protect rights and interests.

One of the important areas in the implementation of this strategy is the creation of mechanisms for resolving commercial disputes based on the following principles:

- joint planning, construction, and profit-making;
- maintaining openness and inclusiveness, fairness, convenience, and efficiency;
- active participation of foreign experts of the countries participating in this project, specializing in the field of national and international law;
- respect for the rights of participants engaged as experts to resolve commercial disputes.

The goal of this process is to create a widely recognized method and institutions for resolving cross-border commercial disputes arising in the implementation of this strategy (Frolova et al., 2018). The active application of international treaties and customs for the settlement of disputes, taking into account the peculiarities of legislation, judicial practice, and the legal culture of the parties, is consolidated.

The most obvious forms of protection of rights that require changes are judicial, arbitration, and mediation, but the greatest emphasis is placed on arbitration and mediation, as recognized methods in international practice, to create a stable, fair, transparent, and predictable business environment under the rule of law.

The Supreme People's Court of the People's Republic of China decided to establish international commercial courts, as well as a committee consisting of international commercial experts, as well as a convenient, fast, and inexpensive "single" dispute resolution center that will provide high-quality and effective legal services to parties from participating countries.

The first International Commercial Court appeared in Shenzhen Municipality, Guangdong Province, and the Second International Commercial Court in Xi'an Municipality, Shaanxi Province, to hear cross-border commercial disputes.

The tasks of the international expert committee will include the resolution of disputes through the mediation procedure, where the committee itself will act as a mediator; in addition, it will give an expert opinion to the courts in resolving disputes complicated by a foreign element (Inshakova et al., 2020).

As part of the implementation of this program, all persons involved in this process must perform certain tasks in resolving cross-border disputes:

- major arbitration centers in China should also actively participate in the dispute resolution process, create collaborations with arbitration institutions of the "One Belt, One Road" member countries to create arbitrations;
- judicial authorities should assist, including taking measures to preserve property and evidence, as well as enforce arbitral awards after the procedure of verification

by national courts, ensure the implementation of mediation agreements concluded in international mediation centers;
- national mediation institutions, which have a good international reputation, also receive state support in conducting mediation procedures in the dispute resolution process;
- law firms are also called upon to participate in mediation and legal aid procedures;
- "smart" courts should also be used to create a legal database, determine foreign law, and provide services to judges to make correct and consistent decisions;
- non-governmental organizations are supported in establishing mechanisms for the prevention and resolution of international commercial disputes.

An important role in the implementation of the strategy is staffing, and attention is drawn to the need to cultivate international lawyers, legal experts specializing in international trade with knowledge of a foreign language (Frolova, 2019).

In July 2018, the Regulation of the Supreme People's Court on a number of issues related to the establishment of international commercial courts, whose competence includes their creation, came into force.

It should be noted that for the first time, the use of audio-visual transmission technologies and other information means of communication in the commission of procedural actions, for example, in the process of collecting evidence, organizing cross-examination and other actions, is legally established. Digital technologies will greatly simplify the dispute resolution process, as well as make it faster and less costly (Hunter, 2019).

In addition, the Supreme People's Court established a committee consisting of international commercial experts, competent international commercial mediation institutions, international commercial arbitration institutions, and international commercial courts to jointly create a dispute resolution platform that combines mediation, arbitration, and judicial proceedings, and to form a "single" international mechanism for resolving commercial disputes.

The International Commercial Court maintains a system for resolving international commercial disputes between the parties through a dispute resolution platform that integrates mediation, arbitration, and litigation in the manner they deem most appropriate.

Creating a single platform for dispute resolution is the most effective and successful way, as it provides the parties with a choice of procedure, making the process transparent, accessible, and inexpensive.

The mediation procedure can be conducted via videoconference, since this procedure is the most flexible from the point of view of its legal nature, and the legislation provides the greatest freedom for the implementation of this procedure since it does not imply binding actions for the parties, except in cases where the parties themselves agree on it.

As for the platform itself, following the applications submitted by the relevant institutions, and after analyzing the activities, for example, taking into account the number of international commercial cases considered, international influence and the level of informatization, the "single window" of the diversified mechanism for resolving international commercial disputes included the China International Trade and Economic Arbitration Commission, the Shanghai International Economic and Trade Arbitration Commission, the Shenzhen International Arbitration Court, the Beijing Arbitration Commission, The China Maritime Arbitration Commission, the Mediation Center of the China Council for the Promotion of International Trade, and the Shanghai Commercial Mediation Center.

In accordance with the standards of the "smart court," the Supreme People's Court of the People's Republic of China will continue to pursue a policy of strengthening the informatization of the "universal" diversified mechanism for resolving international commercial disputes, optimizing the online functions of the dispute resolution platform, promoting the connection of mediation, arbitration and litigation, as well as resolving international commercial disputes impartially, efficiently, and conveniently (Mamychev & Sklyarova, 2020).

According to the rules of the dispute resolution procedure of the International Commercial Court of China (CICC), most of the proceedings can be performed electronically; so according to article 5, the parties can submit materials to the court through the judicial platform on its official website, and in case of any difficulties, the party can submit materials in the following ways:

(1) by e-mail;
(2) by mail;
(3) by personal representation; and
(4) by other means recognized by the court.

If a party submits materials by the means provided for in paragraphs (2) and (3), it provides as many copies of paper documents and duplicates as there are persons involved in the case, as well as recorded on CD-ROMs or other portable storage devices.

The parties to receive court materials from the court and the parties may specify an e-mail address, but only if such receipt is confirmed.

As for the conduct of court meetings, as a general rule, they are held via videoconferencing. If it is not advisable to hold a meeting via videoconference, the parties and/or their representatives are notified of their participation in the meeting.

A similar rule applies to a pre-trial hearing, which may be conducted by videoconferencing or by other means that the court deems appropriate.

When opening a court session via videoconference, except for cases when the court recognizes that a party was unable to participate in the court session online promptly due to a network failure, equipment damage, power outage, or for other reasons, it is considered to have refused to appear in court, and if the party left the court during the court session without permission, it is considered to have left the courtroom during the court session.

In 2019, the Regulations of the Supreme People's Court on the further Provision of judicial services and Guarantees by the People's Courts under the Belt and Road Initiative of the Supreme People's Court of the People's Republic of China were adopted, which consolidate the development of information technology, drawing attention to the trends of the Fourth industrial revolution, improving judicial policy in the field of e-commerce, the introduction of technologies: blockchain, artificial intelligence, and the 5G information network, encouraging innovation in new digital techniques, new types of businesses and models caused by digitalization, network and intelligence, promoting networking, and promoting the construction of the digital Silk Road.

In addition, it is pointed out that the achievements in the creation of "smart courts" should be used, and that research institutes, law organizations, industrial associations, and chambers of commerce should be supported in conducting research and training of personnel specializing in the legislation of the relevant "One Belt, One Road" member countries, to create a legal database and judicial practice in which foreign laws were applied, to increase the transparency of the rules, guidelines, and compliance with the laws of the respective countries, as well as reducing and preventing legal risks.

The need to strengthen the guidelines for international commercial dispute resolution is emphasized, a statistical file system should be established, the websites of international commercial courts should be effectively used to publish typical cases on an unscheduled basis and adopt official documents on time, and the role of judicial practice should be strengthened in defining rules and guidelines of conduct.

In particular, border areas, key cities, and central areas should explore the possibility of using a bilateral or multilateral regional dispute resolution mechanism. Joint dispute resolution platforms are being established in the areas of cross-border trade, trade centers, and port areas. Information technologies and the online dispute resolution mechanism should be used to optimize dispute resolution methods, as well as reform conciliation and mediation procedures, to address such difficulties in cases complicated by a foreign element as complex document exchange, time-consuming process, slow notarization and insufficient knowledge of foreign legislation, as well as to achieve effective and inexpensive resolution of international commercial disputes (Rusakova & Inshakova, 2021).

In 2020, the Supreme People's Court adopted a Notice of Issuing Guidance Opinions on the Services and Guarantees of the People's Courts to further expand openness, which enshrined the implementation of promoting the integration of foreign trials and Internet court services: to meet the needs of the new open economic system, the results of the creation of "smart courts" must be fully commercialized, and the application of advanced technologies such as big data, cloud computing, blockchain, artificial intelligence, and 5G, in a trial with a foreign element, should be expanded (Rusakova et al., 2019). To provide effective, convenient, and inexpensive judicial services to foreign parties, a platform of judicial services for the participation of foreign parties is being created. With the consent of the parties concerned, information technology is used to facilitate the participation of Chinese and foreign parties in court proceedings. The rules of online legal proceedings in cases complicated by a foreign element should be studied and improved, and they should be considered with greater use of information technology. Internet courts and other courts with better information systems should be encouraged to introduce new service modes and platforms, conduct pilot programs, gain experience, and create new rules for the trial of cases complicated by a foreign element, build platforms, adopt judicial rules governing the use of technologies and manage cyberspace.

The need to strengthen the creation of "smart courts" in all aspects was emphasized: it is necessary to support the further application of information technologies such as the Internet, big data, cloud computing, artificial intelligence, blockchain, and 5G networks in judicial work; in order to make judicial work more "smart," for this purpose, efforts will be made to strengthen the research and development of a "smart" judicial system, as well as the creation of appropriate standards and regulations, improve the functions of mobile micro-courts, and accelerate the creation of an online trial service platform with an integrated trial process. The use of the "smart" assistant system in the process of reviewing cases should be strengthened to improve the quality and efficiency of judicial work. It is necessary to accelerate the creation of a transparent online platform for the judicial system and the creation of a more open, dynamic, transparent, and user-friendly judicial system. The mechanism for managing and applying big data technologies should be optimized, and the center for "big data decision-making" should be supported.

It should be noted that China has already established three Internet courts in Hangzhou, Beijing, and Guangzhou, which are competent to hear civil and administrative cases as a court of the first instance at the level of city courts arising from the use of the Internet, namely disputes:

(1) from online purchase and sale agreements, provision of online services, as well as for small financial loans;
(2) related to copyright infringement on the Internet;
(3) violations of personal rights on the Internet;
(4) related to the liability of the manufacturer of goods under online sales contracts;
(5) disputes about domain names on the Internet;
(6) administrative disputes arising in connection with the management of the Internet.

Huangzhou Internet Court has launched a mobile court service for the convenience of users, which is available 24 h 7 days a week. Such a regime has a serious advantage over legal proceedings, which are conducted in the usual mode. The creation of Internet courts has proved its effectiveness in the process of resolving disputes related to the Internet.

10.5 Conclusion

Since the introduction of electronic trading platforms, the process of resolving international commercial disputes has become much easier and more efficient. Participants do not have to think in advance how the dispute will be resolved, in case of its occurrence, which simplifies the process. Moreover, the platform offers the parties a choice of the dispute resolution method, both judicial and alternative mechanisms, based on parity, transparency, and inclusiveness. The main task of any of them is to maintain partnership relations, so almost everyone involves consultation or reconciliation.

The introduction of modern digital technologies creates the necessary conditions for a faster and more effective search for communication mechanisms in the field of entrepreneurial activity. The implementation of the digital agenda increased the number of electronic transactions and, consequently, an increase in the capital raised (Tarakanov et al., 2019).

Acknowledgements This work was financially supported by the Grant of the President of the Russian Federation No. NSH-2668-2020.6 "National-Cultural and Digital Trends in the Socio-Economic, Political and Legal Development of the Russian Federation in the twenty-first century".

References

Artemyeva, Y. A., Ermakova, E. P., Ivanovskaya, N. V., Protopopova, O. V., Rusakova, E. P., Sitkareva, E. V., & Frolova, E. E. (2019). *The financial disputes resolution in the Asia-Pacific region*. Infotropik Media.

Dudin, M. N., Frolova, E. E., Protopopova, O. V., Mamedov, A. A., & Odintsov, S. V. (2019). Study of innovative technologies in the energy industry: Nontraditional and renewable energy sources Entrepreneurship and Sustainability. *Issues, 6*(4), 1704–1713.

Frolova, E. E., Polyakova, T. A., Dudin, M. N., Rusakova, E. P., & Kucherenko, P. A. (2018). Information security of Russia in the digital economy: The economic and legal aspects. *Journal of Advanced Research in Law and Economics, 9*(1), 89–95.

Frolova, E. E. (2019). Legal regulation of green finance in China: Successes and development challenges. *Eurasian Law Journal, 7*(134), 103–108. (In Rus.).

Hunter, S. (2019). *China's innovative internet courts and their use of blockchain backed evidence*. Accessed February 15, 2021. https://conflictoflaws.net/2019/chinas-innovative-internet-courts-and-their-use-of-blockchain-backed-evidence/

Inshakova, A. O., Goncharov, A. I., Inshakova, E. I., & Tymchuk, Y. A. (2020). Digital technologies for alternative methods of resolving conflicts: The prospects of application in Russia and other BRICS countries. In Inshakova, A. O, & Bogoviz, A. V. (Eds). *Alternative methods of judging economic conflicts o the national positive and soft law* (pp. 129–142). Information Age Publishing Inc.

Mamychev, A. Yu., & Sklyarova, S. A. (2020). Digitalization and robotization of agriculture in modern China: Main priorities, directions of innovative policy of the state. *Advances in Law Studies, 8*(S5), 139–155. (In Rus.).

Matytsin, D. E. (2021). Internet-investing as a remote algorithm of the retail investment financing. In *Modern global economic system: Evolutional development vs. revolutionary leap*. Lecture notes in networks and systems. Springer Nature.

Rusakova, E. P., & Inshakova, A. O. (2021). Industrial and manufacturing engineering in digital legal proceedings in the Asia-Pacific region: A new level of quality based on data, blockchain, and AI. *International Journal for Quality Research, 15*(1), 273–289.

Rusakova, E. P., Frolova, E. E., Zankovsky, S. S., & Kupchina E. V. (2019). Problems of implementation of leadership in dispute resolution of the BRICS countries (on the examples of the Russian Federation, China, India). In *6th International conference on education, social science and humanities* (pp. 754–759).

Tarakanov, V. V., Inshakova, A. O., & Dolinskaya, V. V. (2019). Information society, digital economy, and law. In: Popkova, E.G. (Ed.) *Ubiquitous computing and the internet of things: Prerequisites for the development of ICT*. Studies in computational intelligence (Vol. 826, pp. 3–15). Springer Science + Business Media.

Chapter 11
Using Artificial Intelligence in Dispute Resolution

Elena P. Ermakova and Evgenia E. Frolova

Abstract The authors investigate the features of the legal regulation of using artificial intelligence in dispute resolution. It was revealed that (1) To date, no generally accepted definition of "artificial intelligence" has been created; (2) Artificial intelligence can be divided into two levels—low-level (collection of materials, legal expertise, review of contracts, and alike) and strong-level (independent dispute resolution, robot judge); (3) The use of artificial intelligence of the first level (low-level) occurs everywhere—in courts, commercial arbitration, in mediation, and others; cases of using artificial intelligence of the second level (robot mediator) are rare; (4) However, if in the process a human judge is replaced by a robotic judge, then the result in this process (court decision) will be based solely on the algorithms used in the AI system program; therefore, the power of a court decision will not be in the hands of the judge, but in the hands of the programmer; in addition, if the decision is based solely on algorithms, then the goal of ensuring fairness will not be achieved in all cases; (5) The introduction of AI technologies in combination with a human judge may be more successful: (a) a robot can act as a court clerk; (b) the AI can manage the analyzed documents before the decision is made and answer the questions asked by the judge during the final decision; (c) the robot can perform the functions of an expert and present an expert opinion to the court based on the evidence presented during the trial.

Keywords Artificial intelligence · Dispute resolution · Legal services · Robot judge · Robot arbiter · Robot mediator

---

E. P. Ermakova (✉) · E. E. Frolova
Department of Civil Law and Procedure and International Private Law, Peoples' Friendship University of Russia (RUDN University), 6 Miklukho-Maklaya Str., 117198 Moscow, Russia
e-mail: ermakova_ep@pfur.ru

E. E. Frolova
e-mail: frolova_ee@rudn.ru

E. E. Frolova
Leading Researcher, Vladivostok State University of Economics and Service, 41 Gogol St., 690014 Vladivostok, Russia

A. O. Inshakova and E. E. Frolova (eds.), *Smart Technologies for the Digitisation of Industry: Entrepreneurial Environment*, Smart Innovation, Systems and Technologies 254, https://doi.org/10.1007/978-981-16-4621-8_11

JEL Codes F36 · G15 · G21 · K24 · O16

11.1 Introduction

In 2019–2020, law firms and government regulators globally faced unprecedented challenges in terms of blocks of questions and policy proposals to regulate artificial intelligence and automated systems (Goncharov et al., 2019a, b). Legislators were faced with difficult questions about how and when to regulate artificial intelligence (AI) and outline a new legal framework in this area. Attorneys for Gibson, Dunn & Crutcher LLP noted that "the 2019 scientific debate was defined by the chasm between extensive treatises outlining general ethical principles designed to control and mitigate AI risks, on the one hand, and calls for targeted limits on the specific use of certain "narrow" subject-oriented AI products and applications, on the other hand" (Lyon et al., 2020). Chris Price, the CEO of EY Riverview Law, said in early 2020: "We plan to spend $2 billion on innovative legal technology over the next 18 months. That is why it will be difficult for other firms to compete with us" (Suttie, 2020).

AI companies continue to find ways to develop technologies that solve labor-intensive tasks across industries to improve speed and accuracy. In legal science, AI has already found its way: the growing interest in applying AI in law and judicial practice gradually changes the legal profession. Law firms and companies looking to make the best decisions for their businesses in the AI age use AI research and advisory services to assess where AI can optimize workflows.

11.2 Materials

The scientific basis of the article is formed based on scientific works of Russian and foreign scientists. Among the works of Russian authors studying the legal regulation of artificial intelligence in the provision of legal services, it should be noted the textbook of Artemyeva et al. (2017) and scientific articles: Biryukov (2019), Frolova et al. (2020), Goncharov et al. (2019a, b), Khudyakova (2020), Matytsin (2021), Ponkin and Redkina (2018), and Rusakova et al. (2021a, b).

General issues of regulation of artificial intelligence in the field of legal services in foreign countries are disclosed in the works of foreign researchers: Apostolova (2020), Bertolini et al. (2020), Bird et al. (2020), Blyth and Ives (2020), Brudner (2019), Du and Yu (2019), Engstrom and Gelbach (2020), Fischer (2018), Hilborne (2019), Lodder and Zeleznikow (2013), Loebl (2019), Lyon et al. (2020), McGinnis and Pierce (2019), Morgan and Reed (2019), Nikolova (2019), Pichai (2018), Potapeyko (2018), Suttie (2020), Toews (2019), and Turner (2018).

The empirical base is provided for by national legislation, reflecting the characteristic aspects of the legal regulation of artificial intelligence in Russia. In particular,

Decree of the President of the Russian Federation of 10.10.2019 No. 490 "On the development of artificial intelligence in the Russian Federation."

11.3 Methodology

The scientific development of the content of this chapter of the monograph is carried out on the basis of the general scientific method of historical materialism. General scientific methods of cognition are used: dialectical, hypothetical-deductive method, generalization, induction and deduction, analysis and synthesis, and empirical description. The study also used private science methods: juridical-dogmatic, statistical method, method of comparative legal analysis, and other.

11.4 Results

11.4.1 On the Concept of Artificial Intelligence in Jurisprudence

Artificial intelligence (AI) is the ability of a system to perform tasks that typically require human intelligence. This concept is often associated with systems imbued with capabilities associated with "intelligence," such as learning, planning, and generalization. No generally accepted definition of "AI" has been created.

American scientist John McCarthy, the father of the concept of "artificial intelligence," said in 1956 that it is still impossible to determine which computational operations are considered intelligent, and people have not figured out their intelligence. He gave this definition of AI: It is "the computational component of the ability to achieve goals."

Typically, experts distinguish between narrow and general AI. Narrow AI can perform a specific task, such as translating between languages. General AI will have the same cognitive abilities as the human mind and will solve various problems. All existing AI applications, without exception, are considered narrow AI. Machine learning allows systems to learn without being explicitly programmed. Based on algorithms and vast training datasets, systems learn to recognize patterns that have not been previously identified. The "knowledge" obtained in this way can also be applied to new data (Fischer, 2018).

As a Belarusian researcher noted, "Now the prevailing opinion is that artificial intelligence is the ability of a system to create, in the course of self-learning, solutions to problems of a certain complexity and to perform them" (Potapeyko, 2018). Its structure consists of three main elements: a database, a "settler," and an intelligent interface that allows communication. The goal is to make it capable of intelligent reasoning and actions using devices and programs. The scholar noted that there were no criteria for what could be called "rationality" in the case of machines."

In 2019, attorneys for Herbert Smith Freehills, Charlie Morgan, and Rebecca Reed noted "the term artificial intelligence (AI) is used very loosely and is often said to refer to systems that can "think." In short, AI is the use of advanced analytical and logical commands to interpret events and automate decisions and actions" (Morgan & Reed, 2019).

The authors of the article agree with the opinion of the Russian researcher P. N. Biryukov: "We may consider that AI corresponds to a set of concepts and technologies, and not to the sphere of life in which it is represented and functions. So there are many definitions of AI, and each will be correct in its way, depending on the area of application" (Biryukov, 2019).

11.4.2 The Normative Definition of the AI

The Decree of the President of the Russian Federation of 10.10.2019 No. 490 "On the Development of Artificial Intelligence in the Russian Federation" together with the "National Strategy for the Development of Artificial Intelligence until 2030" consolidated the normative definition of AI:

(a) AI is a complex of technological solutions that allows simulating human cognitive functions (including self-learning and finding solutions without a predetermined algorithm) and when performing specific tasks, obtaining results comparable, at least, to the results of human intellectual activity. The complex of technological solutions includes information and communication infrastructure, software (including those that use machine learning methods), processes, and services for data processing and finding solutions;
(b) AI technologies are technologies based on AI, including computer vision, natural language processing, speech recognition and synthesis, intelligent decision-making support, and promising artificial intelligence methods.

From the authors' point of view, the legislative definition of AI is rather heavy construction. Many Russian and foreign authors offer more straightforward definitions of AI.

The latest documents published by the European Commission in 2019 noted that "Artificial Intelligence (AI) refers to systems that exhibit intelligent behavior by analyzing the environment and taking actions—with some degree of autonomy—to achieve certain goals" (Ponkin & Redkina, 2018). Several authors have already shortened this EU Commission definition to the following: "AI systems are systems that exhibit intelligent behavior" (Bird et al., 2020).

In our opinion, this definition, proposed by the European Commission, is too vague and indefinite. Many things can be summed up under this definition, for example, outdoor surveillance cameras that react to movement and, at first glance, demonstrate "intelligent behavior" (Frolova et al., 2020).

The definition of AI proposed by Norton Rose Fulbright in 2018 seems to be quite successful and understandable. However, from the above Decree of the President of the Russian Federation (2019), this definition only lists AI technologies. On the other hand, the decree states that the mere AI is a complex of technological solutions):

AI is a field of computer science that includes:

(1) machine learning;
(2) natural language processing;
(3) speech processing
(4) robotics;
(5) machine vision (Fulbright, 2018).

Google CEO Sundar Pichai wrote, "In essence, artificial intelligence (AI) is a computer program that learns and adapts." (Pichai, 2018). For now, we will focus on these simple definitions from Norton Rose Fulbright and Sundar Pichai, and we will reveal some points related to AI and not always clear to lawyers.

11.4.3 On Machine Learning and Algorithm

Law is information technology: the code that regulates public life was emphasized by American scientists (McGinnis & Pierce, 2019). It explains why the law, in many ways, is particularly conducive to the use of AI and machine learning. "Machine learning and law operate according to strikingly similar principles: they both look to historical examples to derive rules that apply to new situations" (Toews, 2019). Controlled machine learning refers to providing an algorithm's input and output, a sequential operations system (under specific rules) to solve a problem. The algorithm then learns the relationship between them and can make predictions on the training data. People who control machine learning can adjust it until the algorithm reaches acceptable performance in predicting the results (Brudner, 2019).

11.4.4 AI Includes the Study of Human Intelligence

The term "Artificial Intelligence" implies that technology has the human ability to perform tasks that require human intelligence. "This topic is intuitively interesting because one day, our chaotic, frantic human disputes will be determined by the calm, all-knowing logic of an honest and neutral person making decisions electronically. While the concept may be attractive, this leap from human justice to e-justice is enormous" (AMLEGALS, 2020).

Skeptics start to wonder how AI can be used in legal science—an industry where human intelligence is a significant contributor. However, many understand the value of AI in this area—it enhances human intelligence, providing previously unseen capabilities that allow lawyers to make informed decisions.

There are several reasons why the arrival of a robot judge is inevitable. People create billions of disputes every year; soon, there will be tens of billions. This escalation of disputes does not tend to decrease (Goncharov et al., 2019a, b). All disputants want to resolve their problems immediately, placing undue burdens on their country's judicial system. On the other hand, people's faith in judges and arbitrators' ability to be honest and impartial is diminished by this overwork. Along with these developments, computers continue to become more dominant and more deeply integrated into our daily lives. Thus, it goes without saying that if current trends continue, computers will one day be better at handling disputes between people than people themselves.

Therefore, both practice-oriented research is needed, such as creating computer applications that perform tasks that require human intelligence, and fundamental research, such as the provision of knowledge about the law in a computer-readable form.

At the intersection of AI, on the one hand, and law, on the other, lies the area dedicated to the use of advanced computer technologies for legal purposes (Lodder & Zeleznikow, 2013).

11.4.5 Text Analytics, Machine Learning, and Natural Language Processing

The authors of this monograph agree with the opinion of Professor David Engstrom and Professor. Jonah Gelbach that the most critical factor determining the trajectory of legal technology development is likely to be technical, and it proceeds from an inevitable fact: the basis of law is "language." As a result, many legal technical tools depend on text analytics and, in particular, the group of machine learning methods, previously called natural language processing—NLP (Engstrom & Gelbach, 2020).

Deep learning NLP machines make a language computationally driven by transforming words, sentences, documents, or, in a legal context, entire cases into unique vectors called "embeddings." Each vector can be thought of as an arrow from the origin to the point representing the element of interest in a sizeable n-dimensional space, and its magnitude is a function of the presence of words, case references, indexing concepts, or other characteristics.

Once this vast vector space has been constructed and human-annotated labels have been attached to teaching materials (words, sentences, documents, cases), a complex machine learning model can mathematically manipulate vectors using large numbers (billions) calculations to modeling the relationship between them.

With sufficient data and processing power, an AI system's output can solve several legal problems, such as identifying relevant (dispute) or privileged documents, past court decisions that may be controlling, or winning arguments in a case (dispute).

11.4.6 Essential Features of AI

Young Russian scholar Khudyakova E. A. highlighted the essential features of AI: (1) the ability to imitate human intelligence; (2) the ability to self-study based on previous experience; (3) the ability to make decisions autonomously, i.e., solves problems without a predetermined algorithm; (4) lack of life, i.e., artificial intelligence is a set of technological solutions (Khudyakova, 2020).

11.4.7 Artificial Intelligence Levels

Many people believe that AI means the intelligence of a machine that performs any intellectual task and even better than a human can do. This belief is not valid (Matytsin, 2021). Modern AI systems are still far from this level of sophistication. As we found out, AI systems are composed of several methodologies (machine learning, natural language processing, speech processing, robotics, machine vision) and have been deployed in various applications. AI covers a wide range of technologies.

Like many experts in this field, we believe that one should distinguish between AI levels in jurisprudence. As P. Potapeyko wrote: "A classification into "auxiliary," "augmented," and "autonomous" intelligence is proposed. The auxiliary one is, for example, a GPS navigator. Augmented one "allows people to do things that they could not otherwise." Autonomous intelligence "allows machines to operate independently (drones)" (Potapeyko, 2018).

In a study initiated by the EU Parliament in 2020, Andrea Bertolini and her colleagues noted that artificial intelligence technologies could be roughly divided into "Strong" and "Weak" "Strong" AI means that AI systems have the same intelligence as humans or even surpass them. "Weak" AI is focused on solving specific problems using the methods of mathematics and computer science, as a result of which the developed systems are capable of "self-optimization" (Bertolini et al., 2020).

We subscribe to the above point of view on the division of AI into strong and weak, shared by most legal experts (Morgan & Reed, 2019). In this sense, we can also talk about the levels of AI use. The first level of AI use involves helping to resolve a dispute (weak AI). The second level (strong AI) is AI dispute resolution (robot judge, robot arbiter, robot mediator).

11.4.8 The First Level of AI

The first level of AI is already working effectively in the Russian Federation and abroad. The following categories (areas) of legal activity can be distinguished where AI is already applied:

1. Collection of materials or legal expertise (collection of factual data, assessment of the legal situation). For example, such platforms include Kira Systems Software, LEVERTON platform (a branch of the German Institute for Artificial Intelligence), eBrevia software, and others.
2. Review of contracts: platforms and software of this type include analysis of contracts based on AI—"RAVN" from "iManage" by "M & A Due Diligence Robot," "LitIQ," and "LegalSifter" software.
3. Collection and presentation of evidence (use of e-Discovery tools in the US and UK).
4. Forecasting the outcome of the judicial (arbitration, mediation) process. The following platforms and software in this category can be cited: Document search and forecasting system "Intraspexion," Ravel Law's Judge Analytics System, Legal analytics platform "Lex Machina" based on LexisNexis Company; "Premonition" litigation database.
5. Automated work with court documents: "PerfectNDA" software; "Brainspace" tool; in Russia, it is the systems "GAS Pravosudie," "Moy Arbiter."
6. Electronic billing of attorney salaries: software for legal pricing "Brightflag"; cloud-based legal practice management tool "Smokeball."

11.4.9 The Second Level of AI

The second level of AI is independent dispute resolution. An example of implementing this idea in practice can be considered the first robot mediator developed by the Canadian company iCan Systems, which resolved a dispute in a state court—the High Court of England and Wales (Hilborne, 2019; Nikolova, 2019). However, the use of robotic mediators is voluntary and limited to specific categories of disputes.

Models of robot judges and robot mediators were proposed in the past. In 2005, American scientists A. Lodder and D. Zheleznikov presented a three-stage model for resolving AI disputes online. The model's essence was as follows: first, the negotiation support tool (robot) must provide feedback on the dispute's likely outcome if negotiations fail. Second, the tool should try to resolve any existing conflicts using methods of argumentation or dialog. Third, for those not resolved at the second stage, the tool should use decision analysis methods and compensation/compromise strategies to resolve the dispute (Lodder & Zeleznikow, 2013).

In 2020, Australian scholars Toby Blyth and Maddison Ives wrote "The Australian Federal Court has developed an AI-powered machine learning concept to help divorce parties divide up marital property. The spouses' property division is usually discussed in detail by family lawyers; however, the Federal Court conducts an examination using a special order—"FCA Consent Order AI Application." The experts (Blyth & Ives, 2020) noted that cost reduction is likely to make this function attractive to the parties to the dispute."

In January 2019, Chinese researchers Meng Yu and Guodong Du concluded in their publication that the AI technology most needed by Chinese judges is "automatic generation of judgments," i.e., an AI system that reads the case files extracts the critical information and then automatically generates court orders based on criteria for evaluating similar cases. In 2018, some courts in Fujian Province (PRC) tried to use this technology to resolve certain disputes with similar circumstances, such as credit card debt (Du & Yu, 2019).

In June 2019, the Beijing Internet Court launched an Online Judicial Services Center featuring an AI judge with gestures, facial expressions, and a voice that mimics a human's voice. Of course, there are limitations: an AI judge is designed to help judges manage cases but not decide cases on the merits—at least for now (Apostolova, 2020). It is no exaggeration to say that China's Internet courts are characterized by the use of advanced digital technologies, such as blockchain, electronic evidence, voice and visual individualization of trial participants, etc. (Rusakova & Frolova, 2021; Rusakova et al., 2021a, b).

At the Online Dispute Resolution Forum held on October 29–31, 2019, in Williamsburg, Virginia, USA, Dr. Anyu Lee, a Chinese expert, presented the concept of China dispute resolution and stated that the only way to move forward is the resolving minor cross-border disputes by robotic judges, robotic arbiters, and robotic mediators and ensuring their solutions by the social credit system. According to A. Lee, soon, the first advanced robots will be able to speak several languages, know the laws of different jurisdictions, and analyze a large volume of court orders, which will allow them to make precise and consistent decisions. He noted that besides that, judges, referees, and mediators should start training their robots to compete with other robots shortly (Loebl, 2019).

Some lawyers support the views of their Chinese colleagues. As D. Engstrom and I. Gelbach emphasized, many of these publications are intensely futuristic: they are full of predictions by robolawyers, robot judges, or even a possible state of legal singularity when machines can accurately predict the results of cases before they are brought to court (Engstrom & Gelbach, 2020). However, we shall note that relatively recently (less than 20 years ago), the legal community was also wary of online dispute resolution, which is now developed everywhere and has already become commonplace for most jurisdictions. Every country in the world has legislation on online dispute resolution, consisting of one or more laws (Artemyeva et al., 2017; Rusakova et al., 2021a, b).

Jacob Turner, an American lawyer and author of "Robot Rules: regulation of Artificial Intelligence," wrote: "To qualify as AI, a computer program must not have a consciousness like human consciousness. AI should also not resemble anything like the multipurpose or "general" intelligence normally found in humans. Instead, most modern AI examples are "narrow" in nature, meaning that they are only suitable for one task. A narrow AI system may be able to translate all of Shakespeare's works into Chinese, but it cannot make a cup of coffee" (Turner, 2018).

11.5 Conclusion

Digitizing dispute resolution is certainly not new. Indeed, most commercial arbitration and litigation nowadays take place mainly by electronic means of communication, and electronic filing in courts around the world has become more common in recent years. However, there may be more fundamental changes that underlie the dispute resolution process as artificial intelligent software capabilities continue to improve (Morgan & Reed, 2019). If a robotic judge replaces a human judge, then the result in this process (judgment) will be based solely on the algorithms used in the AI program system. Consequently, the power of a court decision will not be in the hands of a judge but the hands of a programmer. This issue is the most controversial one in the application of the second level of AI. Currently, all known AI use cases are associated with the fact that the parties agree to this procedure voluntarily. What may happen if such a procedure is mandatory?

Besides, if the decision is based solely on algorithms, ensuring fairness will not be achieved in all cases. When making a decision, a human judge considers various factors, such as the income and wealth of the parties, and alike, which may not be taken into account (or taken into account formally) when using an AI-supported system in the decision-making process. Many experts believe that the introduction of AI technologies in combination with a human judge may be more successful. For example, a robot can act as a court clerk. The AI can manipulate the analyzed documents before deciding and answer the judge's questions during the final decision. The robot can act as an expert and present an expert opinion to the court based on the trial evidence.

Acknowledgements This work was financially supported by the Grant of the President of the Russian Federation No. NSH-2668-2020.6 "National-Cultural and Digital Trends in the Socio-Economic, Political and Legal Development of the Russian Federation in the twenty-first century".

References

AMLEGALS (2020). India: AI & its effects on arbitration. *Mondaq Social Media*. 23 June 2020. Accessed August 3, 2020. https://www.mondaq.com/india/arbitration-dispute-resolution/956956/ai-its-effects-on-arbitration

Apostolova, K. (2020). The dawn of a new era: Arbitration in the age of AI and digitalisation. *Freshfields Bruckhaus Deringer LLP*. Accessed August 3, 2020. https://www.freshfields.com/495e22/contentassets/

Artemyeva, Yu. A., Ermakova, E. P., Kovyrshina, N. A., & Rusakova, E. P. (2017). Ways to resolve disputes in multi-system legal order (Tutorial, M. (Ed.), p. 423). Infotropic.

Bertolini, A. et al. (2020). Artificial intelligence and civil liability. Legal affairs. Policy Department for Citizens' Rights and Constitutional Affairs. Directorate-General for Internal Policies. July 2020. Accessed September 11, 2020. https://www.europarl.europa.eu/RegData/etudes/STUD/2020/621926/IPOL_STU(2020)621926_EN.pdf

Bird, E. et al. (2020). The ethics of artificial intelligence: Issues and initiatives. European Parliamentary Research Service. Scientific Foresight Unit (STOA). PE 634.452—March 2020.

Accessed November 19, 2020. https://www.europarl.europa.eu/thinktank/en/document.html?reference=EPRS_STU(2020)634452
Biryukov, P. (2019). Artificial intelligence and "predicted justice": foreign experience. *Lex Russica, 11*(156), 79–87. Accessed September 15, 2020. https://cyberleninka.ru/article/n/iskusstvennyy-intellekt-i-predskazannoe-pravosudie-zarubezhnyy-opyt
Blyth, T., & Ives, M. (2020, January). Artificial intelligence and the rise of online dispute resolution. *Insights*. Accessed July 12, 2020. https://www.cbp.com.au/insights/insights/2020/january/artificial-intelligence-and-the-rise-of-online-dis
Brudner, A. (2019, August). *United States: Law in the age of artificial intelligence*. Accessed July 31, 2020. https://www.mondaq.com/unitedstates/new-technology/838022/law-in-the-age-of-artificial-intelligence
Decree of the President of the Russian Federation. (2019, October). On the development of artificial intelligence in the Russian Federation. Accessed November 8, 2020. http://www.kremlin.ru/acts/bank/44731
Du, G., Yu, M. (2019, January). Why are Chinese courts turning to AI?. Diplomat Media Inc. Accessed July 12, 2020. https://thediplomat.com/2019/01/why-are-chinese-courts-turning-to-ai/
Engstrom, D., Gelbach, J. (2020, March). Legal tech, civil procedure, and the future of American adversarialism. University of Pennsylvania Law Review. Accessed August 8, 2020. https://papers.ssrn.com/sol3/papers.cfm?abstract_id=3551589
Fischer, S. (2018, February). Artificial intelligence: China's high-tech ambitions. CSS Analyses in Security Policy CSS. N0. 220. Accessed November 12, 2020. https://ethz.ch/content/dam/ethz/special-interest/gess/cis/center-for-securities-studies/pdfs/CSSAnalyse220-EN.pdf
Frolova, E., Ermakova, E., Protopopova, O. (2020). Consumer Protection of Digital Financial Services in Russia and Abroad. Artificial Intelligence: Anthropogenic Nature vs. Social Origin. In: Popkova, E.G., Sergi, B.S. (Eds.). Springer Nature Switzerland AG. ISC Conference - Volgograd 2020, 76–87. URL: https://www.springer.com/gp/book/9783030393182 https://doi.org/10.1007/978-3-030-39319-9_8
Goncharov, A., Dolinskaya, V., Inshakova, A., & Kalinina, A. (2019a). Corporate conflicts in the system of public relations, Law, and Legislation of the Russian Federation. In E. G. Popkova (Ed.) *"Conflict-free" Socio-economic systems* (pp. 75–88). Emerald Publishing Limited.
Goncharov, A. I., Dolinskaya, V. V., & Trofimov, Y. V. (2019b). Electronic justice: Stages of establishment and issues of development in Modern Russia. In E. G. Popkova (Ed). *Ubiquitous computing and the internet of things: Prerequisites for the development of ICT*. Studies in computational intelligence (Vol. 826, pp. 315–324). Springer Science + Business Media.
Hilborne, N. (2019, February). Robot mediator settles first ever court case. *Legal Futures*. Accessed June 15, 2020. https://www.legalfutures.co.uk/latest-news/robot-mediator-settles-first-ever-court-case
Khudyakova, E. (2020). On the issue of the legal status of artificial intelligence. *Issues of Russian Justice, 5*, 366–373.
Lodder, A., & Zeleznikow, J. (2013). *Artificial intelligence and online dispute resolution. Online dispute resolution theory and practice*. Eleven International Publishing. Accessed June 10, 2020. https://www.mediate.com/articles/ODRTheoryandPractice4.cfm
Loebl, Z. (2019, December). Can a robojudge be fair? *Kluwer Arbitration Blog*. Accessed August 3, 2020. http://arbitrationblog.kluwerarbitration.com/2020/02/06/could-an-arbitral-award-rendered-by-ai-systems-be-recognized-or-enforced-analysis-from-the-perspective-of-public-policy/?doing_wp_cron=1596446412.4099969863891601562500
Lyon, H. et al. (2020, February). 2019 Artificial Intelligence and automated systems annual legal review. Gibson, Dunn & Crutcher LLP. Accessed July 31, 2020. https://www.gibsondunn.com/2019-artificial-intelligence-and-automated-systems-annual-legal-review/
Matytsin, D. E. (2021). Internet-investing as a remote algorithm of the retail investment financing. In: *Modern global economic system: Evolutional development vs. revolutionary leap*. Lecture notes in networks and systems. Springer Nature.

McGinnis, D., & Pierce, R. (2019). The Great undermining: How artificial intelligence changes the role of lawyers in the provision of legal services. *Actual Problems of Economics and Law, 13*(2), 1230–1250. https://doi.org/10.21202/1993-047X.13.2019.2.1230-1250

Morgan C., & Reed R. (2019). Dispute resolution in the Era of big data and AI. Herbert Smith Freehills. Accessed June 9, 2020. https://www.herbertsmithfreehills.com/latest-thinking/dispute-resolution-in-the-era-of-big-data-and-ai

Nikolova, E. (2019, September). Can AI outperform lawyers in dispute resolutions? *The Corporate Law Journal*. Accessed August 3, 2020. https://www.thecorporatelawjournal.com/technology/can-ai-outperform-lawyers-in-dispute-resolutions

Fulbright, N.R. (2018). Artificial Intelligence the ethical and legal implications. Accessed August 3, 2020. https://www.nortonrosefulbright.com/-/media/files/nrf/nrfweb/imported/ca---artificial-intelligence---slideshare.pdf?la=en-ca&revision=8e5c86af-cf7f-425f-858c-d5f83593d2ef

Pichai, S. (2018, June). Artificial intelligence at Google: Our principles. Google Official Blog, Russia. Accessed November 24, 2020. https://russia.googleblog.com/2018/06/google.html

Ponkin, I., & Redkina, A. (2018) Artificial intelligence from the point of view of law. *Bulletin of the Peoples' Friendship University of Russia Series: Legal Sciences, 22*(1), 91–109.

Potapeyko, P. (2018, June 14). The development of artificial intelligence in the EAEU: A view from Belarus. *Eurasia.Expert*. Accessed August 20, 2020. https://eurasia.expert/razvitie-iskusstvennogo-intellekta-v-eaes-vzglyad-iz-belarusi/

Rusakova, E. P., Inshakova, A. O., & Frolova, E. E. (2021a). Legal regulation of Internet Courts in China. Modern global economic system: Evolutional development vs. revolutionary leap. *Lecture Notes in Networks and Systems, 198*, 1515–1521.

Rusakova, E. P., Frolova, E. E., & Arzumanova, L. L. (2021b). Challenges of the judicial systems of the Russian Federation and People's Republic of China in the era of the pandemic. Modern global economic system: Evolutional development vs. revolutionary leap. *Lecture Notes in Networks and Systems, 198*, 1541–1549.

Rusakova, E. P., & Frolova, E. E. (2021). Procedural aspects of proof in China's internet courts: Opportunities for receiving BRICS jurisdiction. Modern global economic system: Evolutional development vs. revolutionary leap. *Lecture Notes in Networks and Systems, 198*, 1598–1605.

Suttie, H. (2020, January). The Next Decade of Legal Services: Embracing Law Market Change. Attorney at Work. Accessed August 12, 2020. https://www.attorneyatwork.com/next-decade-legal-services-part-1-law-market-change/

Toews, R. (2019, December). AI will transform the field of law. *Forbes*. Accessed September 11, 2020. https://www.forbes.com/sites/robtoews/2019/12/19/ai-will-transform-the-field-of-law/#3069931a7f01

Trustworthy AI–Brochure (2019, September). European Commission. Accessed November 19, 2020. https://ec.europa.eu/digital-single-market/en/news/trustworthy-ai-brochure

Turner, J. (2018). Responsibility for robots. *Thomson Reuters*. December 18, 2018. Accessed November 26, 2020. http://disputeresolutionblog.practicallaw.com/responsibility-for-robots/

Chapter 12
Current Problems of Digital Justice in the BRICS Countries

Ekaterina P. Rusakova and Evgenia E. Frolova

Abstract The research goal of this article is to identify the existing problems of implementing digital legal proceedings in the BRICS countries. It is proved that the most serious obstacles to achieving this goal are: (1) The lack of legislative regulation of the use of digital technologies for all civil proceedings; (2) The lack of technical equipment of courts throughout the territory of these countries; (3) The lack of coordinated work of judicial digital platforms or the lack of the ability to carry out certain procedural actions on it; (4) The unsatisfactory attitude of citizens to digital justice; (5) The low standard of living of the majority of the population of these countries. It is concluded that the development of digital justice occurs differently in these countries, and despite the identified problems, some of them have achieved great success. The current situation around the world related to coronavirus infection has played an impetus in some countries for the active integration of modern technologies, such as blockchain, artificial intelligence, and others, into the civil process. Digital technologies used in the process of protecting rights and legitimate interests have proven their effectiveness, but only within individual countries. It is proved that the ongoing changes in the judicial process radically change the approaches to the administration of justice.

Keywords BRICS · Civil procedure · Platform · Court · Artificial intelligence · Blockchain · Technology · Digital justice · Russia · China · India · South Africa · Brazil

---

E. P. Rusakova (✉) · E. E. Frolova
Department of Civil Law and Procedure and International Private Law, Peoples' Friendship University of Russia (RUDN University), 6 Miklukho-Maklaya Str., 117198 Moscow, Russia
e-mail: rusakova_ep@pfur.ru

E. E. Frolova
e-mail: frolova_ee@rudn.ru

E. E. Frolova
Leading Researcher, Vladivostok State University of Economics and Service, 41 Gogol St., 690014 Vladivostok, Russia

Department of Theory and History of State and Law, Far Eastern Federal University, 10 Ajax Bay, Russky Island, 690922 Vladivostok, Russia

A. O. Inshakova and E. E. Frolova (eds.), *Smart Technologies for the Digitisation of Industry: Entrepreneurial Environment*, Smart Innovation, Systems and Technologies 254, https://doi.org/10.1007/978-981-16-4621-8_12

JEL Codes K40 · D18 · K38 · L50

12.1 Introduction

The implementation of digital agendas in the countries under consideration has long been a major challenge. The stages of achievement are different, as the success of the implementation of digital technologies depends on many factors, including economic ones (Dudin et al., 2019).

Within the framework of economic cooperation between the BRICS countries, various acts concerning cooperation in these areas have also been adopted (Rusakova, 2018). As practice shows, despite the different levels of high-tech development, the creation of digital justice has not been achieved by any of the countries studied, although China has achieved the greatest success, which created the world's first Internet courts. India, which is characterized by the presence of large IT companies, cannot boast of implementing its results in the judicial system (Gaivoronskaya et al., 2019). Other countries are gradually integrating technology into various types of legal proceedings.

The legal problems of establishing new communication mechanisms for participants in the judicial process hinder the achievement of digital legal proceedings (Goncharov et al., 2019). The current situation in the world with COVID-19 has pushed all countries to more actively searching for a way out of the challenges, one of which is to ensure uninterrupted access to justice (Rusakova & Inshakova, 2021). It is proved that the transition of the judicial form of protection of the right to digital is not just a goal, but an existing reality of the modern world.

12.2 Materials

The basis of the regulatory framework of the study is made up of various acts regulating the procedure of digital civil proceedings: The Constitution of South Africa of 1996; acts of the Supreme Court of Appeal of South Africa; Decision No. 313 of the Brazilian Council of Justice of 2020 and the provisions of the state courts in its implementation; acts of the Supreme Court of India and judicial precedents of the highest courts of India; notice by the Supreme People's Court of Strengthening and Regulating the Online Litigation Work during the Period of Prevention and Control of the COVID-19 and a number of others; Resolution of the Presidium of the Council of Judges of the Russian Federation No. 439 "On Approval of the Concept for the Development of Court Informatization until 2020" of 2015; Resolution of the Presidium of the Council of Judges of the Russian Federation No. 457 of June 1, 2015 "On the creation of a pilot zone of GAS "Justice" on the basis of the courts of general jurisdiction of Moscow"; Resolution of the Presidium of the Supreme Court of the Russian Federation of March 18, 2020 No. 808 of April 8, 2020 No.

821 in order to counteract the spread of a new coronavirus infection in the Russian Federation.

The doctrinal positions that formed the theoretical basis of the study were studied thanks to scientific works devoted to the phenomena of digital (electronic, virtual, network) development of judicial proceedings, which have become a reality and a necessity, based on generally recognized values that are mediated by law: Goncharov et al. (2019), Matytsin and Rusakova (2021), Rusakova et al. (2019), Rusakova and Inshakova (2021), and Tarakanov et al. (2019).

Objective factors and prerequisites for the formation of the need for creating digital legal proceedings are studied in Barros and Schiller (2020), Broodryk (2019), Damle and Anand (2020), Hunter (2019), Pinheiro (2020), Rocha (2019), Vaid (2020), and Whitear (2020).

The impact of the digital economy on the state of civil society was reflected in the works of such authors as Dudin et al. (2019), Frolova et al. (2018), Gaivoronskaya et al. (2019), and Tarakanov et al. (2019).

The civil procedure law of the Russian Federation, China, and the BRICS member countries was studied by the authors in monograph "Resolution of private law disputes in the BRICS countries," 2018, as well as in the monograph "Resolution of financial disputes in the Asia–Pacific countries," 2019 and in the scientific work of Inshakova.

12.3 Methods

The study used the method of comparative analysis, which was used to compare the development of digital legal proceedings in the BRICS countries, as well as the methods of causal and systemic analysis, which allowed us to identify the main problems and difficulties in integrating digital technologies into the civil process of Brazil, Russia, India, China, and South Africa. Based on dialectical methods of cognition of the process of digitalization in these countries, directions for improving this process were formulated. Based on the results of the analysis, an assessment of the prospects for the creation of digital justice in these countries is provided.

12.4 Results

12.4.1 South Africa Experience

The process of implementing online courts has recently become increasingly important for the judicial system. The conceptual idea is to create a single judicial platform, which all participants in the process can have access to from anywhere in the world.

The concept of online courts is currently being implemented in the High Courts of South Africa, which consists of an end-to-end electronic system where litigants can file documents, manage the case and evidence online from any place and time, without being physically present in court.

It should be noted that the types of civil proceedings that can be considered online are also limited, for example, appeals against decisions in civil and criminal cases, absentee decisions, divorce proceedings, proceedings in an arbitration court, and others.

Online legal proceedings minimize the presence of persons involved in the case in court, which is now extremely important in the era of the coronavirus pandemic; cloud technologies allow you to process a huge amount of information and extract the necessary data, as well as quickly eliminate inaccuracies and prevent judicial errors, for example: when checking the identity of claims.

In the courts, the process of moving the case will occur automatically, which will reduce the flow of paper and court deadlines, as well as increase the efficiency of the judicial system. It will be possible to carry out procedural actions online on a special platform in real time.

One of the advantages of such a system is the rapid verification of documents in electronic form, as well as the ability to make various requests for information on the case, which will be received via SMS messages to mobile devices or e-mail.

The elements that make up the Court Online at present are:

- "Front-End Portal (FE)," which is accessed via the Internet, for any computer or mobile device of a law firm registered in the "Court Online" system»;
- "Workflow application," which is located on the ships' computers and serves the internal routing and workflows in the ships;
- "Case management application," which is downloaded to the computers of the courts and serves the completion of the process of registering cases, generating case numbers, further processing of cases at all stages of the proceedings and monitoring in real time, as well as the overall progress in the completion of cases.

On January 10, 2020, Judge-President Dunstan Mlambo issued a Directive regarding the full implementation of the online trial evidence submission process in the case management app. The Gauteng Division of the High Court of South Africa, which has offices in Pretoria and Johannesburg, has implemented e-case technology in the proceedings, as well as a trial management system.

In a recent South African case, when the High Court allowed witnesses to give evidence by video link, the judge noted that South Africa was lagging behind the rest of the world because there was no legal framework governing remote trial. Professor Dr. Omphemetse S. Sibanda criticized the courts for not moving "at full speed" to e-justice in this time of crisis and noted that only "turtle steps" are being taken in this direction. We have never needed this framework and support for the concept of remote hearings more acutely than we do now. While the global COVID-19 disaster is unprecedented and new and causes untold suffering, it has provided a valuable opportunity for the justice system to fully embrace and support the use of technology to continue to provide essential services to the people of South Africa, namely the realization of the right to a fair settlement of disputes by the courts, a fundamental constitutional right. It can only be restricted if there are no other reasonable means of achieving the purpose behind the restriction of the right. The health and safety of the participants in the trial could be ensured by holding remote hearings. Unfortunately, this is not done in the maximum number of cases, although it would be difficult (Whitear, 2020).

Thus, most court cases are dealt with in the usual way, although the current epidemiological situation could have a more effective impact on the process of creating digital justice in South Africa. Moreover, in his research work, Professor T. Broodryk emphasized the reverse process, when numerous practical directives restricting access to the courts were published in connection with the nationwide isolation. Consequently, the constitutional right of access to justice was severely restricted. This means that access to justice is currently unavailable to a significant portion of the South African population (Broodryk, 2019). The courts considered the possibility of filing class actions to be the most appropriate option, but this option did not solve the problem, but only delayed its solution.

In addition, this type of proceeding does not solve the problem of ensuring the right to judicial protection, but only allows a group of persons who have violated similar rights and legitimate interests to protect their rights by filing not several lawsuits in court from each plaintiff, but one lawsuit. But in any case, the judicial representative from this group will need to apply to the court and be present in the process.

An analysis of the civil procedure legislation shows that the process of ensuring that the parties can submit evidence to the court online is currently underway.

12.4.2 The Brazilian Experience

The following disadvantages are the characteristics of the Brazilian judicial system: the high workload of the courts and long deadlines. The number of appeals to the courts is constantly increasing, and the period associated with the coronavirus has only aggravated this situation.

On March 19, 2020, the Brazilian Council of Justice issued Decree No. 313, regulating the entire judicial system, which established the duty of the courts to guarantee the implementation of basic judicial services, as well as the transition to remote work. Some provisions have been adopted that implement this regulation by other states: Regulation CSM No. 2550/2020, Ato Normativo n. 7/2020, and others.

According to Decree No. 313, courts at all levels were required to provide the following minimum: I-to divide cases into judicial and administrative, depending on the priority and order of urgency; II-to technically ensure the direction and publication of judicial and administrative acts; III-to establish interaction with lawyers, lawyers, public defenders, prosecutors, and the judicial police, primarily remotely and in exceptional cases in person; IV-to provide payment services, institutional security, communications, information technology, and health; and V-urgent jurisdictional measures.

The face-to-face presence of the parties, lawyers, and interested parties should be replaced by a remote presence with the help of available technical means. Courts should ensure that magistrates, civil servants, and court staff work remotely when preparing decisions and sentences, minutes, holding virtual meetings, and performing other administrative activities.

By April 1, 2020, the Council of justice of Brazil established the extraordinary digital platform, which can be made in various legal actions online and conduct hearings (Barro & Schiller, 2020).

In his article, a Brazilian lawyer Rafael Marques Rocha gave the following data that starting from April to March was rendered remotely about 3 million judicial decisions in Brazil, where almost a million in the courts of the state of são Paulo (Rocha, 2019).

The effectiveness of the judicial system was tested directly by practicing lawyers, who were able to participate in court sessions via videoconference during the closure of the courts and their transition to remote work.

The judge of the Court of Appeal, Maria Teresa Gazine, stated that the courts currently consider all cases from consumer protection disputes to family disputes, as well as appeals, exclusively on the basis of the court's electronic platform (Pinheiro, 2020). However, to conduct an online hearing in civil cases, the consent of all participants in the process is required, and only in six of the forty cases, the parties applied for consideration of the person case; in all other cases, the cases were considered remotely. At the same time, the judges do everything possible to consider the dispute; so during the period of restrictive measures, such digital platforms as Zoom, Skype, Google Hang Out, and WhatsApp were actively used to communicate with the parties to the dispute, although it is still unclear how much the use of such funds is legal. The process of proof and the proceedings themselves must meet certain requirements, which must be followed by all participants in the process and strictly observed. The use of such means of communication is not regulated by law and, therefore, cannot be a permissible remedy from the point of view of the law (Tarakanov et al., 2019).

12.4.3 The Indian Experience

Back in 2005, India launched a project to create electronic vessels, which involves several stages. The process of digitalization of the judicial system is led by a specially created Electronic Committee under the Supreme Court of India.

Currently, the third stage is underway, which provides for the complete rejection of paper media in the highest courts of India and the states, as well as the creation of the first virtual court in Delhi and the active introduction of artificial intelligence technologies into the judicial process.

However, in practice, the exact opposite is happening, when for the first time on June 1, 2020, the entire panel of the Supreme Court of India held its first-ever paperless hearing. It was a rare sight to see three judges sitting in a virtual courtroom with laptops instead of bulky folders. Lawyers made presentations via video links and judges typed notes (Vaid, 2020).

Most lawyers support the development and preservation of electronic (virtual) courts in India since they were created as emergency courts.

According to the president of the Supreme Court Bar Association, Dushyant Dave, it is necessary to combine online and ordinary proceedings; but analyzing the existing practice of digitalization of the judicial system, he identified several problems: these are technological shortcomings of the system and the inability to create a single judicial platform, due to the huge number of courts in territories where there is no Internet, as well as the low level of knowledge of computer technologies of most participants in the proceedings. According to statistics for June 2020, 32.7 million cases were pending in Indian district courts, such a huge flow of court cases complicates the process of switching to digital format.

The stated achievements for each stage of creation do not correspond to reality, so far the electronic committee has not resolved the following issues:

1. Management of archived data that has already been submitted to the courts;
2. The definition of what constitutes a case, that is, it is not clear how to search for cases;
3. Technological guidelines for data quality control, there is no single standard for how data should be uploaded and what requirements should be met;
4. Data quality reviews, no data protocols available;
5. Institutional mechanisms, inefficient support for system users (Damle & Anand, 2020).

Despite all the shortcomings of the process of creating e-justice, the legal community hopes to continue implementing this project, which should have a positive impact on the entire judicial system of India: to improve the efficiency of legal proceedings and the quality of services provided.

12.4.4 China's Experience

Examining digital legal proceedings in the People's Republic of China, it should be noted that this process was fully implemented based on three Internet courts: Hangzhou Internet Court, which was established on August 18, 2017; Beijing Internet Court, established on September 9, 2018; Guangzhou Internet Court, established on September 28, 2018. It should be noted that the entire judicial system is actively being transformed into a digital standard of activity.

The legal regulation of the new courts is carried out in accordance with the procedural legislation, the explanations of the Supreme People's Court of the People's Republic of China, which are flexible regulators of rapidly changing public relations arising in the field of judicial proceedings. Issues related to the dispute resolution procedure were settled by the Regulation on Certain Issues of Consideration of Cases in Internet Courts No. 1747 of September 3, 2018, of the Supreme People's Court of the People's Republic of China.

It should be noted that initially, Internet courts were created for online trading platforms, where the parties make various transactions in real time, during the execution of which various disputes arise between the parties that require a quick and transparent solution. Access to this service on trading platforms is carried out in one click, which greatly simplifies the procedure for applying for legal protection. Disputes that may be the subject of consideration of these courts are somehow related to actions carried out on the Internet: from contracts of sale, lending, protection of copyright and related rights, and others.

Blockchain technologies are actively used by judges in the process of proof, and the data obtained in this way does not cause doubts among judges, which can significantly reduce the time of the trial, since the parties do not have to search for evidence, since all information is stored on this platform (Kalinina et al., 2019).

However, there are also a number of problems in the activities of Internet courts; for example, the Hangzhou Internet Court has been criticized for its lack of impartiality, since it is technically supported by the Internet platform of Alibaba and its subsidiaries, whose interests are traced in most disputes (Hunter, 2019).

One of the innovations of these Internet courts is the ability to consider disputes by robots with artificial intelligence technologies. At the same time, as practice shows, the parties to the trial do not object to the participation of a robot judge in the process. The first visual robot judge was a female judge wearing a black robe. Especially for skeptics of the robotic trial, representatives of the Internet court explained that a human judge can actively influence the course of the process, and all actions take place under his strict control.

In the era of coronavirus infection, the task of switching to digital justice was faced by all parts of the judicial system, so the Supreme People's Court of the People's Republic of China published explanations on the work of the courts in the period of COVID-19 No. 40 of the year 2020. In which, it was emphasized that all higher people's courts should actively and effectively coordinate and direct lower courts to facilitate online trials, develop document templates, regulate relevant court procedures, and strengthen leadership over lower courts so that online judicial activities are regulated, unified, lawful, and orderly. Any process based on the judicial platform must comply with the following order: "person, case, and account."

In order to comply with the legality of digital legal proceedings, online court services were provided to the persons involved in the case through electronic means, such as China Mobile Micro Court, China Judicial Process Information Online, the unified national judicial platform, e-mail, and instant messaging services.

The analysis allows us to conclude that China has responded more adequately to all the challenges associated with restrictive measures around the world since the idea of digitalization of all areas of the population's life has long been systematically implemented.

12.4.5 Russian Experience

The process of digitalization of the judicial system in the Russian Federation lasts long enough and is being implemented gradually. There are many reasons: technological, financial, social, and other (Frolova et al., 2018).

The first attempt to transform the judicial system was undertaken in 2001 with the adoption of the Federal target program "Development of the judicial system of Russia" in the years 2002–2006. During this period, the main goals of the reform of the judicial system were announced, namely the development of the material and technical support of the judicial system and its information support (Matytsin and Rusakova, 2021).

In 2015, the Decree of the Presidium of the Council of Judges of the Russian Federation No. 439 "On the approval of the Concept of the development of informatization of courts until 2020" as well as the Decree of the Presidium of the Council of Judges of the Russian Federation of June 1, 2015, No. 457 "On the creation of the experimental zone of GAS "Justice" based on the courts of general jurisdiction of Moscow" were adopted, and then the concept of informatization of activities was developed in relation to the Supreme Court of the Russian Federation.

Initially, the possibility of performing procedural actions in electronic format concerned the procedure for filing a claim, but then, almost three years later. It became possible to download the appendices to the claim.

The law establishes the possibility of holding court sessions via videoconferencing, which is regulated, and we are talking about the presence of the party in the courtroom, which is located in the vicinity of its location.

During the period of restrictive measures related to coronavirus infection, the Decree of the Presidium of the Supreme Court of the Russian Federation and the Council of Judges of the Russian Federation of March 18, 2020, No. 808 of April 8, 2020, No. 821 were adopted in order to counteract the spread of new coronavirus infection in the territory of the Russian Federation, which provided for the courts, if technically possible, to initiate consideration of cases by using video conferencing systems.

The period of restrictive measures demonstrated all the weaknesses and problematic issues of the judicial system; so at the very beginning, the proceedings in most cases were postponed.

Subsequently, in most civil cases, the judges switched to the digital model of the dispute resolution procedure as much as possible. But it is too early to talk about fully digital justice, while it is possible to state the possibility of performing certain procedural actions in electronic format.

12.5 Conclusion

A study of the judicial systems of the BRICS countries pointed to a number of problems that these countries faced in the process of implementing digital subpoenas in this area (Rusakova et al., 2019).

The situation that has developed in the COVID-19 investigation has demonstrated all the weaknesses of the judicial form of protection of rights. The lack of legislative regulation of the use of digital technologies in legal proceedings during the period of remote work has led to a violation of the terms and procedure for the consideration of disputes, and all this has become the norm for many courts. Holding hearings with the use of messengers in many countries was the only way to somehow resolve the dispute. But such actions may in the future be the basis for filing complaints to higher authorities and what seemed to be a way out might soon become a dead end.

It should be noted that many countries have not been able to switch to a new format of interaction, due to the lack of technical support and financial capabilities (Artemyeva et al., 2019).

The technical equipment of ships in many countries, such as South Africa, Brazil, and India, is so weak that it is not necessary to talk about the introduction of any new technologies: blockchain, cloud technologies. It can be stated that in countries with large territories such as Russia, China, and India, it is not always possible to access the Internet, and therefore, the transition to electronic judicial support is impossible; so in such regions, the only way to obtain judicial protection is to proceed in the usual manner.

The last problem is the poor training of personnel in the context of the development of modern technologies when often the judges themselves do not know how to carry out a procedural action in an electronic format, what can we say about the ordinary population.

However, it is safe to say that after COVID-19, no judicial system in the world will remain the same, since society must be ready at any time to respond to global challenges.

Acknowledgements This work was financially supported by the Grant of the President of the Russian Federation No. NSH-2668-2020.6 "National-Cultural and Digital Trends in the Socio-Economic, Political and Legal Development of the Russian Federation in the twenty-first century".

References

Artemyeva, Y. A., Ermakova, E. P., Ivanovskaya, N. V., Protopopova, O. V., Rusakova, E. P., Sitkareva, E. V., & Frolova, E. E. (2019). *The financial disputes resolution in the Asia-Pacific region*. Infotropik Media.

Barros, A. C., Schiller, A. (2020). *Brazilian Courts and COVID-19: A forced switch of paradigm*. Accessed February 15, 2021. https://www.lickslegal.com/covid-19/brazilian-courts-and-covid-19-a-forced-switch-of-paradigm

Broodryk, T. (2019). *COVID-19, the courts and access to justice*. Accessed February 15, 2021. https://juta.co.za/documents/810/Issue_9_Talking_Points__The_Courts_and_access_to_Justice.pdf

Damle, D., & Anand, T. (2020). *Problems with the e-Courts data*. Accessed February 15, 2021. https://www.nipfp.org.in/media/medialibrary/2020/07/WP_314__2020.pdf

Dudin, M. N., Frolova, E. E., Protopopova, O. V., Mamedov, A. A., & Odintsov, S. V. (2019). Study of innovative technologies in the energy industry: Nontraditional and renewable energy sources. *Entrepreneurship and Sustainability Issues, 6*(4), 1704–1713.

Frolova, E. E., Polyakova, T. A., Dudin, M. N., Rusakova, E. P., & Kucherenko, P. A. (2018). Information security of Russia in the digital economy: The economic and legal aspects. *Journal of Advanced Research in Law and Economics, 9*(1), 89–95.

Gaivoronskaya, Y. V., Miroshnichenko, O. I., & Mamychev, AYu. (2019). The immodest charm of digitalization. *Legal Concept, 18*(4), 40–47.

Goncharov, A., Dolinskaya, V., Inshakova, A., & Kalinina, A. (2019). Corporate conflicts in the system of public relations, law, and legislation of the Russian Federation. In: Popkova, E. G. (Ed.) *"Conflict-free" Socio-economic systems* (pp. 75–88). Emerald Publishing Limited.

Hunter, S. (2019). *China's innovative internet courts and their use of blockchain-backed evidence*. Accessed February 15, 2021. https://conflictoflaws.net/2019/chinas-innovative-internet-courts-and-their-use-of-blockchain-backed-evidence/

Kalinina, A. E., Inshakova, A. O., & Goncharov, A. I. (2019). Polysubject jurisdictional blockchain: Electronic registration of facts to reduce economic conflicts. In: Popkova, E. G. (Ed). *Ubiquitous computing and the internet of things: Prerequisites for the development of ICT*. Studies in computational intelligence (Vol. 826, pp. 205–213). Springer Science + Business Media.

Matytsin, D. E., & Rusakova, E. P. (2021). The strategy of quality management in industry 4.0 and formation of a cognitive economy based on industrial and manufacturing engineering in the Russian Federation and countries of the EU. *International Journal for Quality Research, 15*(4).https://doi.org/10.24874/IJQR15.04-03

Pinheiro, L. B. (2020). *Courts across Brazil increase productivity during coronavirus outbreaks*. Accessed February 15, 2021. https://latinlawyer.com/article/1225298/courts-across-brazil-increase-productivity-during-coronavirus-outbreak

Rocha, R. M. (2019). *The remote work of Brazilian courts during the COVID-19 pandemic*. Accessed February 15, 2021. https://www.lexology.com/library/detail.aspx?g=e322a55a-569d-4388-a4ea-1d013944b156

Rusakova, E. P. (2018). *Resolution of private law disputes in the BRICS countries* (p. 302). RUDN University.

Rusakova, E. P., & Inshakova, A. O. (2021). Industrial and manufacturing engineering in digital legal proceedings in the Asia-Pacific region: A new level of quality based on data, blockchain, and AI. *International Journal for Quality Research, 15*(1), 273–289.

Rusakova, E. P., Frolova, E. E., Zankovsky, S. S., & Kupchina, E. V. (2019). Problems of implementation of leadership in dispute resolution of the BRICS countries (on the examples of the Russian Federation, China, India). In *6th International conference on education, social science and humanities* (pp. 754–759).

Tarakanov, V. V., Inshakova, A. O., & Dolinskaya, V. V. (2019). Information society, digital economy, and law. In: Popkova, E. G. (Ed.). *Ubiquitous computing and the internet of things: Prerequisites for the development of ICT*. Studies in computational intelligence (Vol. 826, pp. 3–15). Springer Science + Business Media.

Vaid, D. (2020). *How coronavirus is propelling the rise of online courts in India*. Accessed February 15, 2021. https://www.dw.com/en/how-coronavirus-is-propelling-the-rise-of-online-courts-in-india/a-53774109

Whitear, N. (2020). *Remote justice COVID-19 South Africa lags*. Accessed February 15, 2021. https://clms.ukzn.ac.za/remote-justice-covid-19-south-africa-lags-behind/

Chapter 13
Smart Contracts in the Digital Economy: Contractual Regulation and Dispute Resolution

Sergei A. Sinitsyn, Maria O. Diakonova, and Tatyana I. Chursina

Abstract This article has been prepared for the research purpose of identifying, disclosing, and justifying certain trends in the development of civil law and procedures in the context of the spread of smart contract practices and the expansion of their spheres of application. The key issue in the conclusion and execution of smart contracts is the issue of the exhaustion of options for the protection of civil rights by the algorithm of the program code, the possibility, and conditions for the parties to exercise their right to judicial protection of their legitimate interests. At the moment, there is no uniform approach to choosing an optimal form for the legal regulation of smart contracts within the system of contract law in modern legal systems or international law, while globalization and digitalization of the economy imply the growth of cross-border transactions. The emergence of smart contracts is due to the development of e-commerce, in which the parties' interactions are carried out electronically instead of in physical exchanges or direct physical contact. Smart contracts gaining popularity in circulation are based on two interrelated elements: First, they eliminate the direct participation of a person in some or all cases of executing the agreement, using an automated code designed for execution without reference to the intentions of the contracting parties after publication; second, they eliminate or reduce the self-control and third-party control of the commitment, they make use of decentralized blockchain technology, as well as provide automatic code execution without the potential intervention of any party. This study examines the content,

S. A. Sinitsyn (✉) · M. O. Diakonova
Department of Civil Legislation and Procedure, Institute of Legislation and Comparative Law, Government of the Russian Federation, 34 Bolshaya Cheremushkinskaya St, 117218 Moscow, Russia
e-mail: synss@mail.ru

M. O. Diakonova
e-mail: dolovamaria@mail.ru

T. I. Chursina
Constitutional Law Department, Institute of Legislation and Comparative Law, Government of the Russian Federation, 34 Bolshaya Cheremushkinskaya St, 117218 Moscow, Russia
e-mail: tatyanachursina@hotmail.com

A. O. Inshakova and E. E. Frolova (eds.), *Smart Technologies for the Digitisation of Industry: Entrepreneurial Environment*, Smart Innovation, Systems and Technologies 254, https://doi.org/10.1007/978-981-16-4621-8_13

conclusion, and validity, protection of rights and legitimate interests of the parties, interpretation, and legal nature of smart contracts.

Keywords Legal nature and contents of a smart contract · Validity and conclusion of a smart contract · Protection of rights under smart contracts · Arbitration of platforms

JEL Codes K41 · O38 · L86 · K10 · K15 · K24 · E65 · K15 · O14 · O25 · O33

13.1 Introduction

The nonlinear growth of technologies and the expansion of their spheres of application in all forms of social interaction have opened up new opportunities for the digital economy in terms of the automated conclusion and execution of contracts (Inshakova et al., 2020a, b). New economic and technical realities have required the renewal of contract law, and the question of the content of the smart contract was urgent.

To date, there has been no uniform understanding of the purpose and scope of smart contracts; the requirements for their conclusion and execution; their validity; the possibility of invalidating them; conditions for the jurisdictional protection of rights under them; or the place of smart contracts within the system of contractual obligations.

The irreversible, rapid, and exponential growth of digital technologies and robotics in the production and exchange of goods, as well as their consumption, have been objective factors in reorienting the organization and structuring of economic activity (Frolova et al., 2018, p. 1236–1242). The resultant modification of economic relations has given rise to the necessity to revise national legislation and to assess its adequacy in these new conditions. In combination with these circumstances, the question of the possibility of applying to smart contracts general provisions on obligations, rules on certain types of contracts, and guarantees of protection for the parties to a smart contract to the extent already provided for by national and international law for civil contracts has become acute (Rusakova et al., 2020, p. 665–673).

Neither in Russian nor in foreign legal literature, can we find uniform or universal approaches to predicting the future of contract law based on convincing evidence and argumentation. An analysis of Russian and foreign legal doctrine reveals a difference of views on paths of development and the future of contract law under new economic conditions and digital reality.

13.2 Materials

The regulatory framework of this study is based on the norms of Russian and foreign civil and procedural legislation, legal doctrine, and practice. The subject of the study was the norms of Russian and foreign law, approaches of law enforcement practice and legal doctrine on electronic transactions and the automated performance of obligations, the conclusion, and validity of smart contracts, the consideration of disputes on smart contracts in arbitration, and competent courts.

In the context of global trends in the development of digital technologies, the current integration and harmonization processes necessary for the globalization of markets and eliminating incompatible differences and insurmountable barriers between and among contemporary legal systems, the authors assert the inexpediency of a study restricted to the development of civil and procedural legislation on smart contracts using the example of just one jurisdiction (Russian law), which explains the involvement and analysis of foreign sources.

13.3 Methods

This research has been prepared based on general (deduction, dialectical analysis, and the intersectoral relations of objects) and specialized (comparative-legal, economic-legal, and social-legal) methods of scientific experimentation, which have allowed us to propose for discussion and present the authors' understanding of the optimal forms of regulating smart contracts.

13.4 Results

The development of civil legislation in order to adapt it to the digital economy involves a formulation and resolution of the following interrelated issues.

1. *Demand for smart contracts in the digital economy*—The convenience of using smart contracts is anticipated to be primarily in reducing the costs of preparing and agreeing on a contract text by the parties and in automated execution. The capabilities of machine-readable law provide for text harmonization smart contracts with mandatory norms of the law and prevention of conflicts of contractual terms and legal norms.

It should be noted that the use of smart contracts requires special knowledge and skills in digital technologies. In this regard, the level of digital literacy of the contracting parties is rarely the same, which in itself raises the question of the actual and legal equality of opportunities of the parties, and the need to provide increased guarantees of protection. This circumstance alone excludes the possibility of separate consideration of smart contracts in isolation from any legislation: the provisions of the law on general provisions for certain types of contracts, general provisions on contracts, transactions, and obligations.

2. *The legal nature of a smart contract*—The term "smart contract" was first developed by Nick Szabo in the mid-1990s; the basic idea behind smart contracts is that many types of contractual provisions can be embedded in them (such as linkage or coupling and delineation (delimitations) of property rights), (Szabo, 1994). However, smart contracts, such as transaction processing systems for daily payments and receipts in financial institutions, have been around for decades. Smart contracts are user-defined programs that follow the rules governing transactions.

A smart contract consists of a program code, a storage file, and an account balance. Typically, a user can create a contract by placing a transaction on a blockchain, although not all blockchains support smart contracts. The program code of the contract is fixed when the contract is created and cannot be changed (Chursina, 2020, p. 42–46).

There is no consensus within legal doctrine about the nature of smart contracts (Inshakova et al., 2020a, b). There are polar different points of view that consider a smart contract: a new form of interaction between the contracting parties; an autonomous form of transaction that does not require external intervention and influence due to the very fact that the parties have chosen digital interaction to determine their civil legal relations; a new institution of contract law, in the absence of regulation of which the norms of the law on obligations and contracts can be applied to the relations of the parties vicariously.

It is obvious that contrasting "smart" high-tech with "stupid" traditional contracts is incorrect. The hypotheses that elevate smart contracts from high-tech tools to the level of quasi-legal entities of a new technological reality seem to be phantasmagoric (Ertman, 2017, p. 89–90). The negative consequences of disseminating such approaches are obvious since the prospects imposed for replacing traditional contracts with smart contracts do not take into account the role of the state and law enforcement agencies, primarily the courts, in maintaining the stability and parity of interests of participants in contractual legal relations. Arguments about the independence and inconsistency of content of the smart contract with the essence and features of the obligation contradict the public interests, legality, stability, and transparency of civic commerce, thereby negatively affecting the interests of an undefined circle of persons. It is possible to expect the development of unscrupulous practices of consistent and often involuntary involvement in the digital environment of persons, most often consumers, most of whom do not have the experience, knowledge, and/or skills for using digital communications (e.g., imposing the usage of new technologies in contractual legal relations).

The essence and values of the law suggest that the protection of the interests of such persons as the "weak side of the legal relationship" implies the creation of additional regulatory protection mechanisms, and not placing rosy and unfounded hopes on the principles of self-regulation and self-organization of digital civic commerce. The clarity of such regulatory guarantees will avoid gaps in legal oversight and prevent the growth of illegal and unfair behaviors in contractual practices. At the same time, the existence of previously established general rules on obligations for regulating contractual legal relations in the digital environment may not be sufficient, and, as they develop, specific legal norms may need to be corrected.

It seems appropriate to consider a smart contract as a technology for concluding and executing contracts as transactions. The contents of a smart contract are fixed using a programming language and codes, which do not allow us to consider a smart contract as a new type or kind of contract. Questions of interpreting smart contracts, determining the rights and obligations of the parties, and the conditions and scope of protection are all resolved by general rules of contract law. It would be correct to say that the spread of smart contracts would significantly simplify the technique of contractual work, but not the meaning or content of contract law, which sets limits on the autonomy of the will of the parties and provides guarantees for protecting the legitimate interests of their participants.

The legislation should provide requirements for testing and verifying the software codes used against violations of the rights of the parties and applicable legislation. With time and the accumulation of e-commerce practices and the smart contracts used in them, the question of separate legal regimes for traditional contracts and smart contracts may arise. Given the complexity of legal regulation of relations on the Internet, the regulatory oversight of smart contracts can be a complex task, which cannot be solved simultaneously.

3. *Dispute resolution from smart contracts: procedural aspects*—Currently, the legislation of modern legal systems does not provide for special mechanisms for the protection of rights under smart contracts, including specifics of judicial procedures for resolving disputes; methodological approaches for law enforcement practice in this regard are only beginning to be developed.

However, in the context of recognizing the right to a fair trial as one of the most important of human rights (Article 6, Convention for the Protection of Human Rights and Fundamental Freedoms), the possibility of protecting the rights of participants in smart contracts in the domestic courts of nation-states cannot be excluded. In addition, the use of smart contracts in court proceedings has serious potential, since information regarding actions performed that are stored in such contract has a high degree of reliability and, as a rule, does not require additional research by the court. At the same time, the involvement of experts or specialists is unavoidable to evaluate the contents of a smart contract (Sinitsyn & Diakonova, 2020, p. 29).

At the same time, taking into account the form in which a smart contract is concluded, it is more effective and realistic to use alternative judicial forms of dispute resolution that arise: i.e., through arbitration or conciliation procedures, both with or without the involvement of an intermediary. In this case, an arbitration, mediation, or other dispute resolution clause can be included directly in the smart contract code. Such stipulations may include:

- automatic adoption of interim measures (e.g., suspension of performance of obligations under a smart contract or blocking of funds);

- rules and deadlines for the creation of arbitration, selection of an intermediary;
- procedures and deadlines for dispute resolution and negotiations;
- procedure for execution of arbitration decisions as reached by the parties to the agreement.

The provisions of a smart contract must necessarily include an agreement on arbitration of the "in-platform" type (network arbitration) or "outside-the-platform" type, and substantive law should determine the potential availability of appeal to the competent court for the interested party.

"Network" arbitration involves the use of technological solutions by which a smart contract automatically enforces the equivalent of a traditional arbitration award under the terms of the smart contract; for example, "this can be done… by the parties providing the smart contract with certain features (for example, cryptocurrencies), which, upon the occurrence of a certain condition, are transferred from one party to the other."[1] Only over the long term is it possible that online arbitration can take place in a completely robotic manner (although such a solution has not yet been effectively implemented), without the need to participate in any actions taken by a third party or by the parties themselves. For example, it has been suggested"the algorithm can resolve a dispute based on an analysis of similar transactions and disputes" (Maxwell & Vannieuwenhuyse, 2018, p. 29). The reality is, at least for the foreseeable future, that there is no technology yet to provide a fully automated arbitration system that both the legal and technology communities can rely on.

Off-chain arbitration is more similar to regular arbitration in that it does not feature automatic execution of decisions. This leads to further complications since Article II of the 1958 New York Convention on the Recognition and Enforcement of Foreign Arbitral Awards requires that the arbitration agreement be "in writing." It is not known whether a decision made under an arbitration agreement contained within a smart contract code can be enforced. Perhaps, this is another reason why the parties should seriously consider whether to enter into an arbitration agreement in parallel or not. In addition, given that smart contracts, after all, agreed conditions have been met, trigger an irreversible transaction, arbitration may need to find new remedies for decisions thus taken. Perhaps, such legal remedies will be supported by a set of new arbitration rules designed specifically to resolve smart contract disputes.

4. *Self-regulation and self-determination of platforms in the resolution of disputes from smart contracts*—Internal dispute resolution based on the blockchain remains radically uncertain and does not create clear advantages over traditional litigation. Nevertheless, and in the absence of full clarity of the regulatory oversight of dispute resolution systems that use blockchain, in fact, they already exist (e.g., OpenBazaar and eBay use moderators and special arbitrators of the platforms for dispute resolution).

Thus, several successful dispute resolution protocols are currently known. For example, platforms such as Kleros, JUR, Aragon Network Jury, OpenCourt, and

[1] URL: https://newtech.law/en/on-chain-and-off-chain-arbitration-using-smart-contracts-to-amicably-resolve-disputes/ Accessed: 06.03.2021.

OpenBazaar offer products that are designed to improve the process of executing smart contracts. These applications are similar in many key aspects, but in a competitive environment, their developers claim unique strategies to encourage juries (jury members and arbitrators—jury), different levels of legal protection, and specialized tribunals. The dispute resolution algorithm on these platforms is generally as follows.

After the contract, the buyer contributes a sufficient amount of funds to the smart contract. This crypto-payment will remain in escrow on the blockchain until: (1) the buyer confirms satisfaction with the seller's performance; or (2) any initiated dispute is resolved. In addition, at the smart contract stage, users determine the number of jurors, a specialized tribunal, and a list of possible future remedies.

Most of these dispute resolution applications also offer users the ability to form a natural language agreement to accompany a code-based smart contract. The level of clarity and completeness of the contract varies depending on the platform, with some applications going to great lengths to give the smart contract traditional legal force. Open court, for example, offers templates for drawing up contracts in natural languages in addition to an agreement in the form of code. In all cases, any buyer who believes that the other party has improperly fulfilled the obligation from the smart contract can initiate a dispute resolution. Sellers usually do not have this option, as they have not deposited a payment into a smart contract. Next, the dispute is resolved by a jury.

These platforms strive for pure decentralization and anonymity, so jury selection relies on crowdsourcing. Moreover, candidates for the jury for resolving disputes from smart contracts are volunteers who deposit cryptocurrency for any amount in the hope of being selected as a juror. Once enough candidates have made deposits, a lottery takes place to form a jury. On certain platforms, the probability of being selected in this lottery is directly proportional to the size of the deposit.

Decentralized jury dispute resolution systems differ from any other settlement process in two key aspects.

Firstly, jurors remain completely anonymous throughout the arbitration proceedings.

Secondly, the blockchain uses a financially stimulated majority voting mechanism: those jurors who do not vote by a majority vote stand to lose some or all of their down payment. The developers point out that, in this way, the jury is not interested in voting that is biased or arbitrary but rather is forced to come to the fairest possible decision. At the same time, none of the existing "network platforms" requires jurors to rely on any legal analysis of the dispute or legal precedents when making their decision. The jurors, isolated from each other, simply vote and offer justifications for their decisions. Following an initial decision, most decentralized platforms offer the parties the opportunity to appeal, with the exact format and cost of appeal varying depending on the platform, usually providing for an increase in the arbitration fee to minimize cases of appeal (Buchwald, 2020, p. 1384–1392).

The full explanation to users of the dispute moderation process is limited to four thesis: "In the event of a dispute, both parties will present their evidence. Reasonable efforts will be made to resolve the dispute in a mutually agreeable manner. If a decision cannot be reached by agreement of the parties, the evidence will be weighed in an arbitral award. If it is not clear who is right, the subject of the request will be divided at the discretion of the moderator" (Sklaroff, 2018, p. 301).

The described method of dispute resolution is devoid of a standard of efficiency since there is no possibility for limiting in advance the range of checks used by arbitrators, who may not respect the body of accumulated experience in resolving similar cases. As a result, a decentralized court decision will become increasingly resource-intensive over time, as the parties try to determine all possible circumstances in the program code. In other words, without the use of the arsenal and means of jurisdictional protection of civil rights, the parties will need to discuss each dispute from scratch, without having any idea of how it might be resolved. Organizational problems are also obvious (Matytsin, 2021). The platform should ensure transparency of the formation and activity of arbitrators: publish reviews of practice, explanations, etc. However, it is also true that the more a decentralized dispute resolution system becomes similar to state courts, the more it kills its popularity and demand.

The advantages of resolving disputes by arbitration of platforms are seen in the observance of confidentiality and delocalization of legal relations under a smart contract, in the absence of clear instructions from the law leading to uncertainty of a nation state's court when accepting a dispute in its proceedings.

Smart contracts will make improvements to contract law not automatically, but only if their regulatory support is clear and the rights of the following are protected. In the absence of such contracts, smart contracts cannot be considered a flexible or convenient form of registration of obligations. At the present stage, the use of smart contracts is often localized by separate areas of economic activity, for example, the management of certain categories of routine and undisputed transactions.

Decentralized dispute resolution systems on platforms are not isolated from legal qualifications. Like any contract, a smart contract may have defects that lead to its invalidity or non-conclusion: illegal actions by one or both parties; the platform; an incorrect reflection of the will of the parties in the program code; the illegality of the purpose or consequences of a smart contract; lack of requisite powers by a party; errors in coding and/or reading the code.

Parties having a dispute over a smart contract may not be restricted in their ability to apply to a nation state's court for its execution, as is guaranteed concerning traditional contracts.

13.5 Conclusion

1. The smart contract as technology and form contributes to the realization of private interests only in the digital economy and commerce. A smart contract cannot be implemented outside of digital technologies and space, which refutes judgments about the prospect of replacing all contractual forms used in circulation with smart contracts.
2. Substantively, a smart contract means the execution of a contract in the form of a program code in the blockchain to ensure the subsequent automatic and

autonomous self-fulfillment of the conditions as laid down in the program. There are no grounds for considering a smart contract as a new classification element within the system of contractual regulation (i.e., kind or type of contract).

3. Smart contracts are a new technical opportunity and a tool for implementing contractual freedoms in the digital economy. Neither smart contracts themselves nor their content can be considered in isolation from the regulation of binding legal relations as established under Russian law, including the principles of performance of obligations, freedom of contract, and protection of binding rights. The general direction of development of legislation in the context of ever-increasing practices of using new digital technologies in the execution and conclusion of transactions should exclude the very possibility of creating a parallel reality unregulated by law, in which general guarantees of the exercise of civil rights and/or ensuring fair and accessible justice do not apply, wherein opportunities for the flourishing of illegal behavior are created.
4. Currently, several successful blockchain-based dispute resolution protocols are known. However, their analysis demonstrates that such dispute resolution mechanisms remain radically uncertain and do not create clear advantages over traditional litigation. In this regard, and taking into account the widely upheld recognition of the right to a fair trial as a fundamental human right, parties who have a dispute over a smart contract cannot be restricted in their ability to apply to a nation state's courts for its execution, as is guaranteed concerning traditional contracts.

References

Buchwald, M. (2020). Smart contract dispute resolution: The inescapable flaws of blockchain-based arbitration. *University of Pennsylvania Law Review, 168*(1369), 1385–1393. (In Rus.).

Chursina, T. I. (2020). Legal basis for regulating blockchain technology and smart contracts in Malaysia. *International Public and Private Law, 6*, 42–46. (In Rus.).

Ertman, M. (2017). Smart rules for smart contracts. *The Journal of Things We Like (Lots)*, 89–90.

Frolova, E. E., Zankovsky, S. S., Dudin, M. N., Zinkovsky, S. B., & Kirsanov, A. N. (2018). Studying concepts of the breakthrough economic reforms in selected developed and developing countries and regions of the world: Economic and legal aspects. *Journal of Advanced Research in Law and Economics, 9*(4), 1236–1242.

Inshakova, A. O., Goncharov, A. I., & Salikov, D. A. (2020a). Electronic-digital smart contracts: Modernization of legal tools for foreign economic activity. *Lecture Notes in Networks and Systems, 91*(3–13), 16.

Inshakova, A. O., Goncharov, A. I., & Ershova, I. V. (2020b). Digital blockchain registration of legally significant stages of complex good's export-import supplies by business entities of the EAEU and BRICS jurisdictions. In: Inshakova, A., Inshakova, E. (Eds.) *Competitive Russia: Foresight model of economic and legal development in the digital age. CRFMELD 2019*. Lecture notes in networks and systems (Vol. 110, pp. 328–336). Springer. https://doi.org/10.1007/978-3-030-45913-0_37

Maxwell, W., & Vannieuwenhuyse, G. (2018). Robots replacing arbitrators: Smart contract arbitration. *International Chamber of Commerce Dispute Resolution Bulletin, 1*, 29. https://www.hoganlovells.com/~/media/hogan-lovells/pdf/2018/2018_12_13_icc_robots_arbitrator.pdf

Matytsin, D. E. (2021). Internet-investing as a remote algorithm of retail investment financing. In *Modern global economic system: Evolutional development vs. revolutionary leap*. Lecture notes in networks and systems. Springer Nature.

Rusakova, E. P., Frolova, E. E., & Gorbacheva, A. I. (2020). Digital rights as a new object of civil rights: Issues of substantive and procedural law. In *Advances in intelligent systems and computing* (Vol. 1100 AISC, pp. 665–673).

Sinitsyn, S. A., & Diakonova, M. O. (2020). Smart contracts: The result of technology development or a new institute of civil law and process? *Legislation, 9*, 25–30. (In Rus.).

Sklaroff, J. M. (2018). Smart contracts and the cost of inflexibility. *University of Pennsylvania Law Review, 166*, 301.

Szabo, N. (1994). *The idea of smart contracts.* http://www.fon.hum.uva.nl/rob/Courses/InformationInSpeech/CDROM/Literature/LOTwinterschool2006/szabo.best.vwh.net/smart.contracts.html

Chapter 14
Remote Digital Technologies for Notary Protection of the Rights of Economic Activities Participants

Alexander V. Begichev

Abstract The article's research goal is to identify patterns associated with the implementation of distance digital technologies in notarial activities to increase the efficiency and attractiveness of the extrajudicial form of protecting the rights and legitimate interests of citizens and legal entities. It is proved that (1) performing remote and distant notarial actions should increase the efficiency of the notary's human rights activities; (2) the procedure has economic attractiveness; (3) the alternative procedure for performing notarial actions using digital technologies complements traditional actions, changing the procedure for their implementation; (4) it is necessary to improve the procedure for performing notarial actions in a remote and distant format and expand the competence of the notary, which uses remote digital technologies. It is concluded that the use of distance digital technologies in notarial activities should ensure an operational framework for obtaining notary services without visiting the notary office. Foreign experience in applying such a procedure has proven its effectiveness due to its organizational and economic attractiveness: The parties could seriously save time and receive notary services in the short term without leaving their location. The state's active position and the notary's interest in this procedure's positive result is a guarantee of efficiency. It is necessary to more quickly implement the remote digital procedure for the notarial protection of the rights and interests of economic participants, both among citizens with disabilities and in the business environment, by expanding the use of advanced information and communication technologies for the most popular notarial actions and simplifying the notarial procedure while implementing them and the comprehensive inclusion of notaries in the process of interdepartmental digital interaction of public authorities. It is substantiated that the use of distance digital technologies in notarial activities seriously changes the idea of a notary as a reliable and operational human rights institution that guarantees the confidentiality and legality of the services provided, the results of which have increased legal force in a trial and enjoy the complete confidence of all participants in economic activity.

A. V. Begichev (✉)
Department of Notaries, Kutafin Moscow State Law University (MSAL), 9 Sadovaya-Kudrinskaya St, 125993 Moscow, Russia

A. O. Inshakova and E. E. Frolova (eds.), *Smart Technologies for the Digitisation of Industry: Entrepreneurial Environment*, Smart Innovation, Systems and Technologies 254, https://doi.org/10.1007/978-981-16-4621-8_14

Keywords Notary · Remote and distant notarial actions · Notarial procedure · Human rights activities · Digital technologies · Electronic interaction

JEL Codes K19 · O33 · O52

14.1 Introduction

Currently, the program "Digital Economy of the Russian Federation" is being implemented, approved by order of the Government of the Russian Federation dated July 28, 2017, No. 1632-r, which is aimed at creating conditions for the development of a knowledge society in the Russian Federation, improving the welfare and quality of life of citizens by increasing the availability and the quality of goods and services produced in the digital economy using modern digital technologies, raising awareness and digital literacy, improving the availability and quality of public services for citizens, as well as security both within the country and abroad.

Reforming the legislation on notaries for the country's new tasks to develop the information society was reflected in notarial practice (Inshakova et al., 2017). Statistics records a steady decline in the once-popular notarial act of certifying copies of documents in the usual documentary framework. In 2020, their total number was about 25%, which was unthinkable ten years ago, when this notarial act gained a significant amount (more than 70%). At the same time, there is an increase in the number of notarial actions using information technologies (Rusakova et al., 2020).

In the context of the development of a digital society, the need for notarial protection of citizens' and legal entities' rights increases since the notary is one of the most reliable legal institutions in the continental legal system. In these conditions, the need for highly qualified personnel also increases (Korsik, 2020).

Electronic notarial services are influential and in demand; they increase the availability of qualified legal assistance and allow participants in civil circulation to receive services at their earliest convenience (Kirillova, 2020).

In the November 2020 issue of the Russian-language edition of the American scientific journal—Scientific American—an article, "The Language of Science," was published in which Lorraine Duston, director emeritus of the Max Planck Institute for the History of Science in Berlin, researched how our use of words has evolved over the past 175 years. She highlighted that among the readers of this magazine on the official website (ScientificAmerican.com), the most popular query in 2020 was the word "coronavirus" (Duston, 2020). On the one hand, this study shows us the concern of citizens for their safety in the face of an unknown and dangerous disease, and on the other hand, the search for opportunities for existence in the new conditions of restriction of freedom of movement. Moreover, one of the main priorities is using modern means of communication to remove these barriers, including implementing the legal protection of their rights and interests.

In this regard, the development of "digitalization" in the provision of notarial services is relevant. This topic's coverage is focused on the concept of "electronic notary," which can be characterized as a set of competence of notaries implemented using modern information technologies (Akhrameeva, 2020). The introduction from December 29, 2020, of several essential novelties in the field of "electronic notaries" brings the implementation of notary services to an entirely new level. Information technologies are rapidly being introduced into our lives, and without them, it is impossible to imagine the current state of civil and economic relations.

14.2 Materials

The basis of the research regulatory framework is the Fundamentals of the Russian Federation legislation on notaries; Federal Law of December 27, 2019, No. 480-FZ "On Amendments to the Fundamentals of the Legislation of the Russian Federation on Notaries and Individual Legislative Acts of the Russian Federation" (the so-called law on digital notaries); Orders of the Ministry of Justice of the Russian Federation dated September 30, 2020, regulating the procedure for performing notarial actions in remote and distant formats.

The doctrinal positions that formed the theoretical basis of the research were studied thanks to research devoted to digital notaries: Andropova (2020), Begichev (2021), Goncharov et al. (2019), Inshakova et al. (2017), Kirillova (2020), Korsik (2020), Matytsin and Rusakova (2021), and Renc and Filippova (2020).

The authors investigated foreign experience in the implementation of digital technologies in notarial activities in the works of Begichev (2020) and Renc and Filippova (2020).

The influence of the digital economy on civil society's state was reflected in the works of such authors as Akhrameeva (2020) and Daston (2020).

14.3 Methods

The study used the comparative analysis method, which helped to compare the experience of using digital technologies for the remote arrangements of performing notarial actions in the Baltic countries (Latvia, Lithuania, and Estonia) and the Russian Federation. It also utilized the methods of cause-effect and system analysis, which made it possible to identify the main patterns and practical results of implementing this procedure and distinguish the standard and different features from other formats of notary services provision. Relying on the dialectical methods of cognition of the remote and distant procedure in the Russian Federation, paths were shown to improve legislation in this area and the process itself by expanding the use of advanced information and communication technologies and the inclusion of notaries in the process of interdepartmental digital interaction of public authorities. Based on the analysis results, we assess the prospects for introducing distance digital technologies into notarial activities and give practical recommendations to improve this procedure's efficiency.

14.4 Results

It must be recognized that the development of information technology directly depends on the country's economic situation. First, creating favorable conditions for doing business contributes to the country's economic prosperity and the growth of satisfaction with the quality of life. Secondly, the possibility of developing business ties and satisfying public interest using digital technologies affects a large number of people, which requires timely and well-developed legislative regulation of these processes.

14.4.1 Prerequisites for the Implementation of Information Technology in the Member States of the European Union

In Europe, information technology in notaries began to be applied gradually due to significant differences in the Member States of the European Community's legal systems and, accordingly, different approaches to the legal regulation in this area.

Even though in Europe several completely different models of notaries are used—Latin (including three main models: German, French, and mixed), Scandinavian, and common law (currently represented only by Ireland)—the legal regulation of the sphere in question is actively developing, acts are being issued, both unifying and harmonizing the legislation of the Member States, which regulate the implementation of various aspects of notaries at the supranational level.

Depending on the specific country of the European Union, the fundamental role of the notary is different. Meanwhile, as we observe at the present stage of development of the European Union, the countries that are part of it do not set the goal of creating a single legal space in the field of notaries, since in general, the unification of European law is limited only to the sphere of commercial law and presupposes the preservation of the identity of individual states in areas of family law, property law, property law, and real estate rights.

The digitalization of public institutions should take place subject to equalizing opportunities for legal entities and individuals under the jurisdiction of national sovereignties while receiving modern electronic notary services (Renc & Filippova, 2020).

One of the European notaries' problems is the right of free movement, choice of place of residence, and work of all EU citizens. This is since states with different histories and legal systems coexist in the European Union and some other factors, mainly political, economic, and social. So, even within the civil law family, it is impossible to talk about the complete uniformity of national law systems.

Some researchers note that the presence of fundamental legislation differences will complicate the creation of inter-notarial systems. Simultaneously, several documents can be distinguished that regulate the implementation of various aspects of notarial activity at the supranational level. First, this is the Regulation of the European Parliament and the EU Council No. 650/2012 "On Jurisdiction, Applicable Law, Recognition and Enforcement of Decisions, Adoption and Execution of Notarial Acts, as well as the creation of a European Certificate of Succession," also known as the Rome-IV Regulation. Adopted on July 4, 2012, it responded to the acute issues of ensuring legal certainty regarding cross-border inheritance cases.

From February 16, 2019, they implement Regulation of the European Parliament and of the Council (EU) 2016/1191 of July 6, 2016, "On the promotion of the free movement of citizens by simplifying the requirements for the submission of certain public documents in the EU and amending Regulation (EU) No 1024/2012." The primary purpose of this regulation is to ensure the free circulation of public documents in the European Union, which are understood as "notarial acts; official certificates that are placed in documents signed by persons in their capacity, such as official certificates fixing the registration of a document or the fact of its existence on a certain date, as well as official and notarial certification of signatures."

Other European Union regulations partially regulate notaries' specific issues, which constitute a transnational block of legal norms that are binding on all participating countries.

In the newest vision of the prospects for the development of transnational legislation, it seems that electronic interaction is designed to accelerate the organizational and operational aspects that allow citizens and legal entities of Europe and Russia to exercise their constitutional rights to information, improve the quality of legal services, and the possibility of prompt participation in economic activity (Tarakanov et al., 2019). This is especially true during the period of global epidemiological disasters, such as the spread of the coronavirus infection COVID-19.

14.4.2 The Experience of the Baltic Countries (Latvia, Lithuania, and Estonia) of Remote Order of Certification of Transactions by Notaries

In Latvia, Lithuania, and Estonia, electronic notaries are developing quite actively, first of all, relying on the traditions of the countries of Latin notaries in European countries. At the Tripartite Council of Notaries of the Baltic States, held in the capital of Latvia, Riga, on September 26–27, 2019, they noted the rapid progress of these countries in the development of electronic notary services (https://notarius.lt).

According to the latest report of the World Bank "Doing Business-2019," reflecting the level of the business climate in the country, in terms of the ease of real estate transactions in the list of 190 countries, Lithuania ranked 4th in the world; Estonia, 6th; and Latvia, 25th. It is pleasant to note that Russia ranks 12th on this list. This report analyzes the costs of real estate transactions, including the salary of a notary, the time, and procedure required for them.

The innovations of notaries for the provision of electronic services for the certification of transactions in these countries allow us to assess the degree of their demand in our country for the reliable protection of the rights and legitimate interests of citizens and legal entities (Begichev, 2020).

Latvia. The main innovation of the Latvian legislation is the ability to carry out the notarization of transactions remotely, not only for citizens located in the Republic of Latvia but also abroad. Anyone anywhere in the world with access to a Latvian notary can complete a transaction. This system has been successfully applied in Latvia since July 1, 2018.

It is important to note that the video consultation framework's performance of notarial actions is carried out in Latvian, English, and Russian.

The algorithm for the provision of remote notary services is, first, that a citizen should register on the official portal of the Council of Sworn Notaries of Latvia www.latvijanotars.lv and log in through the portal Latvija.lv. After that, on the portal www.latvijanotars.lv, in the "Search for a notary" section, one needs to sign up for a video consultation with a sworn notary. Simultaneously, one can choose any notary in Latvia's territory, regardless of the citizen's residence. The client on the website chooses a convenient date and time for an online visit to a specific notary and sends a request. Further, the notary either sends the client a confirmation that at a specific date and time, he will be available for the provision of an online service or will offer a different date or time.

To receive the service remotely, the legislation prescribes that the client must be in a quiet and bright room, and he must have a stable Internet connection, a computer, a web camera, a microphone, headphones, a passport or electronic ID card, as well as access to Internet bank for payment of notary services. If a person needs a consultation with a notary and paperwork, the client must have an electronic signature.

Performing notarial actions remotely not only creates certain conveniences for clients but also allows them to save time.

In the so-called virtual notary office, one can perform absolutely all notarial actions and perform them with several people's participation.

Estonia. On January 30, 2019, the Estonian Riigikogu adopted amendments to the Law on Notarization, which, from February 1, 2020, will enable notaries to perform notaries' acts using videoconference via the Internet without the client being physically present at the notary office.

This innovation allows Estonian citizens and electronic residents residing outside Estonia to perform notarial acts.

To implement the innovation on the remote certification of notarial actions, the Estonian Chamber of Notaries has developed special software (X-Road services for organizing data exchange between the digital environment of notaries and the consul program), which will allow the client to enter the system using a national ID card or electronic certificates, as well as an electronic card of an electronic resident.

To perform a remote notarial act, one must contact the Estonian embassy abroad. The embassy officer identifies the applicant and provides a separate room with a computer equipped with special software for one-to-one communication with the Estonian notary via video link. In the future, when developing safe technologies, the client can enter into video communication with a notary from anywhere in the world.

Lithuania. The Lithuanian notary has adopted the German notary system and modified it following its legislation.

Lithuania has a unified electronic register—eNotaras—through which online interactions between notaries and their clients are carried out, including the use of secure electronic means for notarization of documents.

Under the official information presented on the website www.enotaras.lt, by automating the business processes of a notary, the tasks of the eNotaras information system will be (1) the transfer of notarial services provided to individuals, legal entities, and state bodies, (2) ensuring accessibility and the convenience of notaries services for clients, (3) unification (standardization) of the process of submitting, collecting, and storing information, as well as (4) electronic processing of data required by notaries to fulfill their statutory tasks.

The eNotaras information system consists of an internal and an external portal. The eNotaras information system's internal portal is intended for notaries, where certain notary services are performed, and access to the data necessary for performing notarial actions is available.

The portal of the external information system eNotaras is intended for the notary clients, which allows the client of the notary to prepare for the notarial act in the electronic space. The client can order a notary service and submit the necessary documents to the notary by e-mail. After preparing the documents, the client directly applies to the notary office to perform the notarial act.

14.4.3 Specifics of Performing a Notary Act Remotely in the Russian Federation

The fundamentals of the Russian Federation's legislation on notaries are from December 29, 2020, established a list of notarial actions that can be performed in electronic form, i.e., remotely, without a personal admission of the person who applied to the notary. These actions include certifying the correctness of translation made by a notary (Article 81), transferring electronic documents to individuals and legal entities (Article 86), accepting non-cash funds on deposit (Article 87), depositing non-cash funds (Article 88.1), making an executive note for debt collection in an indisputable manner (Article 89), providing evidence in the form of an examination of information on the Internet (Article 103), and issuing a notice from the register of notifications on the pledge of movable property (Article 103.7).

Also, following the example of French legislation, this law established the possibility of doing a remote deed through the electronic notary system (without the parties' joint presence) with the participation of several notaries acting in the interests of each party (Article 53.1). As follows from Article 434 of the Civil Code of the Russian Federation, an agreement in writing can be concluded by drawing up one document (including an electronic one), which is signed by the parties. The authenticity of the will, in this case, is established and guaranteed by each notary participating in the certification of the deed.

According to Article 34.2 of the Fundamentals of the legislation of the Russian Federation on notaries, the specified notarial actions are subject to mandatory registration in a special register of "notarial actions performed remotely and deeds certified by two or more notaries," which is an integral part of the Unified Notary Information System (hereinafter—UIS) and is maintained in electronic form...

The procedure for performing remote notarial actions is disclosed in Article 44.3 of the Fundamentals of the legislation of the Russian Federation on notaries. The peculiarity of performing this notarial action format is to simplify the receipt of notaries' services by citizens and legal entities using digital technologies.

In particular, the provision of notarial services in a remote digital format meets the requirements for the prompt preservation of electronic evidence when providing them. The possibility of remotely contacting a notary for the provision of evidence of notarial actions does not require additional verification of the citizen's legal capacity from the notary.

The law does not link the possibility of the remote and distant format of performing notarial actions with specific reasons, which means its alternative method and the traditional one—personal participation of the applicant in performing notarial actions.

In the development of the provisions of the legislation on the provision of electronic notarial services, the Ministry of Justice of Russia on September 30, 2020, developed a package of normative acts, consisting of 12 orders (registered on October 5, 2020), which detailed the procedure and rules for performing such notarial actions.

The remote and distant formats for performing notarial actions are carried out according to the general rule for conducting notarial procedures, but with some peculiarities divided into stages.

The first stage To perform a notarial act remotely, the applicant sends to the Federal Notarial Chamber (hereinafter—FNP) through the Unified Information System (UIS), including using a single portal of state and municipal services, an application for performing a notarial act remotely with the attachment of documents established by the legislation of the Russian Federation in electronic form.

The requirements for an application are established by the order of the Ministry of Justice of Russia dated September 30, 2020, No. 223 "On approval of the application form for the performance of a notarial act remotely, as well as requirements for the format of such an application and the formats of documents attached to it in electronic form," according to which the application is submitted in the form of XML—a document created using XML schemas and providing reading and control of the submitted data. Electronic documents attached to the application are submitted in the form of PDF files signed with an enhanced qualified electronic signature of the applicant if such documents come from the applicant or the person from whom the electronic documents originate, or a notary who has certified the equivalence of an electronic document to a hard copy, if the legislation of the Russian Federation does not establish other requirements for the formats of the documents attached to the application.

The validity of the electronic signatures with which the application is signed, as well as the electronic documents attached to the application, is verified utilizing the Unified Information System of Notaries following Federal Law No. 63-FZ of 06.04.2011, "On Electronic Signatures." The ownership of these electronic signatures to the relevant persons is confirmed by a notary who has declared his readiness to perform a notarial act.

The second stage Further, the FNC sends the application and the attached documents in the electronic form to the notary who announced earlier in automatic mode that he was ready to perform the corresponding notarial action remotely.

The third stage The notary who received the relevant application indicates the fee charged for the notarial act's performance, indicating the payment details of the account for the payment. Not later than five business days from the date of payment, the notary shall send the applicant a notarial document in the electronic form, or a reasoned resolution to refuse to perform a notarial act remotely, or a resolution to postpone the performance of a notarial act remotely. Within ten days from the date of the decision to refuse to perform the notarial act remotely, the notary shall return the paid amount. This procedure is regulated by order of the Ministry of Justice of Russia dated September 30, 2020, No. 232 "On approval of the procedure for sending an application for a notarial act remotely, making payment for a notarial act, including using electronic means of payment, and returning to the applicant the amount of payment for performing a notarial act remotely, as well as the interaction between the applicant or his representative, who applied for a notarial act remotely, and the notary."

The applicant has the right to choose where the notarial act is performed remotely by specifying the notarial district, or the Russian Federation's entity.

The application and the documents attached to it that do not meet document format requirements are not subject to acceptance. Also, according to Part 3 of Article 45 of the Fundamentals of the Russian Federation's legislation on notaries, electronic documents submitted for the performance of notarial actions should not contain computer programs or other computer information, deliberately intended for unauthorized destruction, blocking, modification, or copying of computer information or neutralization of computer information protection means.

The interaction between the applicant and the notary is carried out remotely without direct contact using the UIS means.

The fourth stage A notary produces a notarial action in the form of an electronic document in PDF or XML format following the requirements established following Part 3 of Art. 45.1 of the Fundamentals of the Russian Federation's legislation on notaries, by order of the Ministry of Justice of Russia dated September 30, 2020, No. 227 "On approval of the requirements for the format of a notarized document in electronic form." The electronic image of a hard copy is formed in the form of a single file. A notarized document must be scanned with a resolution of 300 dpi (dots per inch) in grayscale, with a color depth of eight bits per pixel. The notarial act with attachments in electronic form is signed with a strengthened qualified electronic signature of the notary in PKCS#7 format (separated electronic signature in DER-encoded format).

Certification of transactions in a remote format with two or more notaries is the most representative notarial action. This is especially true if the parties are located in different constituent entities of the Russian Federation and do not have the opportunity to meet at one notary to sign an agreement directly. This procedure certainly saves time and reduces financial costs (Andropova, 2020).

The fifth stage Upon the request of the applicant who applied for a notarial act remotely, a notarial document made in electronic form is issued in hard copy, taking into account the requirements established by Articles 5.1 and 45.1 of the Fundamentals of the legislation of the Russian Federation on notaries, i.e., must have machine-readable markings to ensure verification of its authenticity (QR-code, which makes it possible to verify the authenticity of the protocol for providing evidence, including through a special application on a smartphone). Such information does not constitute the disclosure of information about the performance of a notarial act and is provided to an unlimited number of persons at any time.

14.4.4 Problems of the Implementation of the Digital Format of Notarial Actions

Despite the importance of performing notarial actions using a remote digital order and the timeliness of the appearance of such an opportunity in Russian legislation, one cannot fail to note some problems associated with the implementation of the tasks facing modern digital notaries.

1. The possibility of remotely contacting a notary for performing notarial actions, according to legislators, does not require a notary to verify the legal capacity, i.e., checking the conformity of the will and the wishes of the applicant. This thesis is not entirely accurate since the applicant's will is expressed directly in the application submitted to the notary.
2. The law does not link the possibility of the remote and distant format of the procedure for performing notarial actions using digital technologies for specific reasons, which means its alternative method along with the traditional procedure for personal appeal and participation of the applicant in the performance of a notarial act.

   The notary performs the notarial act remotely, i.e., without the presence of the applicant, and registers it in a particular electronic register of the UIS, as well as in the register maintained by a notary in a hard copy, which, according to experts, is an unnecessary bureaucratic action that complicates notarial record-keeping.
3. A notary who has received an application for the performance of a notarial act remotely and the documents attached to it in electronic form indicates the fee

charged for its performance, indicating the payment details of the account for the payment.

It should be noted that the modern procedure for calculating the amount of payment for a notarial act to provide evidence is based on a variable component of legal and technical work and depends on the amount of information being examined on the Internet, which is reflected in the number of pages of the inspection protocol. Therefore, the notarial tariff's final amount can be calculated only after the notarial act has been performed. However, this procedure changes the procedure for performing a notarial act remotely. To implement this provision, it is possible to propose to indicate the approximate size of the notarial tariff, and the excess amount after the notarial action should be returned to the applicant since such a return procedure is spelled out in this order. It is also logical to consider the possibility of an additional payment by the applicant if the size of the notary tariff has changed upward, which also does not contradict the legislation's requirement.

In general, it should be noted that according to the rules governing the remote procedure for performing notarial acts, the notarial tariff is collected before the notarial act is performed. This provision differs from the traditional procedure for contacting a notary when the notarial tariff is collected only after the notarial act has been performed. It is proposed to reasonably extend this innovation to all types of notarial actions.

4. The applicant is entitled to choose the place where the notarial act is performed remotely by specifying the notarial district of the Russian Federation's entity. However, the legislator has not resolved the possibility of choosing several notary districts at the same time or choosing a specific notary. The last sentence is relevant due to the freedom of choice of a notary by the applicant based on various factors. General rules make it possible to restrict this choice only based on the subject and territorial competence of the notary but do not limit the choice in terms of a particular notary's professional competence. For example, a notarial act to provide electronic evidence is the most complex notarial act and, as a rule, takes place in regions with the most advanced economic activity. As a result of these circumstances, the circle of notaries performing this notarial act promptly and at a high professional level is minimal. Accordingly, the applicant is interested in being able to choose a notary that suits his interests.

This amendment of the legislation will reduce the time for performing notarial actions using digital technologies to exclude the possibility when an uncertain notary has to delay preparing and performing a complex notarial action and excluding unreasonable refusals to perform those (Goncharov et al., 2019).

5. Having received the application and all the necessary information and documents, the notary proceeds directly to the provision of notarial services. There are no deadlines for many notarial actions performed in electronic form in the Fundamentals of the Russian Federation's Legislation on Notaries. However, it must be understood that notarial action does not require a delay and is always associated with the prompt possibility of protecting citizens' and legal

entities' rights and legitimate interests. Considering the requirement of the order of the Ministry of Justice of the Russian Federation dated September 30, 2020, No. 223 on sending a notarial document to the applicant five days after completing all the preliminary procedures, we can conclude that the term for the performance and execution of the notarial action is limited to this period. This provision carries a self-organizing principle but is not always justified since it does not consider the amount of necessary preparatory actions on the part of a notary, including a request for mandatory information from state bodies using electronic interaction systems due time for its processing. Therefore, given that there is no corresponding mechanism in the legislation, this period should be considered as recommended (Matytsin & Rusakova, 2021).

6. For some notarial actions, representing a lengthy preparatory procedure, the application's content is not enough. It requires the applicant's specific direct involvement in the notarial process and constant interaction with the notary to identify the most optimal way to solve the applicant's problem. We can see such an example in the notarial act of securing electronic evidence, which provides the variable possibility of determining the subject of inspection and the amount of information being examined. All these issues arising in preparing and performing a notarial act need to be detailed to avoid adverse consequences, both for the notary and for the applicant, who will have to pay for the notarial act in a more significant amount than he could have expected. For example, when scanning e-mail, one should specify the letters being examined, limiting the period and the specific recipient/sender of the letters. Besides, not all information on the Internet can be viewed remotely without the applicant's direct participation. It will not be easy to inspect the applicant's corporate mail using a remote-access program (Begichev, 2021).
7. It seems that at the initial stage of implementation, the remote and distant format of notarial actions will contain specific difficulties, including technological reasons. Despite this, over time, this format will get used to, and, at a particular stage of development, it can take a leading position in the provision of notarial services. We would also like to implement electronic functions more actively in Russian legislation, using the experience of foreign legislation, including the legislation of the Baltic countries discussed here. In particular, it seems pretty justified to transfer to notaries the functions of registering rights to real estate (the powers of the registrar) and video communication services, which will have a positive effect on the convenience of performing notarial actions in a remote and distant format.
8. Article 10 of the Fundamentals of the Legislation of the Russian Federation on Notaries provides for an imperative procedure for using the Russian language in notarial proceedings. However, when implementing this principle in a remote and distant format, performing notarial actions faces the difficulty of its implementation—the legislator has not resolved the procedure for attracting an interpreter and remuneration for his labor. These issues require an immediate settlement by the applicant, which is not provided by the current legislation.

9. At the applicant's request, a notarial document made in electronic form is issued in a hard copy, taking into account the requirements established by Articles 5.1 and 45.1 of the Fundamentals of the legislation of the Russian Federation on Notaries.
   In practical application, the requirements of the legislation on the issuance of a copy of the notarial deed to the applicant in hard copy require its specification since the following issues remain unresolved:
   (i) payment by the applicant for such actions;
   (ii) the number of copies of the notarial deed issued to the applicant;
   (iii) duplication of notarial acts in electronic form and hard copy, which may cause third parties' questions in the prevalence of one type over another and some other problems.
10. It seems that at the initial stage, advanced users of information technologies will apply for a remote and distant format of notarial actions, and on an ongoing basis—persons who have multiple interests in using this procedure. Therefore, it is pretty fair to assist government agencies to persons with disabilities to choose an alternative form of performing a notarial act.

In general, it can be emphasized that, despite the identified shortcomings, the current legislation has a huge potential for citizens and legal entities to protect their rights and interests when performing notarial actions using remote digital technologies. It seems that these issues will be settled shortly.

14.5 Conclusion

As a European power, Russia seeks to protect the rights of citizens, relying on modern achievements of research, which are on a par with the global concept of digitalization of economic relations. The implementation of state policy in this area is carried out based on the national program "Digital Economy of the Russian Federation," which forms the principles of the e-government.

Modern technologies in Russian notaries have a relatively short history. At the same time, over the past decade, progress in the formation of electronic notaries allows us to derive a modern formula for notarial activity, which is characterized not only by a stable and reliable instrument for the legal regulation of economic, legal relations but also by a modern, progressive, and operational body that performs state functions.

The attractiveness of using remote digital technologies in notarial activities is achieved by establishing an alternative procedure for contacting a notary to receive notarial services without direct contact with the applicant in a remote and distant format. The legislation "On digital notaries" aims to optimize notarial activities, increasing the demand for notarial protection of the rights of citizens and legal entities in the context of the development of a digital society.

Currently, the legal community positively evaluates these innovations, and the practical implementation of the digital order of interaction for obtaining notarial services, we believe, will confirm the feasibility and relevance of this procedure.

The introduction of digital technologies in notarial activities increases the efficiency, quality, and relevance of this human rights mechanism, and the state support in ensuring the implementation of remote and distant procedures increases their importance for all participants in economic activity and is also of particular importance for the entire civil society.

References

Akhrameeva, O. E. (2020). The notarial mechanism for ensuring digital security of the rights of citizens and legal entities in the context of digitalization. *Notary, 1*, 33–37. (In Rus.).

Andropova, T. A. (2020). Advantages and prospects of electronic notarial system. *Notarial Bulletin, 12*, 29–36. (In Rus.).

Begichev, A. V. (2020). Development of electronic notaries in Latvia, Lithuania, and Estonia. *Notarial Bulletin, 1*, 42–52. (In Rus.).

Begichev, A. V. (2021). The procedure for remote evidence securing by notaries. *Notary, 1*, 3–8. (In Rus.).

Duston, L. (2020). The language of science. In the world of science. *Scientific American, 11*, 7–13.

Goncharov, A. I., Dolinskaya, V. V., Trofimov, Y. V. (2019). Electronic justice: Stages of establishment and issues of development in modern Russia. In Popkova, E. G. (Ed.), *Ubiquitous computing and the internet of things: Prerequisites for the development of ICT*. Studies in computational intelligence, vol. 826, pp. 315–324. Springer Science + Business Media.

Inshakova, A., Goncharov, A., Smirenskaya, E., & Dolinskaya, V. (2017). Modern communication technologies in notification of notarial actions in Russia. *Journal of Advanced Research in Law and Economics, VIII, 7*(29), 2144–2151.

Kirillova, E. A. (2020). The state of the Russian notarial system: The introduction of the digital format. *Notary, 6*, 42–44. (In Rus.).

Korsik, K. A. (2020). Notaries in the federal project "Normative regulation of the digital environment." *Notarial Bulletin, 12*, 2–5. (In Rus.).

Matytsin, D. E., & Rusakova, E. P. (2021). The strategy of quality management in industry 4.0 and formation of a cognitive economy based on industrial and manufacturing engineering in the Russian Federation and countries of the EU. *International Journal for Quality Research, 15*(4). https://doi.org/10.24874/IJQR15.04-03

Renk, Ya. G., & Philippova, O. V. (2020). Characteristic transversal modalities in electronic storage. *Notarial Bulletin, 6*, 5–16 (In Russian).

Rusakova, E. P., Frolova, E. E., & Gorbacheva, A. I. (2020). Digital rights as a new object of civil rights: Issues of substantive and procedural law. *Advances in Intelligent Systems and Computing, 1100*, 665–673.

Tarakanov, V. V., Inshakova, A. O., & Dolinskaya, V. V. (2019). Information society, digital economy, and law. *Studies in Computational Intelligence, 826*, 3–15.

Chapter 15
Protection of Women from Violence and Domestic Violence in the Context of Digitalization

Aslan Kh. Abashidze and Olga K. Goncharenko

Abstract Nowadays, digital economics are rapidly changing the world around us, and artificial intelligence and big data are being actively used. These circumstances, while providing states and their citizens with new possibilities, call for modernization of legal and regulatory framework of various areas of public life. In this article, the authors analyze in what way digitalization affects human rights, and in particular how it affects women's rights to be protected from violence, including domestic violence. Different factors due to which women are more vulnerable to violence have been highlighted. These factors include deep-rooted patriarchal views, physiological features of women, strict legal protection of family and personal life inviolability, marginalization, and society underdevelopment and migration crisis. Since recently, the digitalization of society can be noted as one of these factors. It is stressed that the right to personal data protection is inseparable from the right to personal life protection. The article underlines the role of the Council of Europe, in particular the European Court for Human Rights, in the field of preventing and combatting violence against women and domestic violence, provides examples of the case law in part of violation of confidentiality of electronic mail and virtual harassment. In the conclusion, the authors summarize that in conditions of digitalization, there is an ongoing development of new human rights protection mechanisms on the international level. Nonetheless, the issues of women's private life protection and their right to be protected from all forms of violence require a more in-depth examination.

Keywords Digitalization · Human rights · Protection of women · Domestic violence · Virtual intimidation · Privacy protection · Istanbul Convention

JEL Codes B54 · D63 · F53 · J12 · J16 · K38

A. Kh. Abashidze (✉)
Department of International Law, Peoples' Friendship University of Russia (RUDN University), 6 Miklukho-Maklay St, 117198 Moscow, Russia
e-mail: abashidze-akh@rudn.ru

O. K. Goncharenko
Moscow State Institute of International Relations (MGIMO University), Moscow, Russia
e-mail: goncharenko_ok@bk.ru

A. O. Inshakova and E. E. Frolova (eds.), *Smart Technologies for the Digitisation of Industry: Entrepreneurial Environment*, Smart Innovation, Systems and Technologies 254, https://doi.org/10.1007/978-981-16-4621-8_15

15.1 Introduction

The development of information technology is currently leading to the formation of a new so-called virtual reality. According to V. Zorkin, the fundamental human rights, guaranteed by international legal acts and national legislation, including the right to protection of honor and dignity, personal and family privacy, inviolability of private life, the right to secrecy of correspondence, and others are in need to be articulated in the context of digital reality Zorkin (2018). Some authors note that the application of modern digital technology requires the legislators to define the modalities of data circulation, the rights and obligations of participants of "digital" legal relations, the limits of digital technology application, and so on (Chernysheva, 2019).

Digitalization has affected many areas of social life (Inshakova et al., 2018), specifically the prevalence rate of violence against women, most notably, domestic violence.

According to the UN Secretary-General Report, in 2017, there were 137 women murdered daily by their intimate partner or a different family member (Women, Business, and the Law, 2020). According to other sources, more than a billion women are not protected from sexual violence by their intimate partner or a family member, and almost 1.4 billion are not protected from economic violence. The number of women lacking protection from sexual harassment in the sectors of employment, education, and public spaces is 359 million, 1.5 billion, and 2.2 billion, respectively (GREVIO's (Baseline) Evaluation Report, 2018).

Notwithstanding some progress made in improving national anti-violence legislation (Review and appraisal of the implementation of the Beijing Declaration and Platform for Action and the outcomes of the twenty-third special session of the General Assembly, 2020), there are still many challenges concerning countering violence against women and domestic violence. For instance, one in three states lacks a specific legal instrument, as a result of which legal protection against sexual violence as a form of domestic violence remains ineffective. The matters of defense against economic violence are not regulated in as much as half of the UN member states. Two-thirds of states that have legislation against domestic violence do not apply it to unmarried intimate partners. Criminal punishment for sexual harassment is stipulated in a little over 30% of the states worldwide (GREVIO's (Baseline) Evaluation Report, 2018).

Since the 1980s, the international community has been taking different measures to eliminate discrimination against women, which is fundamentally related to the issue of protecting women from violence and domestic violence. These efforts have led to the adoption of international legal instruments aimed at preventing these phenomena either directly or indirectly. However, more of them, including those accepted at the universal level, are recommendatory.

In contemporary international human rights law, there is only one binding international legal action in this area, which is the special regional Convention of the Council of Europe on preventing and combating violence against women and domestic violence (also known as the Istanbul Convention), adopted on April 7, 2011. This

Convention qualifies some acts as criminal offenses that are subject to the implementation in the national legislation of its member states and establishes a special monitoring mechanism, which has been following the realization of the obligations of the state parties to the Convention.

Today, the subject of violence against women and domestic violence prevention is included in the international human rights agenda; we can see an increase in the number of international mechanisms aimed at their prevention (Inshakova et al., 2020). Naturally, it is impossible to analyze all the trends and issues of protecting women from these offenses in just one article. The authors seek to examine the factors due to which women are more vulnerable to violent acts and to consider the practices of the European Court for Human Rights in regard to this issue with an emphasis on private and family life protection.

15.2 Methodology

Despite the high level of development of the factors influencing the spread of violence against women and domestic violence at the doctrinal level Bell & Naugle (2008), Bettman (2009), Goncharenko (2021), Hunnicutt (2009), Inshakova et al. (2018, 2020), Krug (2002), Leikman (1998), Matytsin (2021) and the efforts made by international organizations to combat these phenomena, the problem of the impact of digitalization on the rate of violence has not been sufficiently studied and has not been included in the national and international agenda.

The methodological basis of the study includes a combination of general scientific (analytical, deductive, dialectical, historical, inductive) and private scientific methods (comparative legal, dynamic, formal legal, interpretative, procedural, statistical).

15.3 Results

The obligations of states, international intergovernmental organizations on the universal, regional and subregional levels endowed with human rights functions, including the prevention and combating all forms of violence against women and domestic violence, follow from the statutory provisions of the United Nations (henceforth—the UN) (the preambles, para.1 of art.1, para. (c) of art. 55 of the UN Charter) and the International Bill of Human Rights, which recognize the inherent dignity of all family members, as well as equal and inalienable rights as the fundamental rights and freedoms of every person without any discrimination.

The universal framework on non-discrimination has been strengthened by international Conventions on the elimination of all forms of racial discrimination and discrimination against women, as well as by regional instruments aimed at human rights and freedoms protection, including the rights of certain vulnerable social groups, which include women and children according to contemporary international human rights law.

Since the commencement of its activities, the UN has been addressing the issues of women as of a vulnerable social group, which is confirmed by the fact of a simultaneous creation of the Commission on the Status of Women by the UN Economic and Social Council (henceforth—the ECOSOC) and the Commission on Human Rights. Subsequently, the issue of ensuring gender equality on the human rights protection agenda becomes an object of constant attention of the international community and acquires a broad scope, including the task of eliminating all forms of violence against women.

Unfortunately, as it is reflected in the UN Secretary-General Report on the review and appraisal of the implementation of the Beijing Declaration and Platform for Action, presented in March 2020 on the session of the Commission on the Status of Women, "violence against women and girls remains widespread" (Ending violence against women and girls, global and regional trends in women's legal protection against domestic violence and sexual harassment, 2018). A specific regional European international legal act on combating violence against women and domestic violence, the Istanbul Convention, has been adopted within the Council of Europe. The Convention was preceded by numerous circumstances, which acted as a premise to its development. The relevant provisions on gender equality and non-discrimination on the grounds of a person's sex of the Statute of the Council of Europe, as well as of the European Convention on Human Rights, being an essential European treaty on human rights protection and promotion, served as a general legal basis. To ensure that states comply with the provisions of the Convention, the European Court of Human Rights (henceforth—the ECHR), which is competent to adjudicate individual complaints and provide advisory opinions on legal issues (art. 47), was established as provided by art.19 of the Convention.

An analysis of initiatives to combat violence against women, including domestic violence, taken on the universal and regional levels, suggests that since the 1980s and to the present day, attempts have been made by the UN and the Council of Europe to eliminate the phenomenon of violence that is consequent to entrenched discrimination against women.

The fact that women are more likely than men to be exposed to violence, including such violence inside a family, can be explained by many factors. The WHO Report on violence and its impact on health suggest the so-called ecological model of violence, according to which a combination of personal, situational, social, and cultural factors can influence violence against women and domestic violence, thereby highlighting the multifaceted nature of this phenomenon (global and regional estimates of violence against women: prevalence and health effects of intimate partner violence and non-partner sexual violence, 2013.).

Such factors include, among others:

- the dominance of men over women in family relations;
- traditional distribution of roles between men and women;
- physical inequality between the genders;

- condemnation of women unwilling to adhere to the roles assigned to them by the society (e.g., those who do not want to have a child or want to work in a trade normally attributed to men);
- ineffective measures to prevent violence against women and domestic violence, or a complete absence of such measures.

First and foremost, such factors are referred to deep-rooted patriarchal views in many states (Bell & Naugle, 2008), the position of domination of men over women in contemporary families, and their commitment to traditional gender roles (Hunnicutt, (2009). Patriarchal views on men–women relationships have significantly influenced the formation of family values in most countries of the world, while existing cultural practices and customs allowed the use of domestic violence (Bettman, 2009).

The entrenched patriarchy has led to a social, economic, and political gap between women and men. Moreover, pregnant women and young mothers who are responsible for raising children oftentimes suffer discrimination in employment, which aggravates their financial situation and makes them dependent on men (Matytsin, 2021). In due time, these views formed the basis of feminist economic theory, studied in detail by such researchers as Gunderson, Brown, Bergmann, and others.

Another factor is the obligation of states to comply with national and international legislation on the inviolability of private life, thereby avoiding interference in family affairs. For the same reason, many states do not transfer the cases of domestic violence from private to public prosecution.

As practice has shown, legal definitions of "violence against women" and "domestic violence" are lacking in many national jurisdictions, thus demonstrating the absence of special legislation aimed at combatting these acts. Consequently, the state authorities are not vested with the relevant mandate and do not have the power to prevent and investigate such cases.

Even though these phenomena are common for developing and developed countries, a social factor should be also taken into account. For example, women with disabilities, those having a low income, belonging to religious minorities, alcohol or drug users are subject to cross-discrimination, which means that they suffer discrimination on several grounds in parallel (Krug et al., 2002).

The international community in its present state faces numerous threats, challenges, and global risks, which are fraught with negative consequences of multiple scales, which in turn would result in general worsening of the condition of humanity. Such consequences include poverty, society underdevelopment, and, to a larger extent, the European migration crisis which creates negative implications for women, making them even more vulnerable and in need of special protection. Migrant women regardless of their legal status as well as women-asylum seekers are at an increased risk of violence and face similar challenges in coping with that risk.

With the spread of the COVID-19 pandemic in the year 2020, international organizations and researchers in the field of violence against women and domestic violence have identified new factors that make women more likely to be exposed to these offenses: a difficult sanitary and epidemiological situation as well as emergency human rights restrictions aimed to reduce the number of cases of infection transmission and to protect the population (Goncharenko, 2021).

Digitalization is yet another factor, which contributes to the increase of the rate of violence against women and domestic violence. It is quite obvious that the spread of digital technologies has created serious problems for ensuring personal privacy and protecting honor and dignity (Tarakanov et al., 2019).

According to Chernysheva (2019), the right to personal data protection derives from the right to privacy. The theft of personal information from personal electronic devices or social networks by cybercriminals is becoming the most common phenomenon and is used for personal or economic purposes to discredit another person, to spread information that inflicts damage to the dignity and business reputation (Martinov, 2018).

In this regard, notable are the recommendations provided by the Group of Experts on Action against Violence against Women and Domestic Violence (GREVIO's (Baseline) Evaluation Report, 2018) in the Baseline Evaluation Report on legislative and other measures giving effect to the provisions of the Istanbul Convention. One of these recommendations calls on the Turkish authorities to legislate the crime of stalking, including stalking on the Internet.

That recommendation looks appropriate given the decision of the European Court of Human Rights in the case of "Buturuga v. Romania," in which the issue of ineffectiveness of domestic violence investigation, breach of e-mail confidentiality, and virtual intimidation of the applicant have been addressed for the first time in the case law of the court. In its decision, ECHR found a breach of art. 8 of the European Convention of Human Rights (right to respect for private and family life), as well as the need to respect the confidentiality of electronic correspondence, thereby recognizing virtual intimidation as a form of violence against women, which can be expressed in the form of interference with privacy on the Internet, infiltration of a computer or other electronic devices, dissemination of information or images related to the victim including one's data (Case of Buturuga vs. Romania, 2020).

The development of international law in the field of protecting women from violence and domestic violence, including in the context of digitalization, on the one hand, testifies to the intention of the international community to combat these phenomena. On the other hand, to date, no comprehensive analysis of the relation between the impact of digital reality and the rights of women to protection from all forms of violence has been carried out, and no practical recommendations have been proposed for their protection and elimination of these offenses. It seems that this issue could be reflected in one of the general recommendations of the Committee on the Elimination of Discrimination against Women (CEDAW) or GREVIO.

15.4 Conclusion

The authors argue that there is an obvious correlation between the level of prevalence of violence against women/domestic violence, on one hand, and the societal realities including the digital environment on the other. The latter also serves as a fertile ground and as a framework within which various factors grow and evolve. Despite the progress in identifying the correlated duo of the protection of women from violence and digitalization, we believe there has been inadequately limited in-depth research in this area.

The lack of proper academic research into the role of digitalization as a crucial factor in the field of the prevalence of violence against women and domestic violence explains the lack of so needful new international and national standards in this area.

Nonetheless, it is important to note the current developments of the practice of the European Court of Human Rights on interference with privacy on the Internet in the context of domestic violence, which potentially provides grounds for resolving this issue at the national, regional, and universal levels.

References

Bell, M., & Naugle, A. E. (2008). Intimate partner violence theoretical considerations: Moving towards a contextual framework. *Clinical Psychology Review, 28*(7), 1096–1107.

Bettman, C. (2009). Patriarchy: The predominant discourse and fount of domestic violence. *Australian and New Zealand Journal of Family Therapy, 30*(1), 15–28.

Case of Buturuga vs. Romania. (2020). Application No. 56867/15. http://hudoc.echr.coe.int/eng?i=001-200842 Accessed 28 January 2021.

Chernysheva, Y. (2019). Human rights in a digitalized society. *Psychology and Law, 9*(4), 92–102. https://doi.org/10.17759/psylaw.2019090407

Ending violence against women and girls, global and regional trends in women's legal protection against domestic violence and sexual harassment, 2018. http://pubdocs.worldbank.org/en/679221517425064052/EndingViolenceAgainstWomenandGirls-GBVLaws-Feb2018.pdf Accessed 28 January 2021.

Global and regional estimates of violence against women: prevalence and health effects of intimate partner violence and non-partner sexual violence, 2013. https://apps.who.int/iris/bitstream/handle/10665/85241/WHO_RHR_HRP_13.06_eng.pdf?sequence=1. Accessed 28 January 2021.

Goncharenko, O. (2021). Challenges to the protection of women from domestic violence during the COVID-19 pandemic. *The legal world*, 1.

GREVIO's (Baseline) Evaluation Report on legislative and other measures giving effect to the provisions of the Council of Europe Convention on Preventing and Combating Violence against Women and Domestic Violence (Istanbul Convention) Turkey, 2018. https://rm.coe.int/eng-grevio-report-turquie/16808e5283 Accessed 28 January 2021.

Hunnicutt, G. (2009). Varieties of patriarchy and violence against women: Resurrecting "patriarchy" as a theoretical tool. *Violence against Women, 15*, 553–573.

Inshakova, A. O., Frolov, D. P., Davydova, M. L., & Marushchak, I. V. (2018). Institutional factors of evolution and strategic development of general purpose technologies. *Espacios, 39*(1), 5.

Inshakova, A. O., Goncharov, A. I., & Ershova, I. V. (2020). Digital blockchain registration of legally significant stages of complex good's export-import supplies by business entities of the EAEU and BRICS jurisdictions. In A. O. Inshakova, Inshakova, E. (Eds.) *Competitive Russia:*

Foresight model of economic and legal development in the digital age. CRFMELD 2019. Lecture Notes in Networks and Systems (Vol. 110, pp. 328–336). Springer. https://doi.org/10.1007/978-3-030-45913-0_37

Krug, E. G., et al. (2002). *World report on violence and health*. World Health Organization.

Leikman, L. (1998). From theory to practice. *Crisis Center for Women: Experience of Creation and Work, 12*.

Martinov, A. V. (2018). *Human rights in the digital age*. Materials of the II International Conference "Human rights protection in Eurasia: Exchange of Best Practices of Ombudsmen", pp. 224–232.

Matytsin, D. E. (2021). Internet-investing as a remote algorithm of retail investment financing. In *Modern global economic system: evolutional development vs. revolutionary leap*. Lecture Notes in Networks and Systems. Springer Nature.

Review and appraisal of the implementation of the Beijing Declaration and Platform for Action and the outcomes of the twenty-third special session of the General Assembly, 2020. https://undocs.org/en/E/CN.6/2020/3. Accessed 28 January 2021.

Tarakanov, V. V., Inshakova, A. O., & Dolinskaya, V. V. (2019). Information society, digital economy, and law. In E. G. Popkova (Ed.), *Ubiquitous computing and the internet of things: Prerequisites for the development of ICT/Studies in computational intelligence* (Vol. 826, pp. 3–15). Springer.

Women, Business and the Law. (2020). *Report of the World Bank*. https://openknowledge.worldbank.org/bitstream/handle/10986/32639/9781464815324.pdf Accessed 28 January 2021.

Zorkin, V. (2018). Law in the digital world. *Russian Newspaper*, 7578(115).

Chapter 16
Freedom of Expression—A Double-Edged Right That Continues to Divide Peoples Across the Globe on How Best to Frame Its Scope and Limitations—An Organization of Islamic Cooperation (OIC) Perspective in Times of Digitalization

Aslan Kh. Abashidze and Marghoob Saleem Butt

Abstract Freedom of speech is the foundation of democracy, which ensures the exercise of other human rights and freedoms by individuals and whole groups. Today, this right is recognized in many international acts at the universal and regional levels. Within the framework of this article, the author studies the specifics of the implementation of the right to freedom of opinion and their free expression in Islam. In Islam, suppression and repression are not acceptable. Accordingly, Islamic teachings uphold freedom of expression and recognize the role of critical thinking in empowering human beings as an honorable and responsible creation of God. The Quran upholds the right to freedom of religion and expression. It, however, teaches that one should express oneself through gentleness, courtesy, calmness, and respect. A multilayered approach to promoting freedom of expression while combating hate speech must start from deconstructing the narratives of hate speech and conducting a proper analysis to understand the processes that underpin the rise of hate speech in modern societies. At the core of hate speech construction and normalization is the intellectual legitimization of ethnic, religious, and national superiority, which justifies hatred, discrimination, and violence toward specific individuals and groups. Such legitimization is based on false narratives/ideas that tend to reduce a given ethnicity or religion to a devaluing identity based on nature/biology or cultures such as linking

A. Kh. Abashidze (✉)
Department of International Law, Peoples' Friendship University of Russia (RUDN University), 6 Miklukho-Maklay St, 117198 Moscow, Russia
e-mail: abashidze-akh@rudn.ru

M. S. Butt
Independent Permanent Human Rights, Commission (IPHRC) of the Organization of Islamic Cooperation, Department of International Law, Peoples' Friendship University of Russia (RUDN University), 6 Miklukho-Maklay St, 117198 Moscow, Russia

A. O. Inshakova and E. E. Frolova (eds.), *Smart Technologies for the Digitisation of Industry: Entrepreneurial Environment*, Smart Innovation, Systems and Technologies 254, https://doi.org/10.1007/978-981-16-4621-8_16

violence with Islam, cultural or biological inferiority for Blacks and Jews, and subsequently, their exclusion from and discrimination within a society. These constructed concepts are used by different far rights political parties and social movements' platforms to build negative perceptions against targeted individuals and groups as well as to thwart and antagonize multicultural dynamics in their societies for political gains.

Keywords Digitalization · Human rights · Islam · UN human rights mechanisms · Regional protection of human rights

16.1 Introduction

The development of a peaceful, progressive society is dependent on the enjoyment of the freedom of expression by its inhabitants, which includes the right of every person and all people to hold their own opinions, and to seek, receive, and share information and ideas in different formats (Goncharov et al., 2019). Throughout the history, this right has been recognized by various religions and cultures, and it was instrumental in building scientific knowledge, advancing human civilizations, and fostering understanding among various peoples. Based on its established value, freedom of expression in present times is duly recognized and protected under various United Nations (UN) treaties, regional conventions, and national constitutions and legislation, across the world. This established principle of international human rights law finds its place among the core principles needed to promote and secure human rights-respecting societies.

In Islam, suppression and repression are not acceptable. Accordingly, Islamic teachings uphold freedom of expression and recognize the role of critical thinking in empowering human beings as an honorable and responsible creation of God. The Quran upholds everyone's right to freedom of religion and expression. It, however, teaches that "one should express oneself through gentleness, courtesy, calmness, and respect," such as when Allah commands the prophet Mohammed (peace be upon him) to "*call the people to the way of your Lord with wisdom and goodly and kind exhortation, and argue with them in the most pleasant and best manner*" (Verse 125, Surat An-Nahl). Accordingly, Islamic philosophy considers the freedom of expression as a tool or means to build up love, tolerance, social harmony, and understanding among different members of society to ensure a peaceful coexistence (Jallow, 2015).

At no point, we find expressions or edicts suppressing freedom of expression in Islam. Contrarily, the Quran promotes and urges people to openly discuss all matters, including questioning or rejecting any idea based on evidence and logical arguments. It even challenges them to put forward any evidence or argument that might be used to counter the divine wisdom of the Quran. Islam does not only respect but also demand freedom of expression. This is understood in the context that many Quranic injunctions, such as the order to enjoin good and to forbid evil, are impossible to comply with if people are not free to speak and act freely (Bhat, 2014). Many actions that are deemed praiseworthy in Islam, such as speaking up the truth, providing sincere advice, spreading knowledge, and helping to lift injustices, would remain an illusion if one does not have the right to express oneself openly and freely. As narrated in the authentic hadith, the Prophet Mohammed (peace be upon him) said, "The best form of jihad is to tell a word of truth to an oppressive ruler" (narrated by Abu Dawud and At-Tirmidhi), which is an ultimate illustration of how the Prophet valued freedom of expression.

Islam also guarantees the right of everyone to express opinions and to argue for him or her without any kind of intimidation or fear, to the extent that even the devil has been allowed to speak openly. Its best example is provided in Quran (Surah Al-Hijr Verses 29–33) by describing a conversation between Allah and Iblis (devil), where after creating the human being (Adam) from mud, Almighty asks all angels to prostrate to him and all except Iblis (devil) willingly obey the command. Iblis not only defies the command but also responds to Almighty that he was not going to prostrate himself to a human being, created from sounding clay of altered black mud. This conversation, which goes on between Allah and the devil, is an ultimate Quranic illustration of the Islamic perspective on freedom of expression.

If the devil is given the freedom to speak up and allowed to propagate and distract people with his mischiefs/views, then surely Islam upholds the value of freedom of expression and grants this right to all people without discrimination. That said, Islam does not consider freedom of expression to be an absolute right. Knowing well the harm that can be caused by the unchecked nature of this right, Islam does place certain limits on its exercise, which relates to defamation, blasphemy, or creating social disorder, etc. (Bhat, 2014). In line with this spirit, Islam acknowledges peoples' right to exercise freedom of expression as long as it is exercised responsibly without compromising and affecting the freedom and dignity of other people. The Islamic moral constraints on this freedom are built-in safeguards that prevent injustice, abuse, wickedness, and strife, which can lead to abusive or offensive language in the name of criticism and freedom of speech (Kamali, 1998).

16.2 Methods

The methodological basis of the study includes a combination of general scientific (analytical, deductive, dialectical, historical, inductive) and private scientific methods (comparative legal, dynamic, formal legal, interpretative, procedural, statistical).

16.3 Results

Under the international law framework, the legal understanding, standards, and definition of freedom of expression have been evolving for many decades. In its very first meeting in January 1946, the UN General Assembly (UNGA) passed a resolution that recognized the freedom of information as a fundamental human right and "the touchstone of all the freedoms to which the UN is consecrated" (UN Res 59(1), 1946). This landmark resolution also provided an early definition of freedom of expression as "*Freedom of information implies the right to gather, transmit and publish news anywhere and everywhere without fetters. As such it is an essential factor in any serious effort to promote the peace and progress of the world*" (UN Res 59(1), 1946).

In December 1948, the UNGA adopted the Universal Declaration of Human Rights (UDHR), which is considered as the *Megna-Carta* of the present-day international human rights law. Article 19 of UDHR provides a comprehensive definition of this freedom by stating: "*Everyone has the right to freedom of opinion and expression; this right includes freedom to hold opinions without interference and to seek, receive and impart information and ideas through any media and regardless of frontiers*" (Universal Declaration of Human Rights, 1948, Art. 19). Ever since, all relevant international human rights treaties, experts, and reports from various human rights bodies have expanded on the definition of this right and how it relates to its practical exercise. Accordingly, freedom of expression is recognized under various terms, such as in Articles 19 and 20 of the International Covenant on Civil and Political Rights (ICCPR), *Article 22 of the Cairo Declaration on Human Rights in Islam,* Article 32 of the Arab Charter of Human Rights, Article 10 of European Convention on Human Rights, Article 13 of American Convention on Human Rights, Article 9 of African Charter on Human and Peoples' Rights, and Article 23 of the ASEAN Human Rights Declaration.

However, the framers of this right were also aware of the pitfalls of such a freedom, which could be used in a counterproductive manner, such as incitement to hatred, defamation, discrimination, stereotyping, and violence. Accordingly, as with most human rights, limitations and restrictions apply to the freedom of expression under special circumstances. A careful balance was created to ensure that the freedom is used responsibly without causing harm to others. These restrictions are also included in various legal frameworks that stipulate how the right can be legitimately and responsibly exercised. Under all conventions and declarations mentioned above, freedom of expression is defined in specific terms, which declare that it is not an "absolute right," the exercise of which is subjected to special duties and corresponding responsibilities, based on "avoidance of harm to others" to ensure societal cohesion. Specifically, these measures seek to draw a line between legitimate freedom of expression and illicit forms of expression such as hate speech, incitement to hatred, discrimination, and violence.

To guide the application of the treaty-based rights discussed above, Special Rapporteur and other functionaries have been established to investigate and report on issues related to freedom of expression. For instance, the UN Special Rapporteur on Freedom of Opinion and Expression is an independent expert appointed by the Geneva-based UN Human Rights Council (HRC) to investigate and report on freedom of expression (HRC Res 7/36, 2008, 3). Other organizations such as the

Inter-American Commission on Human Rights (IACHR) and the African Commission on Human and Peoples' Rights (ACHPR) have also established similar mandates to gather relevant information on violations of the right to freedom of expression wherever it occurs to coordinate with various stakeholders, including governments, and to make recommendations and provide suggestions on how to better protect freedom of expression in general and where violations occur, in particular.

Unfortunately, however, the focus of these Special Rapporteurs' mandate has remained at the promotion aspect of this right by reporting and criticizing the limits imposed on this freedom. There had been no effort on part of these mandate holders to point out, criticize, or suggest remedies for violation, misuse, or abuse of this important right. This one-sided and biased perspective has led to many ills in the society, whereby extremist elements have used this uncharted freedom to stereotype, defame, discriminate, and incite hatred against minorities as well as other vulnerable segments of society, which have led to the acts of manifest discrimination and violence including the loss of life.

16.3.1 Difference and Limits Between Freedom of Expression and Hate Speech/Incitement to Hatred

There is a consensus that freedom of expression is an important human right, the exercise of which is crucial for the peaceful and progressive development of each society. Yet, varying the pitfalls attached to its irresponsible use, majority of the countries in the world have domestic laws that censor, restrict, or limit certain types of expressions, including speech that incites violence and hatred or threatens their social cohesion, ethics, national security, and identity. Notwithstanding the importance of this important right, there are two distinct approaches for its exercise. Some advocates of free speech prefer an open marketplace of ideas, with no restriction on any form of expression, believing that the best way to counter hate/harmful speech is encouraging more freedom of expression enabling an open and transparent discussion of ideas that challenges the validity and substance of all expressions (Walker, 2018). Others, however, argue that restrictions on freedom of expression are vital for the protection of other fundamental human rights, including rights to privacy and security, as well as to ensure the protection of minority communities and vulnerable segments of the society from the harm that hate speech may cause (Walker, 2018).

Different approaches to draw the line for acceptable freedom of expression can be seen around the world (Inshakova et al., 2020a, b). Even in countries like the United States of America, which has strong constitutional protection for free speech and vigorously makes efforts to promote the same around the world, there are many restrictions on freedom of expression, such as those against speech that incites "imminent violence/lawless action" and those that censor obscenity (these restrictions have been imposed by the US Supreme Court in various cases such as Brandenburg vs. Ohio, 395 US 444, para. 447 1969 and Jacobellis vs. Ohio, 378 US 184 1964). Based on historical considerations, each country has a varying tendency for prohibiting different forms of speech and the expression of certain opinions. For example, some European countries have passed laws in accordance with a Council of Europe that decided to make it a punishable offense not only to incite hatred but also to publicly deny crimes of genocide (e.g., the Holocaust) or war crimes.[1] On the other hand, some Muslim countries have imposed restrictions and harsher punishments including the death penalty, etc., on expressions that amount to crimes of apostasy, blasphemy, or other statements that threaten the spiritual cohesion of society, national security, and identity (Theodorou, 2016).

In recent years, in the context of fighting terrorism, many countries have excluded the glorification of terrorism from freedom of expression. Accordingly, legislations were introduced by these countries in their criminal code to create a new offense of advocating or promoting terrorist behaviors, otherwise referred to as the glorification of terrorism, which includes all expressions and reckless communications that seek to inspire violent negative behavior against others (Walker, 2018). Among the other laws that can be cited as restricting the absolute freedom of expression are those referred to as anti-hatred laws, to restrict the publication and public expression of messages intended to incite hatred toward members of particular groups. In other words, they prohibit propaganda that incites hatred and represents a violation of social cohesion, national interest, and the stability of the society as a whole (Walker, 2018).

Discussions of the link between freedom of expression and protection against discrimination are necessary and legitimate. When does the promotion of freedom of expression turn into the protection of hate speech and incitement to violence? And, conversely, when does protection against discrimination turn into oppression of freedom of expression? Despite the existence of national restrictions on the freedom of expression in almost every country, a common understanding of how to define and apply these restrictions at the international level is yet to be achieved (Matytsin, 2021). As discussed above, the right to freedom of expression enjoys fairly broad protection in international law. However, conflicting interpretations of the concept by each country have prevented building consensus on its threshold and where it turns into hate speech and incitement to hatred, thus needing criminalization. Specifically, issues related to defamation, incitement, and discrimination based on religion have constituted a major point of disagreement among key members of the international community.

Furthermore, the recent rise of populism and its use of offensive language against people from specific races or religions are a case in the point that highlights the need

[1] European Union, "Council Framework Decision 2008/913/JHA of 28 November 2008 on combating certain forms and expressions of racism and xenophobia using criminal law," Official Journal of the European Union, L 328/55, 6.12.2008. A European Commission report found in 2014 that most EU member states had not yet correctly implemented the rules designed to tackle racist and xenophobic hate crimes: European Commission, Report from the Commission to the European Parliament and the Council on the implementation of Council Framework Decision 2008/913/JHA on combating certain forms and expressions of racism and xenophobia using criminal law, COM/2014/027, January 27, 2014.

to differentiate hate speech from freedom of expression. The absence of clear terms in this regard has intensified the political rhetoric and communication strategies of right-wing populist parties that exploit media platforms to create hatred and spread fear about immigrants and religious minorities, a manifestation of which can be seen in increased cases of Islamophobia and antisemitism (Wodak, 2015). The misuse of freedom of expression in this way does facilitate prejudice against minorities and normalize racist and xenophobic discourse and rhetoric, which favors an ethnic or racial interpretation of social, economic, and political problems and immigration, and increases acts of religious intolerance, hatred, and discrimination against minorities.

16.3.2 The Organization of Islamic Cooperation Perspective on Freedom of Expression

The Organization of Islamic Cooperation (OIC) has taken a firm stand and regularly underscores the importance of distinguishing between freedom of expression that is a fundamental human right for all and hate speech that promotes discrimination and intolerance on ethnic, religious, and other grounds. The OIC considers that combating hate speech serves to promote freedom of expression and has been accordingly leading efforts to raise the international awareness at all levels of the gravity and dangers of hate speech, which constitutes a threat to the values of pluralism, global peace, and harmony.

As part of these efforts, from 1999 to 2010, the OIC submitted a resolution to the UN Human Rights Council for "combating defamation of religions," to highlight and convey its concern over the emergence of new manifestations of intolerance and hatred against Islam and Muslims in different parts of the world. Though adopted by consensus in the first two years, a sharp division over the legal and conceptual grounds on the concept of "defamation of religions" by the Western countries prevented reaching an international agreement on this resolution.

In this long fight against discrimination based on religion, which is particularly fueled by the misuse of the freedom of expression concept—hate speech, the OIC has adopted a consensus-building approach at the international level with the adoption, in 2011, of the United Nations Human Rights Council Resolution 16/18 (Human Rights Council Res 16/18, 2011) on "combating intolerance, negative stereotyping, and stigmatization of, and discrimination, incitement to violence, and violence against persons based on religion or belief." This resolution calls upon all member states to foster religious freedom and pluralism, to ensure religious minorities are properly represented, and to consider adopting measures to criminalize incitement to imminent violence based on religion or belief. Since 2011, this resolution has been annually adopted with a consensus by both the UNGA and HRC, which is reflective of its importance and the need for universal cooperation for its effective implementation.

The Independent Permanent Human rights Commission (IPHRC), which is the principal human rights organ of the OIC, has also clarified that, whereas freedom of expression is one of the key factors for the creation of modern-day inclusive, tolerant, and multicultural societies; hate speech motivated by racism, xenophobia, and intolerance, coupled with impunity for perpetrators to create a climate of fear and social exclusion of the targeted persons and groups, which is anathema to the ideals of pluralism and democracy. Hence, the need is to use this freedom responsibly to ensure the protection of the rights of others, respect the right of privacy and personal dignity, and maintenance of sociocultural harmony (IPHRC Outcome Document, 2015). IPHRC has also emphasized the crucial role of media in both promoting and combating hate speech. It, accordingly, urges the media to exercise caution and avoid perpetuating stereotypes/incitement to hatred that ultimately leads to discrimination and violence against targeted individuals and groups.

16.3.3 Istanbul Process

To ensure effective implementation of HRC Res 16/18, the OIC launched the Istanbul process initiative, in June 2011, in partnership with the USA, the EU, and some other interested countries which aim at ensuring a transparent, sustained, and structured process of engagement to follow up on the implementation of Resolution 16/18, and to provide a framework for a better understanding of stakeholder's different views, interests, and concerns and accordingly shaping an appropriate plan of action based on positive and effective good practices shared in the meetings held under the said process.

Resolution 16/18 constitutes, by its consensual character and the principles established for its implementation, major historic progress in terms of clarification and quest for a balance between freedom of expression and prohibition of incitement to racial, religious, or nationalistic hatred. The hallmark of the resolution, i.e., its action plan provides for a range of balanced measures that not only cater for positive actions such as reaching out to minorities, training of officials, and intercultural dialog, but also call for stricter actions such as the need for the criminalization of acts of incitement that lead to imminent violence. However, the difficulties and hurdles encountered in its implementation, since its adoption, call for a critical reflection on their root causes, the commitment of parties involved, and the modus operandi to be adopted by the Istanbul process for active and measurable progress on all parts of the action plan.

16.3.4 Challenges to Protect Freedom of Expression While Combating Hate Speech and Incitement to Hatred

In an environment infused with populist discourses focused on xenophobic narratives and incitement to hatred against minorities to make electoral gains, it has become a daunting challenge for policymakers and academics to define the narrow threshold between legitimate space for freedom of expression and the dangerous zone of hate speech. Equally potent is the threat of giving cushion to blasphemy laws that have been misused by extremist elements to silence opposition to tyranny, political and religious differences. As discussed above, the right to freedom of expression enjoys fairly broad protection in international law. However, as it is not an absolute right, it must be balanced in consideration of other rights, such as the right to privacy and the right to non-discrimination and to protect the dignity of others.

Governments have the double obligation to protect freedom of expression from one side, as well as to restrict its practice in certain circumstances from the other side, provided they take into account the guidance for restrictions as outlined in Article 19 of the ICCPR, mainly to guarantee the respect of the rights or reputations of others, the protection of national security or public order, or public health/morals, or as necessary in the case of a national emergency as purported in Article 4 of the ICCPR (Bresner, 2015), as well as indicated in general recommendation No. XV on Article 4 of International Convention on the Elimination of All Forms of Racial Discrimination by ICERD Committee, which clearly states that "prohibition of the dissemination of all ideas based upon racial superiority or hatred is compatible with the right to freedom of opinion and expression" (CERD, General recommendation No. 35, 2013).

The rise of far-right movements in recent years represents a real security threat that must be taken seriously. Indeed, the prospects seem to be worrying if the current trend of rising hate speech and identity politics continues in the future. In this regard, what may distinguish a healthy practice of freedom of expression from a toxic one is the resolve of governments to take significant steps toward protecting the achievements in keeping the balance between freedom of expression and fighting hate speech, which undermines people's security and increases the risks of triggering violence in society. It has become apparent that the global politics of hate unleashed by right-wing populist politicians and regimes in the USA, Europe, and beyond have galvanized white supremacist ideologies, promoted discrimination based on racial and religious identities, as well as exacerbated such movements and practices across the globe (CERD, General recommendation No. 35, 2013). These discriminatory practices were manifested through open bans on people from specific religious or racial groups (in the USA), discriminatory practices against migrants (in Europe), and denial of social, political, and economic rights to the minorities (in India and Myanmar), etc.

Ironically, hate speech narratives against ethnic and religious minorities by the far-right extremists give undeserved legitimacy to the narratives of extremist voices on the other side of the spectrum, such as the Islamic State of Iraq and Syria (ISIS) and other religious fanatics. This irony reflects the reality that hate speech does nurture extremism and creates a vicious cycle of hatred, and corresponding consequences that range from discrimination to violence against minorities, as well as threaten not only the peace and security of affected societies but also the peaceful coexistence among the comity of nations. Similar to terrorism recruitment online, right-wing extremism has developed an ability to use the Internet and social media to propagate its cause, in a determined effort to be transnational, benefiting from a free ride on some of globalization's advantages, such as the ease of global travel and communication (Wark, 2019). These abilities are highly efficient in motivating lone wolves through hate speech to take action anywhere in the world and are well demonstrated in the cases of Norway and Christchurch attacks and their Manifestos.

The social dynamics that are being created through the use of hate speech in the political landscape and media, benefiting from the gray area between freedom of expression and incitement to hatred, are real global challenges, which encourage the passive voices of right-wing extremists to take the center stage and be more proactive about their radical perspectives to change society, and to undo achievements of modern societies that recognize equal rights for all (Inshakova et al., 2020a, b). The current rise in far-right extremism represents a cultural cold war over the values of multiculturalism in modern democracies. It threatens societal peace and security and may turn into violent acts, including terrorist attacks, similar to the cases of Norway and New Zealand-Christchurch attacks. The ability of radical voices to misuse the freedom of expression to promote ethnocentric supremacist values reflects the success of contemporary right-wing extremists, like their predecessors, in acquiring the façade of legitimacy (Perry & Scrivens, 2016). This provides a greater capacity to connect with the public and enables the radicalization of youth that are looking for meaning outside traditional political parties, which are becoming less attractive to the young generations across the world.

Emerging political platforms, built around the notions of defending an old identity and ensuring its security and protection against "newcomers," are translated into the electoral success and political credibility for political parties which instrumentalize the democratic processes and values, particularly the ideals of freedom of expression, to explicitly champion racist and xenophobic programs all across the world. Incitement to racial, religious, or national hatred constitutes the key mechanism in the service of this edifice. During 2019, far-right political parties in Europe made significant electoral gains, many of which are unprecedented since the Second World War (BBC News, 2019), and made it compulsory for other parties to make political alliances as well as comprises to gain political power. This growing momentum around identity politics has exacerbated the risk of violent acts by inspired far-right extremists against ethnic and religious minorities, not just in European or Western societies, but across the world.

This political instrumentalization of hate speech in the name of freedom of expression must be seen as not only a threat to democracy but also a catalytic for conflicts between groups and countries in multicultural societies. Intellectual legitimization and political instrumentalization work on giving legitimacy to the passage to symbolic violence, whether verbal, physical, or social against targeted minorities.

16.4 Conclusion/Recommendations

In her annual report to the Human Rights Council, the UN Special Rapporteur on Minority Rights has rightly concluded that "although not all hateful messages result in actual hate crimes, hate crimes rarely occur without prior stigmatization and dehumanization of targeted groups and incitement to hate incidents fuelled by religious or racial bias" (HRC Report No. A/HRC/28/64). This is an apt reminder about the need for addressing the misuse or abuse of freedom of expression in a world where fake news is used by politicians, governments, and social groups for myopic sociopolitical and economic gains with no regard for the individual and collective human rights of targeted individuals and communities resulting in inhuman consequences such as wide spread socioeconomic discrimination, violence including genocide.

In light of such disturbing developments worldwide, there is a need to build consensus on finding/establishing the threshold of freedom of expression, where it converts into hate speech and incitement to hatred, thus needing criminalization as provided in Article 20 of the ICCPR. It will also be in line with the collective will of the international community expressed in the universally adopted action plan of the HRC Res.16/18, which calls for "adopting measures to criminalize incitement to imminent violence based on religion or belief." The action plan of Res. 16/18 also presents a comprehensive prevention plan for all ills related to incitement to hatred. Accordingly, it involves a "multilayered approach," which promotes human rights and tolerance, encourages dialog and understanding among different groups, and builds the capacity of national authorities, including security officials as well as media, thus creating an environment conducive to preventing acts of incitement to hatred, which is of vital significance (IPHRC Outcome Document, 2015).

Such a multilayered approach to promoting freedom of expression while combatting hate speech must start from deconstructing the narratives of hate speech and conducting a proper analysis to understand the processes that underpin the rise of hate speech in modern societies. At the core of hate speech construction and normalization is the intellectual legitimization of ethnic, religious, and national superiority, which justifies hatred, discrimination, and violence toward specific individuals and groups. Such legitimization is based on false narratives/ideas that tend to reduce a given ethnicity or religion to a devaluing identity based on nature/biology or cultures such as linking violence with Islam, cultural or biological inferiority for Blacks and Jews, and subsequently, their exclusion from and discrimination within a society. These constructed concepts are used by different far rights political parties and social movements' platforms to build negative perceptions against targeted individuals and groups as well as to thwart and antagonize multicultural dynamics in their societies for political gains.

To address these dynamics at a global level, there is a need to depoliticize the international discourse on the subject by moving away from an ideological debate to a legal, moral, and ethical discussion within the human rights framework (IPHRC Outcome Document, 2015). Governments should promote an intellectual, moral, and ethical strategy to bridge the gap of misunderstanding or "clash of ignorance" by countering the increasingly negative political rhetoric and biased media coverage of and their interpretations about the negative role of minorities and "the other" in "our problems." To this end, human rights education can serve as an effective tool to combat hatred and promote a better understanding of diversity. Hence, the need for its wider application and integration into national educational plans and other relevant national plans of action is to foster universally recognized human rights values to promote a culture of peace, coexistence, and multiculturalism, thus strengthening the resilience of all societies against extremist and intolerant views (IPHRC Outcome Document, 2015).

The conjuncture of events that the world is witnessing today, including the unprecedented level of interconnectedness and tensions, combined with global crises and pandemics, can produce dramatic social changes and create fundamental new realities in the stock of knowledge and its implications like human societies. Such effects are amplified with the strength of social media in quickly disseminating views and forming opinions as well as its misuse by spreading fake news for fomenting hatred and intolerance. States need to pay special attention and raise awareness about this phenomenon, as well as to monitor the misuse of freedom of expression for incitement to hatred, discrimination, and violence to stop the stigmatization of each other particularly the minorities and vulnerable groups and to promote respect for diversity as well as sociocultural and religious sensitivities of different segments of society that are vital for building inclusive, resilient, and pluralistic societies.

References

BBC News. (2019, November 13). *Europe and right-wing nationalism: A country-by-country guide.* https://www.bbc.com/news/world-europe-36130006. Accessed 11 January 2021.

Bhat, A. M. (2014). Freedom of expression from Islamic perspective. *Journal of Media and Communication Studies, 6*(5), 69–77. https://doi.org/10.5897/JMCS2013.0378

Bresner, K. (2015). *Understanding the right to freedom of expression: An international law primer for journalists.* IHRP|JHR. https://jhr.ca/wp-content/uploads/2019/10/Understanding-Freedom-of-Expression-Primer-ENG-web.pdf Accessed 11 January 2021.

Committee on the Elimination of Racial Discrimination. (2013, 26 September). *On combating racist hate speech* (General recommendation No. 35, UN Doc. CERD/C/GC/35).

Goncharov, A., Dolinskaya, V., Inshakova, A., & Kalinina, A. (2019). Corporate conflicts in the system of public relations, law, and legislation of the Russian federation. In E. G. Popkova (Ed.), *"Conflict-free" socio-economic systems* (pp. 75–88). Emerald Publishing.

Human Rights Council. (2008, March 28). *mandate on the special rapporteur on the promotion and protection of the right to freedom of opinion and expression.* UN Res 7/36.

Human Rights Council. (2011, March 24). *Combating intolerance, negative stereotyping, and stigmatization of, and discrimination, incitement to violence and violence against, persons based on religion or belief*. UN Res 16/18.

Inshakova, A. O., Inshakova, E. I., & Deryugina, T. V. (2020a). Information technologies for the formation of the European space of freedom, security, and justice, artificial intelligence: Anthropogenic nature vs. social origin. In *Advances in intelligent systems and computing* (pp. 444–453). Springer Nature.

Inshakova, A. O., Goncharov, A. I., Inshakova, E. I., & Tymchuk, Y. A. (2020b). Digital technologies for alternative methods of resolving conflicts: The prospects of application in Russia and other BRICS countries.In A. O. Inshakova & A. V. Bogoviz (Eds.), *Alternative methods of judging economic conflicts o the national positive and soft law* (pp. 129–142). Information Age Publishing,.

IPHRC. (2015, November 23). *8th session outcome document on freedom of expression and hate speech.* https://www.oic-iphrc.org/en/data/docs/session_reports/8th/8th_iphrc_thematic_debate_outcome_en.pdf Accessed 11 January 2021.

Jallow, A. Y. (2015). Freedom of expression from the Islamic perspective. *Journal of Mass Communication Journalism, 5*(10). https://doi.org/10.4172/2165-7912.1000278

Kamali, M. H. (1998). *Freedom of Expression in Islam. Fundamental Rights and Liberties in Islam series.* Ilmiah Publishers.

Matytsin, D. E. (2021). Internet-investing as a remote algorithm of retail investment financing. In *Modern global economic system: Evolutional development vs. revolutionary leap.* Lecture Notes in Networks and Systems. Springer Nature.

Perry, B., & Scrivens, R. (2016). Uneasy alliances: A look at the right-wing extremist movement in Canada. *Studies in Conflict & Terrorism, 39*(9), 819–841. https://doi.org/10.1080/1057610X.2016.1139375

Theodorou, A. E. (2016, July 29). *Which countries still outlaw apostasy and blasphemy?* https://www.pewresearch.org/fact-tank/2016/07/29/which-countries-still-outlaw-apostasy-and-blasphemy/ Accessed 11 January 2021.

UN General Assembly. (1946, Decenber 14). *Calling of an International Conference on Freedom of Information.* UNGA Res 59(1), UN Doc A/229, A/261.

Universal Declaration of Human Rights. (1948, December). UNGA Res 217 A(III).

Walker, J. (2018, June). *Hate speech and freedom of expression: Legal boundaries in Canada* (Publication No. 2018–25-E). Legal and Social Affairs Division, Parliamentary Information and Research Service. Ottawa.

Wark, W. (2019, June 27). *Canada has taken an important step in the fight against right-wing extremism.* https://www.theglobeandmail.com/opinion/article-canada-has-taken-an-important-step-in-the-fight-against-right-wing/ Accessed 11 January 2021.

Wodak, R. (2015). *The Politics of fear: What right-wing populist discourses mean* (p. 256). SAGE.

Chapter 17
Ensuring the Right to Education of Children with Disabilities as One of the Factors of Inclusive Growth of the State: The Experience of the Russian Federation

Anastasia A. Belousova, Marianna Ilyashevich, and Valentina G. Mikrina

Abstract The article examines current aspects of the realization of the right to education by children with disabilities, including the use of digital and smart technologies, on the example of the Russian Federation. The Russian Federation is a party to international treaties that secure the right to education for all without any discrimination. However, the State, like many others, faces problems in the implementation of international legal obligations in this area. At the same time, access to education is one of the key factors for ensuring the sustainable economic growth of any state. In the modern world with advanced technologies, children with disabilities have all opportunities to receive a quality education on an equal basis with other children. Appropriate measures must be taken at the state level or the level of private educational institutions with state support. The article analyzes the measures taken by the Russian Federation to fulfill its obligations to ensure the right to education for children with disabilities at the present stage. These issues are also relevant in the light of the provisions of the Sustainable Development Goals, according to which quality education is the basis for a decent life and sustainable development.

Keywords Right to education · Inclusive education · Smart technologies · Digital technologies · Children with disabilities · Human rights · UN Committee on the Rights of the Child

A. A. Belousova (✉) · M. Ilyashevich
Peoples' Friendship University of Russia (RUDN University), 6 Miklukho-Maklaya St, 117198 Moscow, Russia
e-mail: belousova-aa@rudn.ru

M. Ilyashevich
e-mail: iliashevich-m@rudn.ru

V. G. Mikrina
Odintsovo Branch of the Moscow State Institute of International Relations of the Ministry of Foreign Affairs of the Russian Federation (MGIMO University), Moscow, Russia
e-mail: v.mikrina@odin.mgimo.ru

A. O. Inshakova and E. E. Frolova (eds.), *Smart Technologies for the Digitisation of Industry: Entrepreneurial Environment*, Smart Innovation, Systems and Technologies 254, https://doi.org/10.1007/978-981-16-4621-8_17

JEL Codes B54 · D63 · F53 · J12 · J16 · K38

17.1 Introduction

The latest decade was marked by the increased efforts of the international community to protect and ensure the realization of human rights. This is motivated by the progressive promotion of democratic ideals, the cornerstone of which is the prominence and value of the human person in particular. Active promotion of humanitarian ideas has replaced a temperate attitude towards human rights, which has stipulated the creation of numerous human rights institutions. International law with its capabilities to adapt and unite versatile approaches in place on the national level certainly plays a key role in that case. It is the obligation to respect human rights, which stimulates an impulse for elaboration and implementation of smart technologies that constitute a factor of the States' inclusive development (Petrova et al., 2020).

Protection of human rights of persons with disabilities is an important aspect of the international system of human rights protection. Development of that institute lies deep within the development of the human society itself, reflecting the long road from the society ignoring, ghosting, and rejecting persons with disabilities, and attempting to "hide" them, to the inclusion of them in each and every aspect of the functioning of the society and the state. Such an approach, certainly, requires accumulation of resources on the part of the state and other subjects of legal relations, adoption of "unpopular" measures, a search of new forms of involvement of persons with disabilities in all spheres of functioning of the society, including the use of new teaching methods and advanced teaching technologies. Here, access to education comes into sharp focus. Obtaining an education is not a whim or an extravagance. It is a process aimed at personal progressive growth, development of cognitive abilities, skills, without which complete realization of each person's inherent potential is impossible. That fact explains the existence of a variety of international legal documents in this area, based on which states adapt their national legislation. Moreover, one of the goals of sustainable development, which the international community strives to achieve by 2030, is "to ensure inclusive and equitable quality education and to promote lifelong learning opportunities for all" (SDGs, Goal 4). This also needs to be done to ensure the implementation of the other sustainable development goals (Abashidze et al., 2020); the sustainable development goals. Official Web site). Lack of access to education makes attempts to overcome poverty, discrimination, to increase the level of economic development and health care, including the fight against especially dangerous diseases, futile. In that view, the key issue is to ensure access to education for children in general and children with disabilities in particular.

Despite the existing international legal acts and adoption of domestic regulations by states with the view of fulfilling their international obligations, the issue of ensuring even access to education for all does not lose its relevance. A mere aspiration toward the elimination of the existing barriers does not contribute to their factual elimination. Successful experiences (Belousova et al., 2017) and effective practices must be relied on while overcoming any complications, by the trial-and-error method. Specifics of a certain state should be considered, while searching acceptable routes to resolve existing issues, since here the legal status of an individual is addressed. All states have a similar goal, which can (and sometimes should) be reached by different means. In light of the foregoing, specific attention has to be devoted to the practice of certain states, which is the Russian Federation within the current research. The topic of the research is relevant in light of, among others, the National Report presented to the United Nations in 2020, containing a review of the measures, adopted by the government to reach the Sustainable Development Goals 2030.

17.2 Methodology

Issues related to the realization of the right to education for all without discrimination are currently in need of complete recognition, taking into account the importance of this aspect as a whole for the effective functioning of society and the state, as well as the development of the sphere of smart technologies that allow to unlock and utilize the inherent potential of any individual.

Within the framework of the current study, general and special methods of scientific knowledge were used, in particular—the methods of analysis, synthesis, generalization, modeling, as well as the comparative–legal method, the statistical method, and the legal interpretation method.

17.3 Results

The Russian Federation is a party to the Convention of 1989, which enshrines obligations of the contracting parties to ensure and fulfill access to education (Convention on the Rights of the Child. Adopted and opened for signature, ratification, and accession by General Assembly resolution 44/25 of 20 November 1989). It also provides that "a mentally or physically disabled child should enjoy a full and decent life" (Convention on the Rights of the Child. Adopted and opened for signature, ratification, and accession by General Assembly resolution 44/25 of 20 November 1989, Art. 23). Convention on the Rights of Persons with Disabilities of 2006 was ratified by the Russian Federation in 2012.

With the view of fulfilling relevant obligations, in light of the concluding observations of the UN treaty bodies and the recommendations of the UN Human Rights Council following the Universal Periodic Review, the Russian Federation exercises a policy directed at improving the situation of children with disabilities in terms of realizing the right to education, including the use of new teaching methods and the introduction of advanced technologies (Makhotin, 2018).

According to the data for 2021, 10,844,254 persons with different disabilities classifications were registered in the Russian Federation (Official Web site of the

Pension Fund of the Russian Federation). In 2016, 64% of children with disabilities aged from 3 to 8 have attended pre-school educational facilities (Attendance at preschool educational institutions by children aged 3–8 years in 2016. Federal State Statistics Service) and 90% of children with disabilities aged from 9 to 15—facilities of the general educational system or studied at home, while being attached to a specific educational facility (Attendance at educational institutions by children aged 9–15 in 2016. Federal State Statistics Service). In 2019, the number of children with disabilities attending preschool educational institutions increased by 6.5% (Attendance of preschool educational organizations by children with disabilities in 2019. Federal State Statistics Service). Noteworthy is that the rates of students with disabilities, graduating from universities, and obtaining an intermediate vocational education are growing each year. In particular, the amount of students with disabilities, that have obtained an intermediate vocational education within 10 years, has doubled: from 4524 in 2008 to 8913 in 2019. Speaking of higher education—the amount rose from 5770 students in 2008 to 7487 in 2018 (Information about the disabled—students enrolled in professional educational programs. Federal State Statistics Service). The data presented seize only information collected from the governmental educational institutions. The Committee on the Rights of Persons with Disabilities in its concluding observations has also highlighted in 2018 the increase in numbers of persons with disabilities, obtaining education in inclusive educational institutions (Committee on the Rights of Persons with Disabilities, 2018). The number of schools with inclusive conditions of education has increased by 8 times—up to 9,339—that is, from 2.5 to 21.4% from their general amount (Opening remarks of the Deputy Minister of Labor and Social Protection of the Russian Federation, 2018).

Among a variety of national legal acts of the Russian Federation concerning access to education for all without discrimination, most prominent are: Federal Law "On Education in the Russian Federation"; Order "On Establishment of the Order ensuring accessibility conditions for persons with disabilities of the objects and services in the education area, and provide necessary assistance to them"; Order "On Establishment of the Order of the Psychological-Medical-Pedagogical Commission"; Letter of the Ministry of Education and Science of Russia of 18 March 2014 No. 06–281 "On Direction of Instructions" (including "Instructions on the educational process involving persons with physical and mental disabilities in educational institutions, including equipment for the educational process"; Letter of the Federal Service for Education and Science Supervision of 16 April 2015 No. 01–50-174/07–1968 "On Enrolment of persons with disabilities," and others. The "Accessible Environment" Federal program for 2011–2020 also contains provisions crucial for the realization of the right to education, which is planned to be prolonged up to 2025 and which is getting active sourcing.

The listed documents contain provisions determining categories of persons falling under the provisions, regulating the order of their enrolment to educational institutions, teaching process, etc. In particular, according to Article 5 of the Federal Law "On Education," the government has to create conditions for persons with disabilities for obtaining education without discrimination (Federal Law "On Education in the

Russian Federation," 2012). In that view, the terminology, applied in the Russian laws, should be explained (2017).

The Convention of 2006, just as other international legal acts, utilizes the notion of "children with disabilities" (Convention on the Rights of Persons with Disabilities, 2006). Noteworthy is that the English text speaks of the protection of "persons with disabilities," whereas the authentic Russian translation of the text, speaks of "disabled persons." The latter is being criticized by the communities of persons with disabilities as outdated and implying clear discrimination. The person comes first, and only then—his characteristics, says a representative of the Regional Public Organization "Perspective" (Culture of communication with people with disabilities. Language and etiquette). The Committee on the Rights of Persons with Disabilities has also highlighted the use of an incorrect notion in its concluding observations in 2018, stating that it does not follow the human-rights-oriented model and thus recommended amending the terminology in use (Committee on the Rights of Persons with Disabilities, Para. 7, 8).

Along with that, Russian national legislation utilizes a variety of notions: "a disabled child" (Federal Law "On social protection of disabled people in the Russian Federation," 1995), "a child with abnormal development," "a challenged child." The second notion was substituted with the latter in 2007 (Federal Agency for Technical Regulation and Metrology, 2007). The key difference between the notions of "a disabled child" and "a challenged child" lies in the question of which governmental bodies will determine the special conditions included in the plan of these persons' education and mentoring: in the case of a "disabled child" this will be determined by the institutions of medical-social expertise, wherein the case of a "challenged child," which has no disabilities—a body within the educational system (Federal Law "On Education in the Russian Federation," 2012, Art. 79). Thus, it appears obvious that according to the Russian legislation, "disabled children" and "challenged children" are two separate, but closely linked categories.

Inclusiveness and equity form a basis for quality education (2030). As formulated by the United Nations Economic and Social Council, "inclusive education" is a process of integration of children into the general educational process irrespective of their gender, ethnic origin, religion, previous academic progress, health conditions, level of development, socio-economic status of their parents and other differences (Inclusive education. UNESCO). As provided by the Federal Law "On Education in the Russian Federation," inclusive education means an education, which provides an equal opportunity for all to obtain the education, while considering diversity, certain educational needs, and one's characteristics (Federal Law "On Education in the Russian Federation," 2012, Art.2 para. 27).

Despite the difficulties present in the Russian Federation, the introduction of inclusive education is a big step forward, since for a long time the governmental policy was focused on the maintenance of disabled and challenged children in boarding schools, while education and mentoring of them was an exclusive prerogative of a special (correctional) educational institution, which resulted in growing inequalities. Thanks to the introduction of inclusive education, the attitude of society toward persons with disabilities is changing to a more positive one, destroying alienation.

One of the contemporary priorities of the Russian Federation is to ensure access and quality of education for persons with disabilities and those challenged in health well-being, considering all peculiarities and specific educational needs, including the provision of assistive and technical means, including new methods and smart technologies.

The process of introducing inclusive education undoubtedly requires the accumulation of significant funding and resources from the state, training of the personnel, etc., which also takes considerable time, especially considering that such an approach is new for Russia. In particular, the UN Committee on the Rights of the Child (2014) has expressed concern in its concluding observations that children with mental disorders in specialized childcare facilities are considered "unteachable" and are deprived of any incentives for their development, and that the proportion of children with disabilities in schools of general education is still low, despite the new Law on Education, which provides for inclusive education. The Committee has also recommended that measures should be applied to provide teachers with training, schools—with the necessary equipment and digital technologies that will allow children with disabilities to study at the same level as the rest of the children (2021). At the same time, the Committee has also highlighted the progress in the implementation of inclusive education in the Russian Federation and welcomed the legislative changes aimed at ensuring inclusive education for children with disabilities in schools of general education.

Analysis of the alternative reports of non-governmental organizations, comments upon the second cycle of the Universal Periodic Review by the UN Human Rights Council (2013), concluding observations of the Committee on Economic, Social and Cultural Rights (2017), Committee on the Rights of Persons with Disabilities (2018, paras. 48–50), following comments can be made: growth of segregation of children with disabilities in education; slow creation of inclusive schools; absence or low number of inclusive schools in most of the cities; access to education by only a half children with disabilities; lack of awareness among parents of children with disabilities regarding the right to inclusive education; poor readiness of schools to teach children with disabilities; lack of attention on the part of the state to the issue of accessibility of transport for children with disabilities, which results in impossibility to get to school; provisions, governing recommendations for home education are not detailed enough; the problem of access to education for children with special needs, as they study either at home or in special schools and are deprived of the opportunity to socialize, and others.

Prominent changes have been introduced in the Russian Federation within the past 5 years with the view of guaranteeing the realization of the right to inclusive education (Abashidze et al., 2018). Still, the solution of the above problems is linked with certain difficulties arising from differences between states in the applicable terminology and approaches. For instance, it is incorrect to label as segregation the support of children with serious mental health problems because the care, support, and education of such children require special conditions, the implementation of which is possible only in special boarding schools. A wide range of disorders in the development of children with disabilities represents another issue, which also requires an individual approach, and which cannot be provided in general educational institutions. Support and education of children with severe mental disorders in boarding schools are carried out first and foremost in the interests of that child. Even the parents of such children, which have an opportunity to support and educate such children at home, often prefer to place them in a specialized institution, where, in their opinion, children will be better taken care of. Such schools and classes provide children with special conditions required for studying; the issue of lack of communication with their peers, who require no special learning conditions, can and should be solved by conducting various kinds of extracurricular activities, creating coteries and classes that could be attended by challenged and children with disabilities together with those, studying in the general educational institutions. The creation of such conditions for communication is one of the steps toward solving the problem of social adaptation of children with special needs. It is, therefore, possible to point out that the existence of boarding schools does not hinder the implementation of inclusive education, but complements it. In this regard, it is more correct to speak not about the "segregation" of children, but of their "adaptation." Moreover, a positive practice exists, where children from such institutions precede their education in general education schools, thanks to individual approaches applied. The government proceeds from the fact that a combination of effective rehabilitation, an adaptation of the environment, and consideration of the personal factor (maximized individualization in the development of measures in a particular case) constitute the key to complete inclusion (Opening remarks of the Deputy Minister of Labor and Social Protection of the Russian Federation, 2018).

17.4 Conclusion

The federal educational standard for primary and general education of students with disabilities was started to operate from September 1, 2016, thanks to which today the education of children with disabilities and challenged children is provided through the creation of special conditions in educational institutions together with other students, and in separate classes, groups or in separate organizations carrying out educational activities following adapted basic general educational programs. The amount of such children attending preschool educational institutions grows each year. Children with disabilities and challenged children receive psychological and pedagogical support at all levels of the education system, provided by speech therapists, speech pathologists, and psychologists (State report on the situation of children and families with children in the Russian Federation, 2017).

Despite the achievement of significant results, the introduction of inclusive education in the Russian Federation certainly faces some difficulties. The effectiveness of the measures, applied by the government in the area under consideration, should be improved with the view of ensuring full realization by the children with disabilities and challenged children of their right to receive affordable and quality education in accordance with the legislation of the Russian Federation and conformity with its international legal obligations. This will create a real basis for more dynamic development of this sphere of education alone, as well as within the framework of regional programs for the development of education.

Current efforts of the Russian Federation are aimed at the development of the educational process using various communications, audiovisual technologies that allow children with varying degrees of disability to get access to the material for its assimilation and practical application (Tarakanov et al., 2020). In general, the creation of an electronic environment is already a reality, but for children with disabilities, this is sometimes the only opportunity to realize their potential; therefore it is important that the state considers this as one of the priority tasks and continues to intensify development in this area.

References

Abashidze, A., Kiseleva, E., Ilyashevich, M., Nikolaev, A., & Belousova, A. (2018). The right to education of vulnerable groups of children in the Russian Federation in the light of the activities of the UN Committee on the rights of the child. *Journal of Advanced Research in Law and Economics, 9*(3), 792–804.

Abashidze, A. K., Inshakova, A. O., Solntsev, A. M., & Gugunskiy, D. A. (2020). Socio-economic inequality as a barrier on the path of sustainable development: “institutional traps” and perspectives of overcoming them. *International Journal of Sociology and Social Policy, 41*(1–2), 62–75.

Attendance at educational institutions by children aged 9 to 15 in 2016. *Federal State Statistics Service*. http://www.gks.ru/wps/wcm/connect/rosstat_main/rosstat/ru/statistics/population/disabilities/#. Accessed 27 February 2021.

Attendance at preschool educational institutions by children aged 3–8 years in 2016. *Federal State Statistics Service*. http://www.gks.ru/wps/wcm/connect/rosstat_main/rosstat/ru/statistics/population/disabilities/#. Accessed 27 February 2021.

Attendance of preschool educational organizations by children with disabilities in 2019. *Federal State Statistics Service*. http://www.gks.ru/wps/wcm/connect/rosstat_main/rosstat/ru/statistics/population/disabilities/# Accessed: 27 February 2021.

Belousova, A., Ilyashevich, M., Kupchina, Y., & Belousov, A. (2017). Students’ perceptions on teaching styles within public international law curriculum: A case of Russia. *Eurasia Journal of Mathematics, Science and Technology Education, 13*(10), 6945–6952.

UN Committee on the Rights of the Child. (2014). *Concluding observations of Committee on the Rights of the Child on the combined fourth and fifth periodic reports of the Russian Federation* CRC/C/RUS/CO/4–5, Distr.: General 25 February 2014. https://tbinternet.ohchr.org/_layouts/treatybodyexternal/Download.aspx?symbolno=CRC%2fC%2fRUS%2fCO%2f4-5&Lang=ru. Accessed 27 February 2021.

Concluding observations on the initial report of the Russian Federation Committee on the Rights of Persons with Disabilities. UN Document CRPD/C/RUS/CO/1, 9 April 2018. https://tbinternet.ohchr.org/_layouts/treatybodyexternal/Download.aspx?symbolno=CRPD%2fC%2fRUS%2fCO%2f1&Lang=en. Accessed 27 February 2021.

Concluding Observations of the Committee on Economic, Social and Cultural Rights. (2017). *Concluding observations on the sixth periodic report of the Russian Federation. Committee on Economic, Social, and Cultural Rights*. UN Document E/C.12/RUS/CO/6, 16 October 2017. Para. 40, 56, 57. https://tbinternet.ohchr.org/_layouts/treatybodyexternal/Download.aspx?symbolno=E%2fC.12%2fRUS%2fCO%2f6&Lang=en. Accessed 27 February 2021.

Convention on the Rights of Persons with Disabilities was adopted on 13 December 2006. UN Document A/RES/61/106. Art. 7. https://www.un.org/disabilities/documents/convention/convention_accessible_pdf.pdf. Accessed 25 January 2021.

Convention on the Rights of the Child. Adopted and opened for signature, ratification, and accession by General Assembly resolution 44/25 of 20 November 1989. https://www.ohchr.org/en/professionalinterest/pages/crc.aspx. Accessed 26 January 2021.

Culture of communication with people with disabilities.*Language and etiquette*. Official website of the Regional Public Organization of Disabled People "Perspektiva". https://perspektiva-inva.ru/language-etiquette. Accessed 25 January 2021.

Declaration of the rights of the child. Proclaimed by General Assembly Resolution 1386 (XIV) of 20 November 1959. https://www.unicef.org/malaysia/1959-Declaration-of-the-Rights-of-the-Child.pdf. Accessed 3 March 2021.

Federal Law. (1995, 24 November). No. 181-FZ (as amended on 29.07.2018). On social protection of disabled people in the Russian Federation (as amended and supplemented, entered into force on 01.01.2019). Art. 1. http://is.mos.ru/uploads/document//2763318f-1548019156.pdf. Accessed 15 February 2021.

Federal Law. (2012, December 29). *On education in the Russian federation* (No. 273-FZ. Art. 5 (1)). http://kremlin.ru/acts/bank/36698. Accessed 27 February 2021.

Inclusive Education. UNESCO official site. http://www.iiep.unesco.org/en/inclusive-education. Accessed 15 February 2021.

Information about the disabled—students enrolled in professional educational programs. *Federal State Statistics Service*. http://www.gks.ru/wps/wcm/connect/rosstat_main/rosstat/ru/statistics/population/disabilities/#. Accessed 27 February 2021.

Information on the number of disabled people in the Russian Federation, federal districts, and constituent entities of the Russian Federation in the context of disability groups, gender, age, and causes of disability. Official website of the Pension Fund of the Russian Federation. www.pfrf.ru

Inshakova, A. O., Goncharov, A. I., Mineev, O. A., & Sevostyanov, M. V. (2017). Amendments to the civil code of the Russian Federation: Contradictions of theory and practice. Contributions to Economics. *Russia and the European Union: Development and Perspectives, XIII*, 147–153.

Makhotin, D. A. (2018). SMART in education: New approach or influence of technology? *Interactive Education, 5*, 13–15.

Matytsin, D. E. (2021). Internet-investing as a remote algorithm of retail investment financing. In *Modern global economic system: Evolutional development vs. revolutionary leap*. Lecture Notes in Networks and Systems. Springer Nature.

National Standard of the Russian Federation. (2007). "Social services for the population. Terms and Definitions". Federal Agency for Technical Regulation and Metrology, Art. 1. http://docs.cntd.ru/document/gost-r-52495-2005. Accessed 15 February 2021.

Opening remarks of the Deputy Minister of Labor and Social Protection of the Russian Federation. (2018, February 27). G.*G. Lekarev during the 19th session of the Committee on the Rights of Persons with Disabilities when considering the initial report of the Russian Federation on the implementation of the provisions of the Convention on the Rights of Persons with Disabilities*, Geneva. https://tbinternet.ohchr.org/_layouts/treatybodyexternal/Download.aspx?symbolno=INT%2fCRPD%2fSTA%2fRUS%2f30484&Lang=en. Accessed 27 February 2021.

Petrova, E. A., Inshakova, A. O., & Kalinina, V. V. (2020). Information technology impact analysis on the dynamics of the structural change in the regional economy. *Lecture Notes in Networks and Systems, 155*, 1033–1044.

Quality education: why it matters. *The Sustainable Development Goals*. Official site. https://www.un.org/sustainabledevelopment/wp-content/uploads/2018/09/Goal-4.pdf. Accessed 27 February 2021.

State Report on the Situation of Children and Families with Children in the Russian Federation. (2017). *Official website of the Ministry of Labor and Social Protection of the Russian Federation*. https://sustainabledevelopment.un.org/sdg4. Accessed 27 February 2021.

Tarakanov, V. V., Inshakova, A. O., & Goncharov, A. I. (2020). Interactive demo online maintenance foreign trade deals of business entities of BRICS jurisdictions. In A. Inshakova & E. Inshakova (Eds.), *Competitive Russia: Foresight model of economic and legal development in the digital age*. CRFMELD 2019. Lecture Notes in Networks and Systems (Vol. 110, pp. 319–327). Springer. https://doi.org/10.1007/978-3-030-45913-0_36

The Sustainable Development Goals. Goal 4. *Ensure inclusive and equitable quality education and promote lifelong learning opportunities for all*. https://sustainabledevelopment.un.org/sdg4. Accessed 3 March 2021.

UN Human Rights Council. (2013). *Thematic list of recommendations on UPR of Russian Federation* (2nd ed. Cycle—16th session). https://www.ohchr.org/RU/HRBodies/UPR/Pages/RUIndex.aspx. Accessed 27 February 2021.

UNESCO. (2016). Education 2030: Incheon Declaration and Framework for Action for the implementation of Sustainable Development Goal 4*: Ensure inclusive and equitable quality education and promote lifelong learning opportunities for all*. https://unesdoc.unesco.org/ark:/48223/pf0000245656. Accessed 5 March 2021.

Part III
Neo-industrial Technologies in the Legal Regulation of Partnership and Competition of Social and Economic Systems of the Russian Federation and Other EAEU and BRICS Countries

Chapter 18
Upgrading Legal Regulation of Integration in the Context of Digital Economy: The Eurasian Economic Union Agenda

Tatsiana N. Mikhaliova

Abstract Digital transformation of day-to-day life, business processes, public administration urges not only the introduction of innovative legal and institutional instruments, but also the creation of the Single Digital Area in the frames of the EAEU, as well. A combination of agreed policy and common policy is considered necessary for the successful realization of multifactored, plurisubject information, and digital strategy of the Union. Development of the digital infrastructure within the common market and the EAEU institutions, cybersecurity, and protection of the processes are on the EAEU-2025 agenda. An increase in the sociotechnical interaction inevitably leads to the expansion of the Eurasian digital agenda: qualitatively, new types of relations (digital reality) demand innovative regulation, diversification of management mechanisms, axiological "reinstallation" and, generally, transfer from information interaction in the frames of the Union, and digitalization as an instrument to the digital model of the EAEU and creation of the Single Digital Area.

Keywords Agreed policy · Common policy · Digital agenda of the EAEU · Digital integration · Eurasian Economic Union · Eurasian Economic Commission · Eurasian law · Harmonization · Single digital area · Information and communicative technologies · Information interaction · Regional integration · Strategy of digital transformation · Unification

JEL Codes F02 · O33 · O19 · K33 · F02 · F62 · F63

T. N. Mikhaliova (✉)
Law Faculty of the Belarusian State University, 8 Leningradskaya St, 220125 Minsk, Belarus
e-mail: MikhaliovaT@bsu.by

A. O. Inshakova and E. E. Frolova (eds.), *Smart Technologies for the Digitisation of Industry: Entrepreneurial Environment*, Smart Innovation, Systems and Technologies 254, https://doi.org/10.1007/978-981-16-4621-8_18

18.1 Introduction

On inclusion of the digital agenda in contemporary integration processes.

Regional integration becomes an active trend of interstate interaction at the end of XX and the beginning of XXI centuries. In particular, it is reflected in an exponential increase in notifications of regional trade agreements to WTO (RTA Database, WTO, 2021). Not all of them are institutionalized organizations with an advanced level of integration, however, in every part of the world, there are integration projects of different levels of intensity and success (Mikhaliova, 2016).

The notion "integration" derives from the Latin "integratio" meaning a provision of integrity, restoration of integrity and, further on, from the initial form of "integer" as something whole, perfect, complete, cohesive. In the Grand Russian Encyclopedic Dictionary, integration is defined as a state of cohesion of different differentiated parts and functions of the system, organism in the whole, as well as a process, leading to such a state (Grand Russian Encyclopedic Dictionary, 2008).

Interstate integration is defined as a process provided with international legal instruments and directed to the gradual formation of interstate economically and, possibly politically common, cohesive (*integro*) area, based on a common market of goods, services, capital, and labor (Velyaminov, 2015). Freedom of movement of these factors reflects economic and legal characteristics of interstate integration provided for in almost all founding treaties of regional integration unions (e.g., para.1 art. 2 of the Treaty on the Eurasian Economic Union, para.2 art. 26 of the Treaty on the functioning of the European Union).

The common market building would not be comprehensive without digital agenda included in the legal and organizational system of integration associations in a contemporary world, as well as it would be ineffective. The pervasion of digital technologies and primacy of informative and communicative element in all current relations influences the formation of the strategy and agenda of integration associations. So, Digital Single Market Strategy was released in the EU in 2014, 7 years after the Lisbon Treaty.

The digital agenda of the Eurasian Union started in 2016, less than 2 years after the signing of the Treaty on the EAEU. It was December 2020, in the First Eurasian Congress, when the Prime Minister of the Russian Federation proposed to add to 4 freedoms one more freedom of movement of information. Currently, a set of activities on the formation of the Single Digital Area is proposed. Worth noting, readiness to the digital transformation of the economy, and social relations were inserted in the founding treaty and other acts from the very beginning. It can be read between the lines and in the context of general and special provisions notwithstanding the fact that directly the Treaty on the EAEU refers only to the concepts of informative provision of integration processes and information interaction.

18.2 Materials

Normative and legal basis for the research comprise of the Treaty on the Eurasian Economic Union, Protocol on Information and Communication Technologies in the frames of the Eurasian Economic Union, a range of program acts, projects, strategies of the Union's development in the context of digital transformation, sectoral acts of the Union's organs.

The theoretical basis of the study on the information and network transformation of economy, society, generally, refers to the works of Castells (2000). The current research is correlated to the theory of interstate economic integration proposed by Velyaminov (2015) and develops some theoretical conclusions made in the monograph of Mikhaliova (2016). The conceptual framework is investigated in an interdisciplinary context, including by referring to the positions of Nikulina and Starichenko (2018), Inshakova et al. (2020), Matytsin, and Rusakova (2021).

Prospects of the creation of the Single Digital Area in the EAEU are stated by the Eurasian officials, e.g., Sarkisyan (2021), therefore intensifying the necessity of the scientific development of the topic.

Articles and presentations of Khabrieva (2018), Khabrieva and Chernogor (2018), Talapina (2018), Ovchinnikhova and Zor'kin (2018), E. Ermakova et al. (2021) are dedicated to the challenges to the law in the digital age.

Scientific research on economic globalization based on the experience of developed and developing countries has been studied (Zankovsky et al., 2020).

The political and legal context of the EAEU digital agenda are analyzed by Shughurov (2020). Peculiarities of information interaction in the EAEU customs sphere are analyzed in the articles by Novikov and Roghozina (2018), Sablina (2018).

The conclusion on the necessity of the common and agreed policy for the realization of the Union's digital strategy is made based on general theoretical approaches on the types and methods of legal regulation in the integration community.

18.3 Methods

Methodology of the research is based on the content analysis of primary and secondary sources, understanding correlations and concept analysis, leading to doctrinal and applied conclusions and suggestions to improve the legal regulation currently in force and to develop a theory of integration and digital cooperation.

18.4 Results

18.4.1 On the Concepts "Informatization", "Digitalization", "Digital Transformation"

In the doctrine, it is rightly pointed out the difference of the concepts of "informatization" and "digitalization", based on the analysis of economic, legal meanings, ontology of information, and digital contexts (Nekrasov, 2018; Kazakov, 2019; Prokudin, 2020). Some researchers use them as synonymous or similar (Shabayeva, 2019). In a foreign doctrine, attention is drawn to the difference between "digitalization" and "digitization" (Wagner and Ferro, 2020). Unlike the former, which means progressive use of digital technologies and digital information leading to the changes in the functioning of things and social context, the latter is only the transfer of the data from analogous to digital format.

A range of scientists points out the inclusion of the issues of informatization to digital processes, a kind of transfer from informatization to digitalization (Nikulina and Starichenko, 2018). A. Paulin, vice versa, estimates digitalization as a process, which precedes the informatization of society (Paulin, 2018). Digitalization, in his opinion, reflects the current development of society, technologies, and the concept of Economy 4.0 as such. At the same time, informatization and following informatized management will be applied broadly in all spheres and sectors in the nearest future, but it is still not.

According to the Belarusian Standard STB 2583–2020, "Digital transformation Terms and definitions" entered into force on March 1, 2021 "digitalization" is a new stage of automatization and informatization of the economic activity and public administration, the process of transfer to digital technologies, which is based not only on the use on information and communication technologies for the production or management tasks but also accumulation and analysis of Big Data with its help and situation synopsis, optimization of the processes and costs, the attraction of new partners, etc.

Digital transformation is a reflection of qualitative revolutionary changes, which manifest not only in particular digital converting, but also in general reform of the structure of the economy, transfer of the centers of added value into the sphere of digital resources and end-to-end digital processes.

Digitalization affects all public and private spheres, therefore we can claim "the digital imperative of development" (Ovchinnikov and Fatkhi, 2018, Ovchinnikov, 2018). Indeed, digitalization and digital transformation become an indispensable part of the state programs, concepts, economic strategies. These processes took a stable place in the development of society, also by the reason of its special significance for the transfer to the 6th technological order.

Doctrinal comprehension of how the productivity and competitiveness of factors of production, economic actors in any form depending on the ability to generate, process, safely, and efficiently use information based on knowledge, began at the end of the twentieth century (Castells, 2000).

Today digital technologies, the translation of business processes online, the Internet of Things, E-government, artificial intelligence, and many other achievements of digitalization have become regular. There are also new tasks of a derivative order, e.g., access to digital public goods, distribution of responsibility for Internet administration and security (UN General Assembly Resolution A/C.3/74/L.11 "Countering the use of information and communication technologies for criminal purposes").

However, the main issue is that the construction of new relationships leads to the most important transformation, namely the emergence of cyber-spatiality. The digital component moves from a factor that mediates reality and influences it, into its determining one. Digital reality as a qualitatively new type of relations is only revealing itself to us (Khabrieva and Chernogor, 2018). Accordingly, a new law is emerging and this is the "law of the second modern", which regulates various kinds of relations in the context of digital reality, Big Data, robotic systems, and artificial intelligence (Zor'kin, 2018).

18.4.2 On Information Interaction in the EAEU

The Treaty on the EAEU was created as a consolidation of the already achieved integration results and as a new stage of development, the transition to an economic union in its classical sense. Nevertheless, one cannot fail to note the significant presence of provisions on information support for integration processes already in the initial text of the Treaty on the EAEU (art. 23). In this article and the Protocol on Information and Communication Technologies and Information Interaction in the Eurasian Economic Union (annex 3 to the Treaty, hereinafter the Protocol on Information Communication Technologies), the principles of information interaction "in the implementation of general processes within the Union" were envisaged, and the policy in the field of informatization and information technologies was brought to the rank of an agreed policy. These provisions require some clarification.

According to clause 2 of the Protocol on Information Communication Technologies, "general processes" are "operations and procedures regulated (established) by international treaties and acts constituting the law of the Union, and by the legislation of the member states, which begin on the territory of one of the member states, and end (change) in the territory of another member state." The list of general processes within the Union is determined by the Decision of the Board of the Commission dated April 14, 2015, No. 29 and includes 76 positions in 18 areas of information interaction, including in the field of interaction of customs authorities, on the protection and protection of intellectual property rights, in the field of circulation of medicines and medical devices, technical regulation; application of veterinary and sanitary measures; transport (automobile) control; production and circulation of agricultural products; competition policy and state (municipal) procurement; ensuring electronic document flow between the member states of the Eurasian Economic Union and Eurasian Economic Commission, and others.

On a number of these issues, for example, on tariff and non-tariff regulation, technical regulation, sanitary and phytosanitary measures, and some others, the Union law establishes a requirement for a unified policy of sectoral regulation. Uniform policy according to Art. 2 of the Treaty on the EAEU presupposes the application by the member states of unified legal regulation, including based on decisions of the Union bodies.

In the advisory opinion dated April 4, 2017, the EAEU Court formulated a legal position that classifies a certain area as a single policy, it is necessary to comply with the following conditions: (1) the presence of unified legal regulation; (2) the transfer by the member states of competence in this area to the bodies of the Union within the framework of their supranational powers.

All issues of information interaction, which in the modern world are often decisive for achieving an effective and operational result, are *de jure* referred to as a coordinated policy in accordance with paragraph 3 of Art. 23 of the Treaty on the EAEU. The agreed policy presupposes the harmonization of legal regulation, that is, the approximation of the legislation of the member states aimed at establishing similar (comparable) regulation (art. 2 of the Treaty on the EAEU).

This discrepancy between the unified standard of substantive regulation and information potential hinders the development of integration. Obviously, in the areas related to a single policy, it is necessary to establish a single policy regarding their information support. Without a Single Information Area based on the unification and standardization of the information process, the exchange of information will not be effective enough (Nemirova & Vinichenko, 2017).

Moreover, there is an understanding of this at the level of implementation of general processes. Thus, information interaction during the implementation (through the integrated information system of foreign and mutual trade) of the general process "Formation, maintenance and use of a unified register of pharmaceutical inspectors of the Eurasian Economic Union" is regulated within the framework of a single policy by adopting the relevant Decision of the Board of the Eurasian Economic Commission dated October 25, 2016, No. 127, approving the rules, regulations for information interaction, a description of the formats and structures of electronic documents and information used for implementation through an integrated information system of the external and internal trade of the relevant common process, the order of accession to this common process.

The process of preparing similar acts in other areas of general processes indicates a general trend of unification rather than harmonization of information interaction. At the meeting to agree on technological documents governing information interaction when ensuring transport (automobile) control at the external border of the EAEU (rules, regulations, etc.), the parties pointed out the need to fix requirements for some information procedures at the level of acts of direct action, for example, entering notifications into the database.

Thus, in practice, those areas of information interaction that requires a unified approach from the point of view of the subject of regulation are also regulated uniformly, based on the norms of the direct action of acts of the Commission. De jure, however, the EEU Treaty still enshrines a general rule on an agreed policy in the field of information interaction. Therefore, there is a need to amend the Treaty on the EAEU by clarifying the provisions of clause 3 of Art. 23 so that with those issues on which a unified policy is being pursued, the unified regulation of information interaction is also indicated.

18.4.3 From Informatization to Digitalization

Unlike "informatization", the term "digitalization" is not used in the Treaty on the EAEU and its annexes. Nevertheless, based on the text of the Treaty, we can conclude that the elements of a digital society are used in the constituent act of the Union, including in terms and definitions. Thus, the Protocol on Information and Communication Technologies regulates many issues of electronic document management, the use of electronic digital signatures.

These aspects are an integral part of building cyber-social accounting systems at any level (Domrachev et al., 2016). The introduction of cyber-social systems represents a new stage in the development of Industry 4.0 and is critical for innovation and competitive advantages (Karlik et al., 2019).

In addition, the Treaty contains another concept of "cross-border space of trust" that is important for building a digital society within the framework of an integration association. The formation of a space of trust is intended for the free exchange of data and electronic documents, the security of information and telecommunication networks, and information security (Inshakova et al., 2020).

The functioning of the cross-border space of trust is ensured following the Concept for the use of services and legally binding electronic documents in interstate information interaction, approved by the Decision of the Council of the Eurasian Economic Commission of September 18, 2014, No. 73, the Strategy for the development of the cross-border space of trust, approved by the Decision of the Board of the Eurasian Economic Commission of September 27, 2016, No. 105, Regulations on the exchange of electronic documents in cross-border interaction of public authorities of the member states of the Eurasian Economic Union between themselves and with the Eurasian Economic Commission, approved by the Decision of the Board of the Eurasian Economic Commission dated September 28, 2015, No. 125, by the Decision of the Council of the Eurasian Economic Commission dated December 5, 2018. No. 96 "On the requirements for the creation, development, and functioning of the transboundary space of trust." Analysis of this segment of documents shows a high level of legal elaboration of the architecture for the construction and functioning of the transboundary space of trust at the supranational level. An important part of the success of this area of information interaction will be the proper implementation of the requirements contained in these acts in the national segments of the integrated information system, especially in terms of data protection and security.

It is worth to be noted, that some definitions in the Protocol on Information and Communication Technologies are broad and complex. For example, "information protection" is regulated as "the adoption and implementation of a set of legal, organizational, and technical measures to determine, achieve, and maintain confidentiality, integrity, and availability of information and its processing facilities in order to eliminate or minimize unacceptable risks for subjects of information interaction.

This definition simultaneously contains the principles of information circulation, the foundations of information security, and the unity of the categories of social (information) and physical (means of processing) context. It seems quite apt to formulate infrastructure components in the Treaty and the Protocol within the framework of building a cross-border space of trust.

We see such a perception of the text of the Treaty and its annexes as laying the basic constants for the further development of the Union in the context of digitalization that is most in line with the interests of the parties and the integration association, since, as noted above, the promotion of market freedoms is impossible without the creation of unified information and communication networks, and the high dynamics of digital relations highlights the combination of flexibility of technical regulation and stability of the fundamentals. Thus, in the context of the development of the concept of digital sovereignty, the principle of sovereign equality of states, laid down in the EEU Treaty as the basic principle of integration, acquires new directions of implementation, but does not change its essence. At the same time, the formation, for example, of common protocols for the security of the use of information and communication technologies, the protection of personal data can and should be carried out promptly and in a unified manner, which requires an active law-making position of the Commission. The idea of the need for the "advanced effect" of law in the era of digitalization and the role of the principles of law in this (Shafalovich, 2020) in integration association has its application.

The Treaty acquires special significance not only in two traditional directions of its interpretation and application—as a constituent act of the organization, institutionalizing the foundations of integration (institutionalizing function of the Treaty), and as a regulator of the single market (regulatory function of the Treaty), but also as an act that determines the vector of interaction in the integration space (predictive function of the Treaty). In this case, the role of the EAEU Court and the dynamic interpretation of Union law increase.

The digital potential of the EAEU is revealed not only through the intensification and transformation of the internal market, but also the possibility of "creating a global logistics corridor between Europe and Asia", in relation to which it will be effective to be comparable with the formats of UNCITRAL and others (Domrachev, 2016). Digital globalization as a new stage of globalization (Golovenchik, 2019) can and should be based on regional unions.

Thus, the Treaty provided for the legal and organizational basis for creation of a Single Digital Space, which is autonomous and secure inside and coherent with external systems. It made it possible to reach a common digital agenda in a short time.

18.4.4 On the Digital Agenda of the EAEU and a Single Digital Space

The adaptation of existing law to new digital relations is possible, but one cannot do without creating new norms (Talapina, 2018). This thesis fully applies to integration law.

It is not entirely possible to agree with the conclusion that "a specific feature of the EAEU model is the allocation of directions and priority initiatives as the basis for the development of public–private partnership projects," an aspect of which "is the political and legal model of integration", enshrined in the relevant digitalization tools (Shugurov, 2020). The software method is used in various forms and other integration associations. For example, the EU also uses strategic planning, the development of common priorities, the method of open coordination (Strategy: the European Commission's Priorities).

What could become innovative from the point of view of legal forecasting and program-strategic planning, especially in the dynamic digital era, is the systemic short-, medium-, and long-term setting (and adjustment, if necessary) of goals and objectives.

At the moment, some key strategic acts on the digital agenda are in force: an action plan for the implementation of the main directions of the development of the single-window mechanism in the system of regulation of foreign economic activity, approved by the Decision of the Supreme Eurasian Economic Council dated May 8, 2015, No. 19, Main directions of the implementation of the digital agenda of the Eurasian Economic Union until 2025, approved by the Decision of the Supreme Eurasian Economic Council dated October 11, 2017, No. 12, the Concept of cross-border information interaction, approved by the Decision of the Eurasian Intergovernmental Council dated August 9, 2019, No. 7.

In the Decision of the Supreme Eurasian Economic Council dated December 11, 2020 "On strategic directions for the development of economic integration until 2025", direction 5 is devoted to the formation of the digital space of the Union, digital infrastructures, and ecosystems, and includes nine main segments of digital transformation: traceability of goods in the EAEU, cross-border trust space, and electronic document management, the integrated information system of the Union, digital ecosystems (including data circulation, personal data protection), digital transformation in the field of intellectual property, e-commerce, external digital agenda, increasing the technical support of digitalization (unimpeded Internet traffic) and improving mechanisms for developing initiatives and implementing projects.

Analyzing these tracks of digital transformation, it should be noted, on the one hand, their complexity (from technical equipment to unified information systems), but on the other hand, their trade and economic nature, in general. The actively developing Union must also be perceived as a new information space (Tarakanov et al., 2020). The emergence of a new information space requires systemic anti-entropic mechanisms, primarily axiological, value attitudes. At the Yalta conference, held annually by the International Affairs magazine, in 2017, the idea was voiced about updating the common historical memory and the importance of managing information flows, for creating a positive image in the mass consciousness, the attractiveness of the EAEU member states, including outside, in the international arena.

From the point of view of digitalization, the issue of expanding the information presence, positive "branding" of the Eurasian integration project is seen in the creation of its domain of the first (top) level. All official sites are currently hosted on the ".org" domain—eurasiancommission.org, courteurasian.org. This first-level domain is used by the most international organizations and generally appears to be convenient and reasonably secure.

However, from the point of view of defining Eurasian interaction as going beyond classical international cooperation to create a supranational integration project, with its development paradigm, strategy of internal and external positioning, i.e., creation of a Eurasian "brand", the transition of Eurasian actors (official union bodies, additional integration non-governmental organizations, legal entities and individuals-residents of the EAEU, etc.) can become a strong argument for the progress of Eurasian integration, the implementation of the Single Digital Area.

The question of registering the top-level domain.EA in ICANN, by analogy with the European Union domain.EU, was discussed in 2014, but the idea was not implemented. Of course, when registering such a top-level domain, it will also be a question of the Union itself being the administrator of such a domain, which will require endowing its bodies with special competence in this area, which is, reaching appropriate agreements between the member states. Accordingly, the EAEU will be able to establish rules for users, including such that sites and resources are located on servers within the territory of the Union, etc.

It should not be forgotten that from the point of view of international economic law, and it is integration associations that can (this is their purpose, in fact) avoid conditions of the most favored nation for third countries in those positions where they provide member states with more favorable conditions for mutual trade (Matytsin & Rusakova, 2021). Moreover, they can impose restrictions on foreign trade based on the interests, threats, and risks for the integration association.

According to T. Sargsyan, Deputy Chairman of the Management Board of the EDB, if at first, the EU limited itself to supporting its own companies and ecosystems, then in the last year it has embarked on the path of building an obvious system of restrictions for digital corporations: the European Data Strategy and the draft act on digital markets 2020 are examples of such a system (Sargsyan, 2020). Similarly, there is every reason in the Eurasian region to combine the development of freedom of movement of information and digital resources with a strategy for protecting the information and digital Union from the outside.

In the EAEU digital agenda, digital transformation is positioned as an integration driver, and therefore, with the further analysis of clause 5.4 of the EAEU Development Strategy in the implementing acts of the Commission, international treaties within the Union, close attention should be paid to cross-industry solutions and the construction of digital ecosystems. This trend is a general characteristic of the digital age, but concerning integration, it is especially important, since it is capable of multiplying (the *spill-over effect*).

The digitalization of various sectors of the economy on the scale of an integration project should be accompanied by the maintenance of transparency and accessibility of integration resources, the creation of unified information and telecommunication networks, and the ability to use them for citizens and residents of the Member States, individuals, and legal entities. With the introduction of the institution of single citizenship of the Union, which, as we believe, is necessary for further building the EAEU; it is advisable to consolidate uniform digital rights and obligations for citizens. At the same time, it is possible to start working out and consolidating digital rights and obligations (we emphasize, only in such an inextricable connection, since great opportunities for digitalization also give rise to great threats) already now at the level of the Commission.

When designing the Single Digital Space, it is necessary to provide for the provision of electronic services by the integration bodies to individuals and legal entities. By analogy with the G2B and G2Px processes (elements of e-government), it is advisable to create digital communications in the procedures between the bodies of the integration association and business, as well as citizens. These will significantly "bring closer" different actors of integration. It will make the integration system a multisubject polycentric network model—the most effective type of building socio-systems in the modern world according to M. Castells.

Thus, the procedures for applying to the EAEU Court should gradually be brought to a digital standard (for example, sending by electronic means of communication any procedural documents, using an electronic digital signature) with a step-by-step reduction to more complex elements of electronic justice (for example, creating electronic offices and administering legal proceedings with the use of technology, the introduction of telecommunication technologies at various stages of the administration of justice).

These and any other issues of digitalization of the activities of the EAEU bodies require careful planning and legal support, starting from the inclusion of relevant provisions in the acts of the Union's primary law and ending with technical legal acts. A uniform and ultimately unified regulation of data circulation in the EAEU is necessary. Minister for Internal Markets, Informatization, Information and Communication Technologies of the Eurasian Economic Commission G. Vardanyan during the panel session "Digital agenda in the EAEU Strategy until 2025", which was held as part of the digital forum in February 2021, said that it is necessary to define clear approaches to the separation of data, what it is advisable to exchange within the framework of integration processes, and what information should be stored exclusively in the Member State, as well as to develop mechanisms for ensuring data security (for example, anonymization) that can help expand the list of types of information for exchange. Assessing this from a legal point of view, within the framework of the competence of the Union and the Member States currently provided for in the Treaty and possible legal instruments of interaction, it could be done in the form of an international treaty within the Union. In future, it is advisable to refer this competence to the sphere of a unified policy and to adopt an appropriate act of direct action of binding legal force—for example, a decision of the EEC.

Back to the beginning of the study, the designation of the main concept of any successful integration as freedom of movement of certain factors, which classically include goods, services, capital, labor force and to which freedom of movement of information is added in the modern era. However, it is possible to do some generalization of a higher order: effective interstate integration is associated with the *freedom of movement of development resources*, and in future, these resources are largely associated with digital transformation, including the creation of socio-digital networks.

18.5 Conclusion

The digital era allows us to bring integration closer to the large population of the Eurasian Economic Union, to make its institutions understandable, to create a truly "area without borders" for trade, communication, security, while preserving national identity and sovereign rights. The expansion of the digital competence of the Union is relevant in areas where legal regulation takes place in the direction of unification. Harmonization of the information interaction in other spheres is also necessary. The great opportunities of the Digital Age also give rise to challenges in security matters, ethics, and protection of all subjects of legal relations. It requires operational regulation with the involvement of expert potential at the Union level, as well as harmonized organizational and legal mechanisms for implementing responsibility in case of violation of the established rules. The digital agenda of the EAEU is wide enough, but these are only the first steps toward digital transformation. The formation of a predictive scientific concept for the development of the Union in the digital era will make the Eurasian integration project effective, long-term, and sustainable.

References

Castells, M. (2000). *Information age: Economy, society, culture* (p. 608). GU HSE.

Digital Cooperation roadmap: Implementing the recommendations of the high-level panel on digital cooperation: Report of the Secretary-General, 74th Session General Assembly, 29 May 2020. https://undocs.org/ru/A/74/821

Domrachev, A. A., Evtushenko, S. N., Kupriyanovskiy, V. P., & Namiot, D. E. (2016). On innovative initiatives of the EAEU member states in the field of building a global digital economy. *International Journal of Open Information Technologies, 4*(9), 24–32.

Ermakova, E., Frolova, E., & Sitkareva, E. (2021). International economic integration and the evolution of the principles of civil procedure. Modern Global Economic System: Evolutional Development vs. Revolutionary Leap. Lecture Notes in Networks and Systems. *Book Series LNNS, 198*, 1589–1597.

Golovenchik, G. G. (2019). *Digitalization of the Belarusian economy in modern conditions of digitalization* (p. 257). BSU Center.

Grand Russian Encyclopedic Dictionary. (2008). *Grand Russian Encyclopedia, 1887.*

Inshakova, A. O., Goncharov, A. I., & Ershova, I. V. (2020). Digital blockchain registration of legally significant stages of complex good's export-import supplies by business entities of the EAEU and BRICS jurisdictions. In A. Inshakova & E. Inshakova (Eds.), *Competitive Russia: Foresight model of economic and legal development in the digital age*. CRFMELD 2019. Lecture Notes in Networks and Systems (Vol. 110, pp. 328–336). Springer. https://doi.org/10.1007/978-3-030-45913-0_37

Karlik, A. E., Platonov, V. V., Krechko, S. A. (2019). Organizational support for digital transformation of cooperation networks and the introduction of cyber social networks. Scientific and technical statement of SpbSPU. *Economic Sciences, 12*(5), 9–22.

Khabrieva, T. Y. (2018). Law in the face of digital reality challenges. Zhurnal rossijskogo prava [*Russian Law Journal*], *9*(261), 5–16.

Khabrieva, T. Y., & Chernogor, N. N. (2018). Law in digital reality. *Zhurnal rossijskogo prava* [*Russian Law Journal*], *1*(253), 85–102.

Matytsin, D. E., & Rusakova, E. P. (2021). The strategy of quality management in industry 4.0 and formation of a cognitive economy based on industrial and manufacturing engineering in the Russian Federation and countries of the EU. *International Journal for Quality Research, 15*(4). https://doi.org/10.24874/IJQR15.04-03

Mikhaliova, T. N. (2016). *Legal regulation of regional economic integration: challenges and prospects* (p. 196). Institut Radiologii [Institute of Radiology].

Nekrasov, V. N. (2018). Innovation, informatization, digitalization: correlation and features of legal regulation. *Voprosy rossijskogo i mezhdunarodnogo prava* [*Questions of Russian and International Law*], *8*(11A), 137–143.

Nemirova, G. I., & Vinichenko, A. A. (2017). *The mechanism for improving the quality of public services in the field of customs in the context of digital transformation* (p. 130). RIO of Russian Customs Academy.

Nikulina, T. V., & Starichenko, E. B. (2018). Pedagogical education in Russia. *Informatization and Digitalization of Education: Concept, Technology, Management, 8*, 107–113.

Novikov, A. B., & Ragozina, N. A. (2018). Legal support for the creation of unified information space of the Eurasian Economic Union (EAEU) in the field of customs regulation. *Juridicheskaja Nauka [law Science], 6*, 39–41.

Ovchinnikov, A. I. (2018). Trends in the development of law in the context of a new technological order. *Philosophija prava* [*Law Philosophy*], *3*(86), 26–31.

Ovchinnikov, A. I., & Fatkhi, V. I. (2018). Law and the digital economy: Main directions of interaction. *Philosophija Prava [law Philosophy], 86*, 128–134.

Paulin, A. (2018). *Digitalization vs. informatization: different approaches to governance transformation*. CEE E/Dem and e/Gov Days, 251–262.

Prokudin, D. E. (2020). *From informatization to digitalization, philosophical analytics of the digital age* (pp. 38–52). Publishing House of St. Petersburg University.

Sablina, M. V. (2018). Problems of information interaction of customs authorities of the Eurasian Economic Union. *Alley of Science, 10*, 715–718.

Sargsyan, T. S. (2021) GOELRO integration plan for the XXI Century. *Rossiya v globalnoi politike, 19*(3). https://doi.org/10.31278/1810-6439-2021-19-3-136-149

Sarkisyan, T. (2021). *Digital sovereignty and digital agenda of the EAEU*. https://eabr.org/press/news/tsifrovoy-suverenitet-i-tsifrovaya-povestka-eaes/

Shabayeva, O. A. (2019). Law in Digital Reality: Problem statement. *Sibirskij Juridicheskij Vestnik [siberian Law Bulletin], 1*(84), 16–20.

Shafalovich, A. A. (2020). Challenges to the law in the era of digitalization (the Case) of the Republic of Belarus). *Teise, 114*, 113–121.

Shugurov, M. V. (2020). Trends and prospects for the development of regional scientific and technological integration in the context of the digital agenda of the EAEU: Political and legal dimension. *Pravo i Politika [law and Politics], 9*, 119–141.

Talapina, E. V. (2018). Law and Digitalization: New challenges and perspectives. *Journal of Russian Law, 2*, 5–17.

Tarakanov, V. V., Inshakova, A. O., & Goncharov, A. I. (2020). Interactive demo on-line maintenance foreign trade deals of business entities of BRICS jurisdictions. In A. Inshakova & E. Inshakova (Eds.), *Competitive Russia: Foresight model of economic and legal development in the digital age*. CRFMELD 2019. Lecture Notes in Networks and Systems (Vol. 110, pp. 319–327). Springer. https://doi.org/10.1007/978-3-030-45913-0_36

Velyaminov, G.M. (2015). International law: experiences. *M: Statute*, p.1006.

Wagner, B., & Ferro, C. (2020). *Governance of digitalization in Europe: A contribution to the exploration shaping digital policy—Towards a fair digital society?* Legal Notice, 45.

Zankovsky, S., Bezbakh, V., Rusakova, E. P., & Inshakova, A. (2020). Social consequences of economic globalization: Experience of developed and developing countries and perspective of optimization. *International Journal of Sociology and Social Policy, 41*(1/2), 211–223.

Zor'kin, V. (2018, 29 May). Law in the digital world. Reflections on the sidelines of the St. Petersburg International Legal Forum. *Rossijskaya gazeta* [Russian Newspaper.

Chapter 19
Digital Technologies of the Bank of Russia for Regulating Investment Relations

Denis E. Matytsin

Abstract The relevance of the study is because the current Russian legislation regulating the legal status of entities that regulate investment relations and professionally provide services in this market contains some contradictions. The purpose of the chapter is to identify these contradictions and substantiate proposals and recommendations for their elimination. Based on statistical observation and analysis of judicial practice, the author examines the licensing requirements for professional participants in the securities market and the current state of organizations engaged in brokerage, dealer, and depository activities. There is an increase in the licensing requirements for professional participants and a natural reduction in the number of organizations engaged in brokerage, dealer, and depository activities. The current legislation regulating the legal status of entities that regulate the stock market and provide intermediary services contains some contradictions. In particular, the information letters of the Bank of Russia, which are recommendatory, are not provided for by the federal legislation on the Bank of Russia, but non-compliance with the procedure contained in these letters is the basis for revoking licenses for professional activities of securities market participants. The study confirmed that the conclusion of contracts between professional participants of the investment market and their clients is carried out mainly in electronic form. The analysis revealed that the infrastructure of the Russian stock market is being improved in the direction of expanded use of digital technologies and acceleration of legally significant actions related to the transfer of ownership rights to securities. It is proved that during the transition from an administrative economy to a market economy, the state set the task of maintaining control in the banking and financial spheres, but at the same time the task of taking measures to reduce state participation in market relations. By giving the Bank of Russia the authority to issue letters of recommendation, the state thereby solves this problem.

D. E. Matytsin (✉)
Department of Business Law, Arbitration and Civil Procedure, Institute of Law, Volgograd State University, Volgograd, Russia
e-mail: Dmatytsin@ya.ru; ppagp@volsu.ru

Department of Civil Law and Procedure, Volga Branch of the International Law Institute, Volgograd, Russia

A. O. Inshakova and E. E. Frolova (eds.), *Smart Technologies for the Digitisation of Industry: Entrepreneurial Environment*, Smart Innovation, Systems and Technologies 254, https://doi.org/10.1007/978-981-16-4621-8_19

Keywords Bank of Russia · Investor · Securities · Broker · Dealer · Manager · Investment adviser · Depository · Register holder · Trade organizer · Clearing organizations · Repository · Commercial and non-commercial organizations · Digital technologies

JEL Codes E58 · K22 · G24

19.1 Introduction

The institution of the central bank is the most important component of the financial system of each state, performing the functions of money issuance, banking supervision, ensuring the financial stability of the state, etc. (Cimon & Garriott, 2019). The question of the place of the Central Bank of Russia in the financial system of the Russian Federation is still debatable, due to the specific ambiguity of the Bank in the system of state organs. By the norm of Article 1 of Federal Law No. 86-FZ of 10.07.2002 "On the Central Bank of the Russian Federation (Bank of Russia)," the Bank performs its functions and powers independently of other federal state authorities, state authorities of the subjects of the Russian Federation, and local self-government organs (Federal law, 2002). This provision logically follows from the current rule of Article 10 of the Constitution of the Russian Federation, which establishes the division of state power into legislative, executive, and judicial. At the same time, the Bank of Russia is essentially a legal entity. According to the norm contained in clause "g" of Article 83 and clause "g" of clause 1 of Article 103 of the Constitution of the Russian Federation, the Chairman of the Central Bank is appointed and dismissed by the State Duma on the proposal of the President of Russia.

The Central Bank of Russia is recognized by most scientists as a mega-regulator of the financial system. This statement is related to the existence of separate but interrelated powers with the Russian Government, for example, to develop and implement policies for the development and stability of the financial market. It should be noted that a key change in the legal status of the Bank of Russia occurred in 2013, due to the entry into force of Federal Law No. 251-FZ of 23.07.2013 (Federal law, 2013), according to which the powers of the Federal Financial Markets Service were transferred to the Bank of Russia. As A.N. Lisachev notes, the abolition of the Federal Service for Financial Markets gave a legal entity that is not part of the system of government authorities the powers of a state body in the field of financial market control (Lisachev, 2015).

19.2 Materials Methods

In the course of the study, some regulations were analyzed. Federal Law No. 251-FZ of 23.07.2013 "On Amendments to Certain Legislative Acts of the Russian Federation in connection with the transfer of powers to regulate, control and supervise financial markets to the Central Bank of the Russian Federation," Federal Law No. 46-FZ of March 5, 1999, "On the Protection of the Rights and Legitimate Interests of Investors in the Securities Market"; Federal Law No. 86-FZ of 10.07.2002 "On the Central Bank of the Russian Federation (Bank of Russia)," Federal Law No. 259-FZ of 02.08.2019 "On Attracting Investments Using Investment Platforms and on Amendments to Certain Legislative Acts of the Russian Federation," Federal Law No. 7-FZ of 07.02.2011 "On Clearing, Clearing activities and the Central Counterparty." The regulations, explanatory letters, and instructions of the Bank of Russia were also studied.

Scientific works of foreign and Russian authors, such as Abramova (2017), Inshakova et al. (2017, 2020), Berzon (2019), Chaldaeva and Kilyachkov (2019), Davydov (2015), David Cimon and Corey Garriott (2019), Gabov (1999a, 2010), Goncharov (2014), Grishaev (2016), Grishin (2019), Inshakova et al. (2019), Nishide and Tian (2019), Lisachev (2015), Loader (2014), Mikova (2012), Samigulina (2017), Matytsin (2021).

The research methodology is based on the materialist worldview and the general method of historical materialism and is focused on the system approach, including general scientific and specific methods of cognition: induction and deduction, subject-object, structural–functional, descriptive and comparative analysis.

19.3 Results

The importance of the Bank of Russia in the sphere of control and supervision of the Russian stock market is manifested in the law-making activities of the mega-regulator; in this regard, the question of the legal force of the Bank's acts is reasonable. According to Article 7 of the Federal Law on the Bank of Russia, the Bank issues regulatory acts in the form of instructions, regulations, and instructions that are mandatory for the entire vertical of government bodies, as well as for all legal entities and individuals. At the same time, the rules for the preparation of such acts are established by the Bank independently. In the field of investor rights protection in the securities market, the Bank of Russia issues regulations that are mandatory for commercial and non-commercial organizations, their officials, and individuals, including individual entrepreneurs. These regulations serve as a tool for stopping and preventing violations of the legislation of the Russian Federation on joint-stock companies and on the securities market (p. 1 and p. 2 of Article 11 of the Federal Law on the Protection of Investors' Rights in the Securities Market) (Federal law, 1999).

At the same time, the Bank of Russia issues, in addition to these regulations, Letters and Information Letters of the Bank of Russia, which are essentially advisory. Formally, by law, these letters do not relate to regulatory acts; however, violation of their provisions is the basis for revoking the license to conduct banking activities (Davydov, 2015). In the first quarter of 2020 alone, the Bank issued more than 20 letters, including four related to its activities on the stock market: Letter of the Bank of Russia No. 55–4-3–1/168 dated 29.01.2020 "On Filling out the Reporting Form 0420418 by Professional Participants of the Securities Market" (Letter from the Bank of Russia, 2020b). Letter of the Bank of Russia dated 13.03.2020 No. 54-3-2-1/645 "On the reporting of the Management company and the specialized depository" (Letter from the Bank of Russia, 2020c). Letter of the Bank of Russia dated 20.03.2020 No. 015–54/2082 "On a set of measures to support the Collective Investment Market" (Letter from the Bank of Russia, 2020d). Letter of the Bank of Russia dated 15.01.2020 No. 55-7-1/28 "On termination of contracts in the financial market." These acts significantly affect the structure of the operational activities of participants in the securities market. For example, the Letter "On termination of contracts in the financial market" reflects a decrease in the use by brokers of the possibility of remote termination of contracts, in particular brokerage service contracts, in connection with which the client's presence in the broker's office or notarization of documents is required for termination of contracts (Inshakova et al., 2017). Because of this, according to the Department of Securities and Commodity Market, the use of such practices by professional market participants leads to infringement of the rights of customers, generates additional time and material costs for customers and generally negatively affects the attractiveness of the financial market (Letter from the Bank of Russia, 2020a). Although this letter is not binding, it is also capable of influencing the participants of the securities market, without additional regulatory regulation.

Regarding the nature and legal force of the letters, we believe that in the transition from an administrative economy to a market economy, the state has set the task of maintaining control in the banking and financial sphere, but at the same time taking measures to reduce the state's presence in market relations. By giving the Bank of Russia the authority to issue letters of recommendation, the state thus solves this problem.

The Bank of Russia's control and supervisory functions in the sphere of securities issuance and circulation, as well as corporate relations in joint-stock companies and some other powers (Article 4 of the Federal Law on the Central Bank, 2002) (Federal law, 2002), are aimed at regulating relations between all financial market participants, including professional participants in the securities market. S. P. Grishaev points out that a characteristic feature of equity securities is that their circulation is carried out through an organized securities market with its specific institutional participants: brokers, dealers, clearing organizations, trust managers (Grishaev, 2016).

Chapter 2 of Federal Law No. 39-FZ of 22.04.1996 "On the Securities Market" discloses the legal status of market participants through certain types of professional activities, among which the following subjects can be distinguished: (1) broker;

(2) dealer; (3) manager; (4) investment adviser; (5) depository; (6) register holder (registrar); (7) trade organizer; (8) clearing organizations; (9) repository.

According to clause 18, clause 1, Article 2 of the Law on the Securities Market, professional participants in the securities market are legal entities that carry out one or more types of activities provided for in Chap. 3 of the said law. In the special literature, it is noted that the reason for the legislator's choice of such an approach to the definition of professional activity lies in the regularity of this activity on the part of the state, and not in terms of its content (Chaldaeva & Kilyachkov, 2019). Therefore, the separate market activity that is subject to regulation by the state is professional. If a legally separate activity is not subject to regulation, although it is such in the content, then it is not considered professional. In this regard, there is both a legal understanding of professional activity in the securities market and an economic understanding of the division of labor in the field of securities trading. From this point of view, the process of securities circulation consists of some independent types of useful activities. The focus of such activities is adjusted to issuers or investors and ultimately takes the form of providing services, the specialization of which depends on the type of professional activity. Thus, professional activity in the securities market (economic understanding) is the provision of services to issuers and investors for the disposal of securities on a paid basis.

It is important to note the key point in understanding the content of the professional activity of market participants, which consists in profitable investment in a certain type of separate services provided to market participants (issuers and investors), in contrast to the latter, whose purpose is a profitable investment in securities and capital raising. Accordingly, the activities of professional participants form the main or main source of profit from the provision of services. So, it is fair to say that the "income" of the issuer is the funds raised in the form of capital, the investor—the net income on the security, and the professional participant—the profit from the sale of its services (Berzon, 2019).

Because professional activity is associated with the provision of services to legal entities and individuals, it can be carried out subject to licensing. Licensing by the state in the securities market is carried out by the control body—the Central Bank of the Russian Federation. Obtaining a license is accompanied by a procedure for applying to the Central Bank of the Russian Federation, characteristics of commercial activities, copies of constituent documents, etc. Moreover, the licensing authority checks the compliance of the license applicant with the following requirements.

Organizational and legal form: a legal entity, trade intermediaries must be commercial organizations, non-commercial organizations can also be non-commercial. The size of the authorized capital: depending on the type of activity, various standards are established in the form of special numerical coefficients multiplied by a constant value—2 million rubles. So, for registrars, a minimum of 100 million rubles (2 * 50) is set, for dealers—3 million rubles (2 * 1.5). In general, the minimum amount of own funds is defined as the product of a constant value (2 million) and the standard of the sufficiency of own funds. Detailed standards are set out in the Instruction of the Bank of Russia No. 4373-U dated 11.05.2017 (Bank of Russia Instruction, 2017a).

Table 19.1 Number of organizations that have a license to carry out professional activities in the securities market as of 12.05.2020 (Bank of Russia, 2020) and as of 16.11.2018 (Berzon, 2019)

Type of professional activity	Number of organizations		Fraction (%)	
	16.11.2018	12.05.2020	16.11.2018	12.05.2020
Total including	1312	1080	100	100
Brokerage activities	380	277	28.96	25.64
Dealer activity	343	308	26.14	28.51
Securities management activities	238	196	18.14	18.14
Depository activities	316	267	24.09	24.72
Register management activities	35	32	2.67	2.96

Source Compiled by the author from these sources

Qualification requirements for personnel: The Bank of Russia sets the minimum number of specialists who have a qualification certificate issued based on passing the basic and specialized qualification exam.

The license is issued for an unlimited period; however, the licensing authority has the right to suspend the license or cancel it, in cases provided for by the Regulation on License Requirements. The suspension of the license entails a ban on carrying out activities until the identified violations are eliminated, and upon the elimination of violations, the Bank of Russia decides on the renewal of the license (Regulation of the Bank of Russia, 2015).

It is possible to quantify professional participants in the securities market by the number of licenses issued by the Central Bank of the Russian Federation. Based on these data, it is possible to understand the diversity and volume of organizations engaged in professional activities at present, and, consequently, to adequately assess the need to build effective legal regulation (Table 19.1).

As can be seen from the dynamics of the data in Table 19.1, over the past two years, the number of organizations that have the right to operate in the securities market has decreased by 17.68%. However, the total number remains at a high level, with the vast majority of participants being credit institutions (banks).

Professional participants have the right to combine several types of activities at once, but due to Article 10 of the Law on the Securities Market, the activities of maintaining the register of securities owners cannot be combined with other types. Moreover, the above-mentioned Instruction of the Bank of Russia No. 4373-U sets restrictions on combining activities and operations with financial instruments.

Let us briefly consider the content of the activities of each of the professional participants.

Brokers. based on a brokerage service agreement, brokers enter into securities transactions on behalf of and at the expense of clients. The essence of brokerage activity is trade mediation between issuers and investors, investors and investors (hereinafter referred to as clients). A brokerage service agreement is a mixed agreement consisting of the elements of an order, commission, and agency agreement, depending on the specifics and will of the parties. In practice, the broker and the client can also enter into separate independent contracts for orders, commissions, or agency services. But the qualifying feature is the participation of the broker in exchange trading on its behalf in the interests of the client and at his expense. The main duty of brokers is to execute clients' orders in good faith and on time. There may be situations when the execution of an order is associated with a conflict of interests between the broker and the client, then if the execution of such an order resulted in damage to the interests of the client, the broker is obliged to compensate for the losses.

Dealers At their own expense and on their behalf, dealers make transactions for the purchase and sale of securities by publicly announcing the purchase and/or sale prices of certain securities with the obligation to buy/sell these securities at the announced prices. The dealer can only be a commercial organization or a state corporation if this is provided for by the relevant federal law. Unlike an investor, as follows from the legal definition of Article 4 of the Securities Market Law, a dealer, when announcing the price of a transaction publicly, undertakes to conclude this transaction with anyone who wants to and does not have the right to refuse any person or change the price condition. In this case, a similar rule applies to a public contract (Article 426 of the Civil Code of the Russian Federation). The investor, in turn, has the freedom to choose the counterparty and the terms of the contract. The public announcement of prices on the market is usually referred to as a quote (Mikhailenko, 2019).

Dealers' activities are divided into professional (in the context of the law) and actual trading, which is carried out by the dealer at his own expense. Organization and support of the liquidity (turnover) of securities the main function of dealers, which is professional, includes the announcement of purchase and sale prices; the fulfillment of obligations; the establishment of the quantitative maximum of securities at the announced prices and terms. Unlike a broker, a dealer operates directly on the market and has great advantages, which is why the legislation on the securities market imposes special requirements on dealers: acting in the interests of the client; executing the client's orders in the first place; timely and reliable communication of information about the market to clients; prohibition of price manipulation. Thus, Article 4 of the Law on the Securities Market establishes the rule that in the event of a dealer's evasion from entering into a contract, a claim may be brought against him for forcing him to enter into a contract and (or) for compensation for losses to the client.

Brokerage and dealer activities occupy a special place in the securities market (Nishide & Tian, 2019); therefore, there are self-regulating organizations of brokers and dealers that develop rules and standards for the implementation of professional activities that are mandatory for participants. Following the norm of Article 8 of Federal Law No. 223-FZ of July 13, 2015, "On Self-regulating Organizations in the Financial Market," membership of a financial organization in an SRO is mandatory, if there is such an SRO (Federal law, 2015). From December 1, 2019, by the requirements of the legislation and the requirements of the Bank of Russia's Directive No.

4585-U of October 26, 2017 (Bank of Russia Instruction, 2017b) a basic standard for brokers has been introduced, which is mandatory for all brokers, regardless of their membership in the SRO. This standard sets out the general principles of professional activity (to act reasonably and in good faith, the priority of the interests of customers over their own, etc.), the rules of interaction with customers and other market participants, etc. (Standard, 2019).

Managers, following the provision of Article 5 of the Law on the Securities Market, the trust management of securities, funds intended for transactions with securities, and the conclusion of contracts (derivative financial instruments). As managers, as a rule, credit organizations (banks) and investment funds, headed by professional economists, lawyers, and other specialists, act as managers (Pham, 2020). At the same time, by paragraph 1 of Article 2 of Federal Law No. 156-FZ of 29.11.2001 "On Investment Funds," a joint-stock investment fund is a joint-stock company whose activity consists in investing property in securities and other investment objects. The certification of investors' rights is, as a rule, an investment unit that does not relate to issue-grade security, but at the same time certifies the financial relationship with the management company and secures the right to a share in the general undivided ownership of the fund (Goncharov, 2014). At the same time, the joint-stock investment fund is prohibited from carrying out other activities. Following paragraphs, 1 and 3 of Article 38 of this law, the management company (JSC or LLC) manages the assets of a joint-stock investment fund only based on the management company's license (Federal law, 2001). In turn, managers in the securities market carry out their activities based on a trust management agreement, and a license is not required if the trust management is related only to securities (paragraph 3 of Article 5 of the Law on the Securities Market). In addition, it is necessary to distinguish the term "management company" in the sense of "management organization," which, per paragraph 1 of Article 69 of the Law on Joint Stock Companies, by decision of the General Meeting of Shareholders based on a contract, a commercial organization may be transferred the powers of a sole executive body following paragraph 1 of Article 53 of the Civil Code of the Russian Federation.

The basis for the performance of the manager's activities is the trust management agreement. The subject of the contract is formed as follows: the client transfers funds for a certain time to invest in securities, and the manager undertakes to manage the totality of the client's property (securities and cash) in the interests of the client or a third party. The client pays the manager a fee related to his expenses. The manager keeps separate records of his property and the property received by the management. Moreover, the legislation establishes a number of restrictions on the activities of the manager. He is prohibited from acquiring securities owned by the manager at the expense of clients' funds; acquiring securities under management at his own expense. It is also prohibited to purchase securities of organizations that are in the process of bankruptcy; to transfer trusted securities as collateral for their obligations. It is prohibited to enter into a bank deposit agreement and an insurance contract with the use of trusted funds. The manager exercises all rights and bears all obligations under the securities, including participation in the general meeting with the right to vote, unless otherwise provided by the agreement. For this purpose, the legislator gave the manager the right to file a claim in court with any claims in connection with the implementation of trust management, including the presentation of which is possible by shareholders and other owners of securities. However, all legal costs are paid at the expense of the property under management.

Registry holders (registrars) According to paragraph 1 of Article 8 of the Law on the Securities Market, the following activities are recognized for maintaining the register of securities owners: the collection, recording, processing and storage of data that make up the register, as well as the provision of such data. Only legal entities are entitled to carry out the activity of maintaining the register, and this may be: the issuer or a non-professional holder; the registrar or a professional holder—a licensed specialized organization operating independently of the issuer. The registrar has the right to maintain registers of an unlimited number of issuers, but the issuer has the right to conclude a contract for maintaining the register with only one registrar. This activity is exclusive of any other type of activity and cannot be combined with other types of activity on the securities market.

Depositories Provision of services for accounting and transfer of rights to non-documentary securities, immobilization of documentary securities, as well as their storage, and in cases provided for by federal law, provision of services for accounting for digital rights. A significant difference between the registrar and the depository is the registrar's participation primarily in the primary market, while the depository's participation is in the secondary market. The registrar provides services to issuers and is responsible for fixing, maintaining, and updating relations between issuers and investors. The depository is mostly responsible for accounting for changes in the ownership rights of the owners of securities after their release into the circulation.

Civil relations between the depository and the user of its services (the depositor) arise from the contract of the deposit account. Under the depositary agreement, the depositary is not entitled to dispose of the deposited securities, except for the execution of the instructions of the depositor; such securities may not be subject to foreclosure on the obligations of the depositary; the depositary is responsible for the custody of the securities. The subject of the agreement consists of the obligation of the depositary to accept for storage and (or) keep records of the rights to securities, the term of the agreement, the procedure for transmitting information about the disposal of assets and payment for the services of a representative.

A. V. Gabov notes the existence of four main points of view about the nature of the depot contract, which has the right to exist (Gabov, 1999a, b): (1) independent non-named contract or (2) provision of special services (Kozyr et al., 1996); (3) mixed contract (Mikova, 2012); (4) storage agreement (Gabov, 2010). Meanwhile, we agree with the position of the authors that the depot account is not a form of security, but a method of accounting for the volume of securities of each holder (Abramova, 2017). And the legal nature of the depot contract depends on the specific content determined by the parties in each case.

There are two types of depositories: settlement and client. The settlement depository provides services only to professional participants of the organized securities market. Its functions are inextricably linked with the activities of clearing organizations and the organizer of trade. In accordance with Federal Law No. 414-FZ of

07.12.2011 "On the Central Securities Depository," the settlement depository exists in the form of a national depository (Federal Law, 2011b), its status is currently held by the Non-bank credit Organization Joint-stock Company "National Settlement Depository,", which is part of the group of companies of PJSC "Moscow Exchange." It is a single settlement depository of the organized market and the parent organization concerning the other depositories.

Client (or custodian from the English custodian—trustee, custodian) depositories provide services directly to the owners of securities (investors), but in some Western countries, their activities are combined with the functions of the registrar (Loader, 2014). In Russia, their activities are differentiated, and depositories can combine only with the activities of a broker, dealer, or manager. The functions of client depositories include:

(1) participation in the settlement of the transaction through the delivery or acceptance of securities and their crediting to the "depo" account;
(2) securities storage: open-without division by customers, closed-storage in registered "depo" accounts;
(3) payment of dividends and coupon income to customers;
(4) support of corporate actions: the exercise of voting rights on shares, amendment of the issuer's charter, liquidation, reorganization, merger and acquisition of the issuer, distribution of net profit, conversion of securities, etc.;
(5) implementation of clients' tax obligations on securities;
(6) provision of information services.

It is the depository form of accounting for the transfer of ownership rights to securities that is the only possible one, in cases where the number of transactions reaches a high value. The movement of cash and securities is synchronized. Today, the market infrastructure is growing at a noticeable rate, largely due to the introduction of digital technologies and the automation of rights accounting algorithms. In many European countries and the USA, the institution of depositories has displaced or is displacing registrars. The activities of depositories were expanded with the introduction of amendments to Article 7 of the Law on the Securities Market by Federal Law No. 259-FZ of 02.08.2019 "On Attracting Investments Using Investment Platforms and on Amendments to Certain Legislative Acts of the Russian Federation," which added accounting for digital rights. According to clause 11 of Article 8 of the said law, the investment platform (Part 5 of Article 11 of Federal Law No. 259) may acquire and alienate utilitarian digital rights, which are recorded by the depository (Federal law, 2019).

Investment advisors—Due to the development of information and digital technologies, investments in securities have become available to a wider range of people, but for effective investment of their funds, citizens need professional advice about investment objects, market conditions, investment ideas, etc. In world practice, it is customary to contact a personal financial adviser (Grishin, 2019). Since there was no independent institution of a financial consultant in the domestic legislation, and their role was performed by banking and brokerage organizations, since December 21, 2018, amendments to the Law on the Securities Market entered into force, in terms of regulating investment consulting activities as an independent professional activity. Investment advisory activities are carried out based on an investment advisory agreement and consist in the provision of advisory services in respect of securities, transactions with them, as well as contracts—derivative financial instruments. This activity does not include providing advice to professional market participants in relation to the services they provide and the financial instruments they issue.

Both legal entities and individual entrepreneurs who are members of the relevant self-regulatory organization are entitled to provide consulting services. Employees of organizations and individual entrepreneurs are subject to qualification requirements and the obligation to comply with the requirements established by the legislation and acts of the Bank of Russia.

Securities can be traded on both on-exchange trading platforms and over-the-counter ones. The division of the market on the basis of the organization is based on the order of transactions: (1) if securities are placed and traded with the organizers of the trade—the market is organized and its activities are regulated by the rules of the organizer of the auction; (2) if buyers and sellers conclude transactions independently finding each other or using the services of brokers, but not in an organized market, the market is recognized as unorganized (Samigulina, 2017). Previously, the Law on the Securities Market referred to the professional activity and the organization of trade. However, in 2011, Federal Law No. 325-FZ of 21.11.2011 "On Organized Trading" was adopted, which regulates the activities of organizing securities trading. A trade organizer following this law is a person who provides services for conducting organized trading on the commodity and/or financial markets based on an exchange license or a trading system license. This activity is carried out based on a contract for the provision of services for conducting organized auctions, and all participants, unless otherwise established by law or contract, pay for these services. There are also restrictions imposed on the organizers—a ban on engaging in industrial, insurance, banking, and other activities related to professional activities in the securities market and some others.

Before January 1, 2013, the Securities Market Law regulated the activity of determining mutual obligations—clearing activities. Currently, the activities of clearing organizations are regulated by Federal Law No. 7-FZ of 07.02.2011 "On Clearing, Clearing Activities and the Central Counterparty." According to paragraph 3, paragraph 1, Article 2 of this law, clearing is understood as the definition of obligations to be performed under contracts, including as a result of the implementation of netting obligations, the preparation of documents that are the basis for the termination or performance of such obligations, as well as their security (Federal Law, 2011a).

The following features of the Bank of Russia's legal regulation of the activities of professional market participants can be distinguished:

- increasing the licensing and professional requirements for participants and as a result the termination of the activities of organizations;
- there is a decrease in the number of professional market participants by 17.68%;

- the list of services is expanding over time, which is reflected in the preparation of new contractual structures that combine more than one element of the contract and meet modern standards;
- the regulatory framework in the field of professional activity of market participants is constantly being improved, there is a high frequency of amendments to the Law on the Securities Market, as well as the adoption of new acts;
- the nature and mechanisms of the Russian stock market infrastructure are changing, which is reflected in the expansion of digital technologies (Elena I. Inshakova et al., 2019) and speeding up the execution of legally significant actions related to the transfer of ownership of securities.

19.4 Conclusions

Thus, it is revealed that the current legislation regulating the legal status of entities that regulate the financial market and provide intermediary services contains some contradictions. In particular, the information letters of the Bank of Russia, which are recommendatory, are not provided for by the federal legislation on the Bank of Russia, but non-compliance with the procedure contained in the letters is the basis for revoking licenses for professional activities of securities market participants. We believe that in the transition from an administrative economy to a market economy, the state has set the task of maintaining control in the banking and financial sphere, but at the same time, the task of taking measures to reduce state participation in market relations. By giving the Bank of Russia the authority to issue letters of recommendation, the state thereby solves this problem. Based on statistical research and analysis of judicial practice, an increase in licensing requirements for professional participants and a natural reduction in the number of organizations engaged in brokerage, dealer, and depository activities in general by 17.68% was established.

It was revealed that the procedure for concluding contracts between professional participants and clients (investors) is changing—more often contracts are concluded in electronic form. In addition, the analysis of changes in legislation in this area allows us to conclude that the infrastructure of the Russian stock market is changing in the direction of expanded use of digital technologies (Inshakova et al., 2019) and accelerate the commission of legally significant actions related to the transfer of ownership of securities.

Acknowledgements The work was supported by Russian Science Foundation (project no. 20-18-00314).

References

Abramova, E. N. (2017). The legal nature of the deposit account contract. *Banking Law, 2*, 48–55.

Bank of Russia Instruction. (2017a, May 11). *On requirements to own funds of professional participants of the securities market* (No. 4373-U). Bulletin of the Bank of Russia, 53.

Bank of Russia Instruction. (2017b, October 26). *About requirements to contents of basic standards of protection of the rights and the interests of natural and legal entities - recipients of the financial services rendered by members of the self-regulatory organizations in the sphere of the financial market uniting brokers, managing directors, depositaries, registrars* (. 4585-U). Bulletin of the Bank of Russia, 103(1937).

Bank of Russia. (2020). *Securities market*. https://cbr.ru/registries/rcb/

Berzon, N. I. (2019). *Securities market: Textbook for academic undergraduate studies*. Yurite.

Chaldaeva, L. A., & Kilyachkov, A. A. (2019). *Securities market: Textbook for academic undergraduate studies*. Yurite.

Cimon, D., & Garriott, C. (2019). Banking regulation and market-making. *Journal of Banking & Finance, 109*, 105653.

Davydov, N. D. (2015). The reason for the revocation of licenses from credit institutions by the Central Bank of Russia. *Problems of the Modern Economy, 28–1*, 141–144.

Federal Law. (1999). *On the protection of the rights and legitimate interests of investors in the securities market* (accepted the State Duma of March 5, 2002, No. 46-FL).

Federal Law. (2001). *About investment funds* (accepted the State Duma of November 29, 2002, No. 156-FL).

Federal Law. (2002). *On the Central Bank of the Russian Federation* (Bank of Russia) (accepted the State Duma of July 10, 2002, No. 86-FL).

Federal Law. (2011a). *On clearing, clearing and central counterparty* (accepted the State Duma of February 7, 2011, No. 7-FL).

Federal Law. (2011b). *On the central depository* (accepted the State Duma of December 7, 2011, No. 414-FL).

Federal Law. (2013). *On amending certain legislative acts of the Russian Federation in connection with the transfer to the Central Bank of the Russian Federation of powers for regulation, control and supervision in the field of financial markets* (accepted the State Duma of July 23, 2013, No. 251-FL).

Federal Law. (2015). *About self-regulatory organizations in the financial market* (accepted the State Duma of July 13, 2015, No. 223-FL).

Federal Law. (2019). *On attracting investments using investment platforms and on amending certain legislative acts of the Russian Federation* (accepted the State Duma of August 2, 2019, No. 259-FL).

Gabov, A. V. (1999). Depositary agreement in the system of compulsory law. *Law and Economy, 7*, 14–17.

Gabov, A. V. (1999). Depositary agreement in the system of compulsory law. *Law and Economy, 8*, 12–16.

Gabov, A. V. (2010). *Problems of civil legal regulation of relations in the securities market* (Doctor of Law thesis), Moscow.

Goncharov, A. I. (2014). Investment stake as a certificate of the right to own share of property in the fund. *Bulletin of Volgograd State University. Series 5, Jurisprudence, 3*(24), 21–27.

Grishaev, S. P. (2016). *Securities: Types and practices of application*. Editorial office of the Russian newspaper.

Grishin, K. A. (2019). *A new professional participant in the securities market*. Investment Advisor. Joint Stock Company: Corporate Governance Issues. https://pravo.cliff.ru/news-legal/aktsionernaya-praktika/aktsionernaya-praktika_2198.html

Inshakova, A. O., Kazachenok, S. Y., & Sevostyanov, M. V. (2017). Legislative approbation of the extended powers of notaries in the sphere of civil turnover of immovable property: The Russian

Federation and the European legal orders. *Journal of Advanced Research in Law and Economics, VIII, 5*(27), 1534–1544.
Inshakova, E. I., Ryzhenkov, A. Y, & Inshakova, A. O. (2019). Neo-Industrialization of the Russian economy: Technological and digital development. In E. G. Popkova (Ed.), *Ubiquitous computing and the internet of things: Prerequisites for the development of ICT*. Studies in Computational Intelligence (Vol. 826, pp. 239–250). Springer.
Inshakova, A. O., Goncharov, A. I., & Ershova, I. V. (2020). Digital Blockchain Registration of Legally Significant Stages
of Complex Good's Export-Import Supplies by Business Entities of the EAEU and BRICS Jurisdictions. In A. O. Inshakova & E. I. Inshakova (Eds.), *Competitive Russia: Foresight model of economic and legal development in the digital age*. Proceedings of the International Scientific Conference in Memory of Oleg Inshakov (1952–2018) (pp. 328–336). Springer.
Kozyr, O. M., Makovsky, A. L., & Khokhlov, S. A. (1996). *Commentary to the Civil Code of the Russian Federation (Part Two)*. International Center for Financial and Economic Development.
Lisachev, A. N. (2015). Issues of mega-regulation of the financial market in the Russian Federation by the Bank of Russia. *Scientific and Methodological Electronic Journal Concept, 6*, 196–200.
Letter from the Bank of Russia. (2020a, January 15). *On termination of contracts in the financial market* (No. 55–7–1/28). http://docs.cntd.ru/document/564161950
Letter from the Bank of Russia (2020b, January 29). *On filling in the reporting form 0420418 by professional participants of the securities market* (No. 55–4–3–1/168). http://docs.cntd.ru/document/564210462
Letter from the Bank of Russia (2020c, March 13). *On the reporting of the management company and the specialized depository* (No. 54–3–2–1/645). http://docs.cntd.ru/document/564485243
Letter from the Bank of Russia. (2020d, March 20). *On a set of measures to support the collective investment market* (No. 015–54/2082). http://docs.cntd.ru/document/564578198
Loader, D. (2014). The role of the clearinghouse, trade repositories, and central securities depositories. In *Clearing, settlement, and custody* (3rd ed., pp. 17–23): Butterworth-Heinemann.
Matytsin, D. E. (2021). Internet-investing as a remote algorithm of the retail investment financing. In *Modern global economic system: Evolutional development Vs. revolutionary leap*. Lecture Notes in Networks and Systems (pp. 1850–1857). Springer.
Mikhailenko, M. N. (2019). *Securities market: Textbook and workshop for academic undergraduate*. Yurite.
Mikova, E. V. (2012). Legal nature of securities turnover and depository activities of a commercial bank. *Historical, Philosophical, Political, and Legal Sciences, Cultural Studies, and ART History. Questions of Theory and Practice, 2*(16), 138–140.
Nishide, K., & Tian, Y. (2019). Brokered versus dealer markets: Impact of proprietary trading with transaction fees. *International Review of Financial Analysis*. https://papers.ssrn.com/sol3/papers.cfm?abstract_id=2470355
Pham, M. H. (2020). In law, we trust—Lawyer CEOs and stock liquidity. *Journal of Financial Markets*, 100548.
Regulation of the Bank of Russia. (2015, July 27). *On licensing requirements and conditions of professional activity in the securities market, restrictions on the combination of certain types of professional activities on the securities market, as well as on the procedure and deadlines for submitting termination reports to the Bank of Russia, related to professional activities on the securities market, in case of cancellation of the license of a professional participant in the securities market* (No. 481-P). Bank of Russia Bulletin, 75.
Standard. (2019). *The basic standard for the protection of the rights and interests of individuals and legal entities - recipients of financial services provided by members of self-regulatory organizations in the financial market, uniting brokers* (approved by the Decision of the Financial Supervision Committee Central Bank of the Russian Federation May 8, 2019). https://cbr.ru/Queries/UniDbQuery/File/90002/20
Samigulina, A. V. (2017). The legal aspect of investor relations in the securities market. *Bulletin of Arbitration Practice, 1*, 22–28.

Chapter 20
Features of the Legal Infrastructure of the Turnover of Investment Objects in the Russian Federation

Denis E. Matytsin

Abstract The relevance of the research presented in this chapter of the monograph is due to the urgent need for the development of legal regulation of the turnover of non-documentary securities and digital objects in modern Russia, which determines the purpose of this scientific work. Millions of Russians, in the face of declining returns on traditional banking products, direct their investments to the stock market, making both independently and through the mediation of professional market participants exchange deals with securities. The study, based on the study of the regulatory framework for regulating the procedure for circulation of equity securities, substantiates the conclusion that in practice, there are three types of legal gaps when transferring ownership rights to shares and bonds. First, there is the potential for falsification of material documents that mediate the transfer of shares from one person's account to another person's account. Secondly, there is uncertainty about the qualification of actions as unilateral or bilateral transactions in some cases related to the disposal of non-documentary securities. Third, there is a lack of interaction between the register of notarized transactions on the alienation of securities and the databases of organizations that maintain the depot account and the register of shareholders. In modern Russia, the development of Industry 4.0 technologies is a challenge to legal science to a new understanding of the state of security on digital investment platforms, to a new view of the legal nature of undocumented securities and related legal phenomena that arise as a result of scientific and technological progress. Based on the identification of three types of legal gaps, the conclusion is justified that to eliminate them, there is a need to improve the legislative tools of the system of accounting and registration of the transfer of ownership rights to non-documentary securities. The author proposes to fix the rules on the mandatory registration of purchase and sale transactions and other transactions that mediate a change of ownership in the distributed registry system, which should be accessible to all entities involved in deals. The

D. E. Matytsin (✉)
Department of Business Law, Arbitration and Civil Procedure, Institute of Law, Volgograd State University, Volgograd, Russia
e-mail: Dmatytsin@ya.ru; ppagp@volsu.ru

Department of Civil Law and Procedure, Volga Branch of the International Law Institute, Volgograd, Russia

A. O. Inshakova and E. E. Frolova (eds.), *Smart Technologies for the Digitisation of Industry: Entrepreneurial Environment*, Smart Innovation, Systems and Technologies 254, https://doi.org/10.1007/978-981-16-4621-8_20

recommendation is justified to verify the authority of a person to dispose of security by activating an electronic digital signature or to use cryptographic protection tools, which will increase the level of law and order and the volume of securities turnover on the market.

Keywords The Constitution of the Russian Federation · Shares · Equity securities · Non-documentary securities · Register of shareholders · Civil Code of the Russian Federation · Multifunctional exchange trading platform · Investments · Securities turnover

JEL Codes G18 · G24 · K12 · K15 · L14 · L24 · L86 · O16

20.1 Introduction

The legal basis for the circulation of non-documentary securities is a set of regulatory legal acts that include the norms of various branches of law. The fundamental ideas come from the constitutional provisions on which civil legislation is based. According to Part 1 and Part 2 of Article 8 of the Constitution of the Russian Federation, freedom of economic space, and freedom of economic activity are guaranteed in the state, and private, municipal, state, and other forms of property are recognized and protected equally. Article 34 of the Constitution of the Russian Federation establishes the right of everyone to freely use their abilities and property for carrying out entrepreneurial and other non-prohibited economic activities. Following paragraph "g" of Article 71 of the Constitution of the Russian Federation, the establishment of the legal basis of the single market is the responsibility of the Russian Federation. In addition, the Russian Federation is responsible for the adoption and amendment of acts of civil, civil procedure, and arbitration procedure federal legislation (The Constitution of the Russian, 2014). It follows that the establishment of the legal basis for the circulation of non-documentary securities is the responsibility of the federal center.

20.2 Materials and Methods

The scientific development of the material is carried out based on a set of normative and doctrinal sources. The article uses federal laws and other normative acts of the Russian Federation. In particular, the following federal laws were studied. The Constitution of the Russian Federation. Civil Code of the Russian Federation. Federal Law No. 142-FZ of 02.07.2013 "On Amendments to Subsection 3 of Section I of Part One of the Civil Code of the Russian Federation." Federal Law No. 514-FZ of 27.12.2018 "On Amendments to the Federal Law on the Securities Market and Certain Legislative Acts of the Russian Federation in Terms of Improving the legal regulation of the issue of Securities." Federal Law No. 39-FZ of 22.04.1996 (as amended on 27.12.2019) "On the Securities Market" and other regulatory legal acts.

Doctrinal sources are represented by scientific publications of domestic lawyers and economists. The works of Russian and foreign authors were used, including Agnessa Inshakova, Goncharov, Elena Inshakova, Ryzhenkov, Sevostyanov, Lungu, Brylyakov, Matytsin, Mariola, Urbano, Bustos, Sady-Kimbaya, Huang, Yuan, Nikiforov, Bogustov, Pochezhertseva, Sukhanov, Tarakanov, Lubyagina, Borodkin, Zenin, and others.

The content of this chapter of the monograph is developed based on the materialist worldview and the general scientific method of historical materialism. General scientific methods of cognition are applied: dialectical, hypothetical-deductive method, generalization, induction, and deduction, analysis and synthesis, empirical description. The study also uses private scientific methods: normative-dogmatic, comparative-legal, structural–functional, etc.

20.3 Results

The main body of legal norms regulating public relations in the field of non-documentary securities, of course, is concentrated within the framework of civil legislation and federal laws adopted following it (Inshakova & Goncharov, 2017). The key category of civil law is the object of civil rights. The term "object of civil rights" should be understood as 'a legally passive material or ideal phenomenon, in respect of which unilateral legally significant actions can be committed" (Objects of Civil Rights, 2019), since there is no legal definition in the current legislation. We also agree with the position of scientists regarding the fact that the objects of civil legal relations are concretized and detailed objects of constitutional legal relations (Lungu & Brylyakova, 2018).

According to Article 128 of the Civil Code of the Russian Federation (hereinafter referred to as the Civil Code of the Russian Federation), undocumented securities are classified as property rights that are part of a group of other property that forms one of the groups of objects of civil rights (Civil Code of the Russian Federation, 1994). Unlike documentary securities, which relate to things and represent a material phenomenon (the object of the material world), due to the absence of a materialized document, non-documentary securities form an ideal phenomenon, as well as non-cash funds and the novel of Russian civil legislation-digital rights (Inshakova et al., 2020).

The generic identity of non-documentary securities is determined by Article 142 of the Civil Code of the Russian Federation from which it follows that the following investment objects are securities. First, documents certifying obligations and other rights, the implementation and transfer of this are possible only upon presentation of such documents. Secondly, the rights of obligations and other rights stipulated in the relevant decision or act of the issuer, the implementation and transfer of which is possible in compliance with the rules and requirements established by Article 149 of the Civil Code of the Russian Federation. Thus, this norm contains the basic generic concept of documentary securities (the first) and non-documentary securities (the second).

The norm of Part 1 of Article 149 of the Civil Code of the Russian Federation reflects the theory of unilateral will, we note only the essence of the concept—the responsibility for the execution of undocumented security lies with the person who issued it. The current rule of law (Part 1 of Article 149 of the Civil Code of the Russian Federation) also applies to persons who have provided security for the performance of obligations under the relevant security. In addition, it should be noted that the Federal Law of 02.07.2013 No. 142-FZ (Federal Law, 2013) Chap. 7 of the Civil Code of the Russian Federation "Securities" was supplemented with the third paragraph. This section highlights the rules governing the main provisions of non-documentary securities: enforcement, transfer of rights, protection of violated rights, consequences of reclamation, and consequences of loss of accounts. Thus the rules on non-documentary securities were separated and placed in a separate pedagogical institute within the broader legal institution of securities.

Further regulation of the circulation of non-documentary securities, as well as the use of the rights that they certify, is carried out by federal laws adopted following the Civil Code of the Russian Federation. From 01.01.2020, Federal Law No. 514-FZ of 27.12.2018 (Federal Law, 2018) introduced amendments to Federal Law No. 39-FZ of 22.04.1996 "On the Securities Market" (hereinafter referred to as the Law on the Securities Market), regulating the direct circulation of securities (Federal Law, 1996).

Among the amendments, it should be noted that non-documentary securities have ceased to be registered. The Law names four equity securities: shares, bonds, an issuer's option, and Russian depositary receipts. According to paragraph 1, part 1 of Article 2 of the Law on the Securities Market, equity securities are characterized simultaneously by the following characteristics: (1) fix the totality of property and non-property rights; (2) are placed in issues or additional issues; (3) have an equal volume and terms of implementation within one issue.

These features allow us to conclude that the characteristic feature of securities in circulation is their issue, i.e., issue. The issuing procedure is a sequence of actions established by the legislation of the issuer for the placement of equity securities (Article 2 of the Law on the Securities Market). Moreover, the issue of equity securities must be registered following the established procedure. Currently, the registration authority is the Bank of Russia, which is guided in this matter by the relevant Regulation (Bank of Russia Ordinance, 2017).

The procedure for issuing securities is divided into the following stages:

1. A decision is made on the placement of securities. As a rule, such a decision is within the competence of the general meeting of shareholders.
2. The decision on the placement of securities is approved by the Board of Directors (Supervisory Board). Or another competent management organ no later than six months from the date of the relevant decision, except in cases when the

placement takes place during the establishment of the company or during the reorganization.

3. The state registration of the issue (additional issue) is carried out by the registering organ. Since 01.01.2020, in addition to the Bank of Russia, the registrar, the exchange, or the central securities depository have the right to register issues in the cases provided for in Article 20.1 of the Securities Market Law. When applying for state registration of an issue, the issuer is obliged to provide a list of the necessary documents specified in Chap. 5 of the Regulation. In addition, these documents can be sent in electronic form on the website of the Bank of Russia or the registering organization using the Internet information and telecommunications network, including through access to the personal account.
4. After the registration of the issue, securities are placed, which consists of transactions aimed at transferring ownership rights to the first owners, making accounting entries on the personal accounts of the first owners.
5. The issuing procedure is completed by the state registration of the report on the results of the issue, which reflects all the necessary information about the placement of securities; the response is submitted no later than 30 days after the end of the placement period.

After the issue has passed, the securities go into circulation. The circulation of securities of the Law on the Securities Market is understood as the commission of several civil transactions aimed at the transfer of the right to securities. The circulation of securities before the registration of their issue as well as before their full payment is prohibited (Article 27.6 of the Law on the Securities Market). At the same time, the moment of transfer of rights depends on whether they are recorded in the register or the depository. In the first case, from the date of making a receipt entry on the personal account of the acquirer, and in the second, from the date of making a receipt entry on the depot account (Article 29).

Currently, the only multifunctional exchange trading platform in Russia is the Moscow Exchange Group. The group includes the central securities depository non-bank credit institution joint-stock company "National Settlement Depository," as well as the non-bank credit institution-central counterparty "National Clearing Center" (Joint-stock Company), which allow for a full cycle of investment services. Individuals are allowed to participate in trading on the exchange by entering into a contract for the provision of brokerage services or opening an account with a brokerage company. Limits on transactions are set and professional and property requirements are imposed on individuals when they are recognized as qualified investors who can independently make exchange deals. These restrictions are established by the state to ensure the safety of citizens from significant property losses. Since civil transactions on the exchange are subject to an increased degree of risk and an uncertain degree of confidence in the issuer. Securities are purchased primarily to improve their financial condition by regularly receiving an increase. That is the mathematical difference between the purchase and sale price, the interest, and the dividends due (Bustos & Pomares-Quimbaya, 2020). As a rule, the purchase price and, accordingly, the sale price at organized auctions are directly proportional to the amount of income received on them. In other words, if there is an increase in the yield on security, then the market price increases, and if it decreases, it decreases (Inshakova & Goncharov, 2017).

The legal regulation of the market and its functioning in modern Russia is quite automated and optimal. At the same time, the development of Industry 4.0 technologies inclines legal science to a new understanding of the state of security in digital investment markets (Sánchez & Urbano, 2019) and also to understand the legal nature of non-documentary securities and, most importantly, related phenomena (from the point of view of law) that arise as a result of scientific and technological progress (Tarakanov et al., 2019).

Here, of course, we are talking about cryptocurrency, tokens (Huang & Zhao, 2017), or using the terminology of the latest Russian legislation—digital financial assets.

In legal science, scientists have repeatedly noted that there is a change in the form and leveling of the signs of a thing in the classical sense. Non-documentary securities, however, have retained the legal regime associated with the exercise of certifiable rights ("rights from the paper"), which are the subject of the law of obligations and corporate law. Accordingly, the provisions of property law as an object of civil rights apply to "rights to paper" (Nikiforov, 2010). However, some researchers believe that non-documentary securities do not fit into the traditional understanding of securities, justifying this by the lack of properties of public credibility, formalism, presentation, and constitutive value (Bogustov, 2012). In this regard, others explained the peculiarity of the non-documentary form as follows: non-documentary security originally appeared as a set of legal facts, forming a complex composition these facts are characterized by different legal nature—binding and public law. Therefore, this composition, for example, according to Z. A. Pochezhertseva, should not be called "the emergence of rights to an undocumented security," but "the emergence of an undocumented security." It follows from this that the special regulation of such a complex composition of legal facts in the market is due to the solution of two sets of tasks by the regulator. The first-a relatively "conceptual" block is associated with maintaining the balance of state regulation and self-regulation. The second conditionally "instrumental" block includes a combination (convergence) of private and public legal means of legal regulation (Pochezhertseva, 2013). However, it should be noted that the legislator chose a completely different approach, reducing the powers of, for example, self-regulatory organizations (SROs), transferring most of the powers to the regulator—the Bank of Russia. Currently, SROs have the right only to train citizens in professional activities in the securities market and take exams, if they are accredited by the Central Bank of the Russian Federation (Article 49 of the Law on the Securities Market).

Examining the legal nature of undocumented securities, E.A. Sukhanov notes that the Law on the Securities Market very unsuccessfully fixed in the domestic legal order an extremely controversial category of "undocumented securities." The Law is subject only to equity securities (shares and bonds) and does not in any way extend its effect to documentary securities. Therefore, the institution of "undocumented securities" is not directly related to the traditional civil law institution of securities:

we are talking about the civil law regime of two different types of objects of civil rights (Russian Civil Law, 2015). We should agree with this point of view, and therefore, in our opinion, the amendment to Article 128 of the Civil Code of the Russian Federation regarding the determination of the place of undocumented securities in the system of objects of civil rights is positive in this regard. Unfortunately, the Law on the Securities Market is still terminologically different from the Civil Code of the Russian Federation, despite significant changes in recent years.

On the other hand, most scientists agree that with the "loss" of shares and bonds of the documentary form, only the way of legitimizing the rights they secure has changed. And the introduction of such a form into circulation is primarily due to the need for economic turnover. According to D. V. Lubyagina, this circumstance, on the contrary, emphasizes the decisive importance of the rights contained in them. Therefore, the development of various structures due to the movement of scientific and technological progress not only does not detract from but also does not change the main function of securities (Lubyagina, 2017). This approach allows us to draw an unambiguous conclusion that in the case of non-documentary securities, the content prevails over the form. Because, first, as historical retrospect shows, securities arose as a result of simplifying the turnover of the obligations enshrined in them. Secondly, the objects of the material world have long ceased to be the only source and repository of information.

The position that not only things are recognized as objects of property rights but also binding rights and disembodied things has long been heard in the literature. The reason for this recognition, as already noted, is technological progress, thanks to which fundamentally new objects begin to exist in civil circulation, including three types of ideal phenomena: undocumented securities, non-cash funds, and digital rights (Borodkin, 2018). However, it should be noted that common law countries are characterized by a proprietary concept of the interpretation of objects of property rights. Thus, intangible objects of intellectual property are objects of real rights. According to I.A. Zenin, the attribution of intellectual property to the result of intellectual activity is due to many historical, economic, psychological, and legal factors (Zenin, 2005).

20.4 Conclusions

Thus, based on the study of the regulatory framework for regulating the procedure for circulation of equity securities, the following conclusion is justified. In the implementation of the transfer of ownership of shares and bonds in practice, there are three types of legal gaps. First, there is the potential for falsification of material documents that mediate the transfer of shares from one person's account to another person's account. Second, there is uncertainty about the qualification of actions as unilateral or bilateral transactions in some cases related to the disposal of non-documentary securities. Third, the lack of interaction between the register of notarized transactions on the alienation of securities and the databases of organizations that maintain the depot account and the register of shareholders. To eliminate these problems, the author substantiates the need to improve the legal instruments of the system of accounting and registration of the facts of the transfer of ownership rights to non-documentary securities. This is achieved by establishing a rule on the mandatory registration of purchase and sale transactions and other transactions that mediate a change of ownership in a distributed registry system, to which all entities involved in transactions must have access. At the same time, it is proposed to verify the authority of a person to dispose of security by activating an electronic digital signature or using cryptographic protection tools. This will increase the level of law and order and the volume of securities turnover on the market.

Acknowledgements The work was supported by Russian Science Foundation (project No. 20-18-00314).

References

Bank of Russia Ordinance No. 4477-U dated 03.08.2017 "On Amendments to the Regulation of the Bank of Russia" 11 August 2014 No. 428-P rev. dated 18.12.2018 "On the standards for the issue of securities, the procedure for state registration of the issue (additional issue) of emissive securities, state registration of reports on the results of the issue (additional issue) of emissive securities and registration of securities prospectuses". Bulletin of the Bank of Russia, 77.

Bogustov, A. A. (2012). *Securities as objects of rights in the civil law of the CIS member states. Comparative legal analysis* (Ph.D. thesis). Moscow. 27.

Borodkin, S. V. (2018). A new approach to the classification of objects of property rights. *Journal of Russian Law, 2*, 6–12.

Bustos, O., & Pomares-Quimbaya, A. (2020). Stock market movement forecast. A Systematic Review. *Expert Systems with Applications, 156*, 113464.

Federal Law. (1996, April 22). No. 39-FZ (as amended on 27.12.2019). *On the securities market* (as amended and supplemented, entered into force on 01.01.2020). Collected Legislation of the Russian Federation, No. 17, Art. 1918.

Federal Law. (2013, July 2). No. 142-FZ *On amendments to subsection 3 of Section I of part one of the civil code of the Russian Federation*. Collected Legislation of the Russian Federation, No. 27, Art. 3434.

Federal Law. (2018, December 27). No. 514-FZ *On Amendments to the Federal 'On the Securities Market' and certain legislative acts of the Russian Federation in terms of improving the legal regulation of the issuance of securities*. Collected Legislation of the Russian Federation, 53 (part I), Art. 8440.

Goncharov, A. I., & Inshakova, A. O. (2019). *Objects of civil rights: textbook for an undergraduate, specialist, and graduate programs* (A. I. Goncharov & A. O. Inshakova, Eds., 2nd ed., pp. 20–23). Yurayt; Publishing House of Volgograd State University.

Huang, T., & Zhao, Y. (2017). Revolution of securities law in the Internet Age: A review on equity crowd-funding. *Computer Law & Security Review, 33*(6), 808.

Inshakova, A. O., & Goncharov, A. I. (2017). Securities as a subject of financial and economic interest and object of civil law relations. *Law—Higher School of Economics Journal, 4*, 97–110.

Inshakova, A. O., Inshakova, E. I., Ryzhenkov, A. J., & Sevostyanov, M. V. (2020). Civil law in the digital economy: Analysis of doctrinal adaptation trends. competitive Russia: Foresight model of economic and legal development in the digital age. In A. O. Inshakova & E. I. Inshakova (Eds.),

Proceedings of the International Scientific Conference in Memory of Oleg Inshakov (1952–2018) (pp. 421–429). Springer Nature.
Lubyagina, D. V. (2017). Influence of the form of security on the content of civil law regulation. *Advocate, 2*, 18–23.
Lungu, E. V., & Brylyakova, E. S. (2018). Constitutional legal relations, and civil legal relations: A comparative analysis. *State Power and Local Government, 7*, 13–18.
Nikiforov, A. Y. (2010). *Uncertified securities as objects of civil legal relations* (Ph.D. thesis). Tomsk, 26.
Pochezhertseva, Z. A. (2013). *Legal regulation of the securities market* (private and public law principles. Ph.D. thesis). Yekaterinburg. 27.
Sánchez, M., & Urbano, A. (2019). Security in digital markets. *Journal of Business Research, 101*, 487.
Sukhanov, E. A. (2015). *Russian Civil Law. A common part. Real right. Inheritance law. Intellectual rights. Personal non-property rights* (4th ed.), Moscow: Statut, 331.
Tarakanov, V. V., Inshakova, A. O., Dolinskaya, V. V. (2019). Information society, digital economy, and law. In E. G. Popkova (Ed.), *Ubiquitous computing and the Internet of Things: Prerequisites for the development of ICT.* Studies in Computational Intelligence (pp. 3–15). Springer Science+ Business Media.
The Civil Code of the Russian Federation (part one), 30 November 1994 No. 51-FZ (as amended on 16 December 2019). Collected Legislation of the Russian Federation, No. 32, Art. 3301.
The Constitution of the Russian Federation (adopted by popular vote on 12.12.1993) (taking into account the amendments introduced by the Laws of the Russian Federation on amendments to the Constitution of the Russian Federation dated 30.12.2008 No. 6-FKZ, dated 30.12.2008 No. 7-FKZ, dated 05.02.2014 No. 2-FKZ, dated 21.07.2014 No. 11-FKZ). Collected Legislation of the Russian Federation, No. 31, Art. 4398.
Zenin, I. A. (2005). *The civil and commercial law of foreign countries: Textbook.* Allowance. Moscow, 92.

Chapter 21
Priorities and Principles for the Development of the Space for the Use of New-Industrial Digital Technologies 4.0 by Foreign Trade Companies of the EAEU and the BRICS Member States

Agnessa O. Inshakova, Alexander I. Goncharov, and Marina V. Goncharova

Abstract The chapter of the monograph reveals the scientific and methodological approach to determining the main (priority) directions of the development of digital technologies in Industry 4.0. These technologies increase efficiency and should be used by economic entities of the EAEU and BRICS jurisdictions in the implementation of foreign trade activities.Two priority main directions of development are justified. Five fundamental principles of the application of digital technologies in Industry 4.0 are defined. These principles contribute to the proper implementation of cross-border export–import transactions between residents of the member states of integration associations with the participation of the Russian Federation. The space for the use of neo-industrial digital technologies 4.0 by foreign trade companies of the EAEU and BRICS member states should be digitized in the Internet portal "Foreign Economic Activity of the Greater Eurasian Partnership online". On this portal, each participant of foreign economic activity can open and operate their account of legal significance. The priorities and principles of the development of the space for the use of new-industrial digital technologies 4.0 by foreign trade companies of the countries participating in the Greater Eurasian Partnership should be fixed in an interstate conventional agreement.

A. O. Inshakova (✉) · A. I. Goncharov
Volgograd State University, Volgograd, Russia
e-mail: ainshakova@list.ru; gimchp@volsu.ru

A. I. Goncharov
e-mail: gimchp@volsu.ru

M. V. Goncharova
Plekhanov Russian University of Economics, Volgograd, Russia
e-mail: reu-kafedra-ekonomiki@mail.ru; goncharova.sofia@gmail.com

A. O. Inshakova and E. E. Frolova (eds.), *Smart Technologies for the Digitisation of Industry: Entrepreneurial Environment*, Smart Innovation, Systems and Technologies 254, https://doi.org/10.1007/978-981-16-4621-8_21

Keywords Foreign economic activity · Priorities · Principles · Digital technologies · Integration associations · Industry 4.0 · Inter-jurisdictional polysubject blockchain

JEL Codes G18 · G24 · K12 · K15 · L14 · L24 · L86 · O16

21.1 Introduction

Industrial technologies have already penetrated almost all spheres of public relations, and foreign economic activity is no exception. Today, it is impossible to imagine foreign trade transactions without the use of digital technologies. To successfully conduct business, participants in foreign trade turnover must be familiar with the nuances of the latest legislative changes, be aware of the introduction of new, innovative digital technologies. The foreign economic activity requires participants to make decisions quickly and accurately, while the use of digital technologies reduces risks, ensures transparency, legality, and business growth. Traditional forms of international trade are gradually being transformed into commodity exchange based on artificial intelligence technologies, which naturally lead to the reorganization of markets, trade turnover, and business models, which are now based on new digital technologies, such as the "Internet of Things", artificial intelligence, blockchain, and Big Data.

We support the increasingly intensive use of foreign trade companies, in particular, digital blockchain technology. Blockchain technology can be used to view not only goods but also legally relevant information for all participants, from manufacturers, shippers to port authorities, and government agencies, through a virtual dashboard. A blockchain is a distributed registry, a complementary database, which is a continuous chain of blocks, the data of this database is stored simultaneously on many computers. Blockchain technologies increase the speed of processing paper documents by 10 times, allowing you to simplify and speed up the work with issued documents (https://www.cnet.com/news/ibm-maersk-tradelens-blockchain).

In 2018, for example, the companies "Maersk" and IBM for the implementation of foreign trade activities for the first time created an innovative blockchain platform TradeLens. The largest ports are PSA Singapore and the Port of Rotterdam, as well as the customs authorities of Singapore, the Netherlands. Saudi Arabia and Australia, the transport and logistics companies Agility, CEVA Logistics, Damco, Kotahi, PLH Trucking Company, WorldWide Alliance (about 94 organizations in total) became participants in this register. The platform is powered by blockchain technology, and over time, it can likely grow into a global trading ecosystem. Participants of the platform can quickly, conveniently, and reliably conclude transactions, exchange the necessary information, as well as carry out the subsequent movement of goods across borders and international trade zones (https://www.cnet.com/news/ibm-maersk-tradelens-blockchain).

The digital technology of electronic (virtual) residency (E-residency) used in Estonia seems to be promising. In particular, the program allows a citizen of any country to create a company and conduct its activities in Estonia remotely. An E-resident can manage their company from anywhere in the world via the Internet, use the services of an electronic bank, pay taxes, and enter into contracts using a digital signature (http://migratetoestonia.com/e-residency-elektronnoe-grazhdanstvo-estonii/).

In the scientific works of foreign colleagues, we do not find a response to the problems of digital technologies in the EAEU and BRICS (Muller & Voigt, 2018; Szalavetz, 2019; Weyer et al., 2015).

21.2 Materials and Methods

The following documents were examined as part of the regulatory framework:

- The Treaty on the Eurasian Economic Union and its Annexes;
- Statement on the digital agenda of the Eurasian Economic Union;
- The decision of the Supreme Eurasian Economic Council of 11.10.2017 No. 12 "On the main directions of implementation of the digital agenda of the Eurasian Economic Union until 2025";
- The main directions of the implementation of the digital agenda of the Eurasian Economic Union until 2025 and their Annexes;
- Moscow Declaration of the XII BRICS Summit;
- BRICS Economic Partnership Strategy until, 2025;
- Order of the Government of the Russian Federation No. 1776-r of 08.07.2020.

The study of doctrinal sources covers the scientific works of some Russian scientists, including these authors: Tarakanov, Inshakova A., Goncharov, Ershova, Kalinina, Inshakova E., Stepanova, Glazyev, Matytsin. The scientific works of a number of foreign authors were also studied, including: Szalavetz, Muller, Voigt, Kagermann, Gilchrist, Weyer, Schmitt, Ohmer, Goreck. At the same time, at the scientific and methodological level, the priorities and principles of the development of the space for the use of digital technologies 4.0 by foreign trade companies of the EAEU and BRICS member countries have not received the attention of researchers until recently.

The content of this chapter of the monograph is developed on basis of the materialist worldview and the general scientific method of historical materialism. General scientific methods of cognition are applied: dialectical, hypothetical-deductive method, generalization, induction, and deduction, analysis and synthesis, empirical description. The research also uses private scientific methods: dogmatic, comparative-legal, hermeneutic, structural–functional, etc.

21.3 Results

The Ministry of Economic Development of the Russian Federation has created a "Single portal of Foreign Economic Information" for Russian exporters in terms of using electronic platforms to promote and sell their goods. Another Russian development, the RSTrade information, and service platform (in Russian, English, Indonesian and Chinese), is aimed at increasing trade turnover between the countries of South and Southeast Asia and the countries of the Eurasian Economic Union, as well as attracting foreign investment in Russian projects. The platform integrates 4 sections: "Goods", "Services", "Investments", "Contract production". In particular, the platform contains catalogs of manufacturers and suppliers of goods and services, investment projects and investors, as well as current engineering developments. The Internet portal is created on basis of an integrated approach to solving problems of interaction between the seller and the buyer: from the moment of placing the goods on the trading platform to the moment of discussing the terms and conditions of delivery, passing customs procedures, etc. The platform makes it possible to resolve legal issues that arise during the execution of transactions in real-time (Inshakova et al., 2020a; b).

In foreign economic activity, digital technologies of Industry 4.0 are a key way to increase the volume, quantity, and content of commodity diversity of foreign trade deals. In our opinion, their effective application should be conditioned by scientifically verified priority (main) directions, in which these technologies should be applied in a concentrated and purposeful manner. In this regard, we highlight the first main direction—the expanded and increasingly intensive use of a complex of new-industrial digital technologies in the space of the Greater Eurasian Partnership (EAEU + BRICS) (Glazyev, 2018). The second main direction of the development of the space for the use of neo-industrial digital technologies 4.0 by economic entities and state bodies (services) of the Greater Eurasian Partnership is the end-to-end tracking of the movement of goods from the release from the assembly line of the manufacturer to the moment of its receipt by the end consumer.

Revealing *the first main direction* it is necessary to point directly to the leading, flagship role in the wide deployment and promotion of innovative and effective digital technologies of Industry 4.0 of the People's Republic of China. Other partner states, along with the ability to acquire, copy, or otherwise legally use Chinese digital developments, have a very healthy and positive interest in overcoming the real gap and digital gap with China and raising the level of their national science and digital technologies to its level. At the same time, given the technological lag behind China, for example, Russia, by about 50 years, it is irrational to spend resources on creating digital technologies from scratch that have already been created and widely used.

In our opinion, the latest technologies and a high degree of digitalization of business processes and public administration, in general, have ensured China's victory over COVID-19 in 2020. In the context of the pandemic, China has used digital technologies in four targeted areas. Firstly, limiting the transmission of coronavirus infection: LBS-services (location-based services), big data analysis, and robotics were used to track and identify cases of increased risk of infection, restrict movement and minimize contact between people with each other. Secondly, informing people anywhere in the country every hour: digital platforms and communication technologies to ensure transparency and availability of all information online have become the basis for nipping in the bud the slightest manifestations of panic in the general population. Thirdly, ensuring a high quality of life in emergencies. Chinese information and communication companies make the most of the latest digital technologies of O2O (from online to off-line) virtual reality, allowing not only to meet the basic needs of life during the period of self-isolation but also to neutralize depression and restrictions associated with prolonged stay in a confined home, in particular, to virtually visit museums, cinemas, theaters, zoos. Fourthly, the recovery of business activity. To do this, China's economic entities used digital technologies as widely and intensively as possible: they worked with potential buyers through Internet channels and broadcast the resumption of operational activities; the work of the companies ' labor collectives was resumed remotely as soon as possible, also via the Internet.

We believe that international agreements that can be agreed upon by all partner states at the conventional public legal level on the leading role of China in creating a holistic digital and technological architecture of a single space for foreign economic activity of the Greater Eurasian Partnership are quite acceptable. This is fully justified by the leading trade and economic and advanced information technology role of China in the XXI century.

Analyzing *the second main direction*, it should be noted the following. By creating new digital technologies that are already functioning in other economic spheres for foreign trade, it is necessary to consistent and persistent achieve completely transparent, end-to-end tracking of the movement of goods from the release from the assembly line of the manufacturer to the moment of its receipt by the end consumer. We believe that over the next 10 years, the process of innovative design, production financing, advertising, selection, purchase and sale, payment, delivery of goods anywhere in the world—the entire process will be mediated by a single remote, end-to-end digital technology applied via the Internet. The consumer will have the opportunity to participate remotely, but personally, in all the above-mentioned stages of creating a product directly for themselves, being continuously connected, tracking the process of creating a product, and interactively filling its functionality with specific, pre-selected options for their consumer preferences. Foreign trade interstate alliances will gradually lose their relevance, but a joint partnership of states in foreign economic activity will be relevant for about 50 years. The prospects for a Large Eurasian Partnership seem quite positive to us since it has a fundamental geographical advantage, namely, the largest continental space of Eurasia on the Planet. In this regard, efforts should be focused on the formation of centers of responsibility within the Partnership for the use of neo-industrial digital technologies: production, settlements, transportation-customs, electronic border, and cybersecurity. These five areas of responsibility in the implementation of foreign economic activity should be perceived and considered as fundamental principles. These are the principles of the development of the space for the use of neo-industrial digital technologies 4.0 by economic entities of the EAEU and BRICS member states.

The *first principle* of the development of digital technologies at the stage of production of goods provides for a deep development of the combination of the highest level of automation and robotization of high-tech conveyor production and continuous online tracking of all its discrete procedures. These online Internet tracking technologies can be used by end-users and investors, government regulators, as well as tax and customs authorities. As part of the implementation of this principle, a preventive information public–private partnership is provided. Since the actual production volumes of goods for export are initially known, buyers and their sources of financial resources are known in advance, prices for the purchase and sale of goods are known in advance, all financial transactions are tracked, it becomes impossible to hide the objects of taxation. It is also impossible to smuggle goods abroad, as a result, the level of criminality of public relations decreases. At the same time, the online multi-sided input quality control of components, spare parts, and other materials supplied for the production of goods is also of great importance. Also, online monitoring and continuous video recording are carried out completely along the entire trajectory of the conveyor movement of the created product, while the operations of specific employees are recorded which is of legal importance for each of them in assessing the degree and quality of labor participation in the production process. Such videos should be stored on the manufacturer's server at all times, which will help to speed up the identification of perpetrators in cases of defects in the production process of goods.

The *second principle* of the development of digital technologies is the intensification of financial settlements between counterparties online. This sphere of foreign economic activity is currently the most advanced and advanced in terms of the use of digital technologies. Remote banking services, the use of electronic signatures in the implementation of money transfers are the very first information and communication technologies, these digital services work quite effectively and are used by the absolute majority of participants in foreign economic activity. The largest commercial banks also offer comprehensive service products for the export and import of goods in terms of the formation of a transaction passport, letter of credit and bill payments, control, and implementation of customs payments. Modern digital smart contract technology should be integrated into these well-established banking technologies. These smart digital technologies provide for remote Internet recording of the execution of the intermediate stages of a foreign economic transaction planned by the parties to the contract, which allows you automatically send legally significant commands, including for the transfer of planned payments (Inshakova et al., 2020a, b). At the same time, we propose to implement the next level of technological integration as we work out the interconnected application in the combination of remote banking services and smart contracts. Namely, to display the actual implementation of foreign trade smart contracts, including monetary settlements on them, on the interstate Internet portal "Foreign Economic Activity of the Greater Eurasian Partnership online" (Tarakanov et al., 2020).

In 2020, the information system "One Window" was put into operation in Russia. It will allow Russian exporters to get online access to services related to the conduct of the foreign economic activity. The "One Window" system digitally presents the services and services required by the exporting company to solve the main business tasks at each stage of the export cycle. To interact with various agencies, the exporter will only need to enter the necessary data into the system once, and then receive the necessary services. The information system "One Window" contains 24 services, including services related to analytics, as well as electronic storage of certificates and licenses. The new order of the Russian Government expands the list of available services in the "One Window" system. In total, 28 services should work in the system (Order of the Russian Government No. 1776-r of 08.07.2020).

The "One Window" system will allow exporters to free themselves from a large paper document flow, to establish electronic communication with various government agencies and organizations involved in the provision of services in the field of export.

At the beginning of 2020, the Eurasian Intergovernmental Council set a task to prepare amendments to the EAEU law in the field of international electronic commerce. After working on the issue at the meeting on October 9, the Commission reported on the results of the work carried out.

The *third principle* of the development of digital technologies is the achievement of complete electronic remote control and continuous Internet tracking of goods moving along the route of a foreign trade transaction, combined with mobile and easy electronic customs procedures. It should be objectively recognized that digital customs technologies, in particular, in the union of the EAEU, are the most advanced in their development. The implementation of the digital agenda in foreign trade activities began with a change in the organizational structure of customs authorities. Declarations for goods were sold only to specialized customs authorities that carry out electronic document management. The first electronic customs offices appeared in 2018. According to the level of implementation of innovative technologies, the customs authorities of the member states of the EEC are recognized leaders: in the countries, the technologies for performing individual operations in automatic mode are legally fixed and used. The implementation of these technologies is carried out on basis of paragraph 3 of Article 82 of the Customs Code of the EAEU. Every day, the information systems of the customs authorities process about 200 thousand customs documents without the participation of officials. For example, in 2020, the EEC Board approved the draft Agreement "On a unified system for identifying participants in Foreign Economic Activity" within the framework of the EAEU. The proposed system will use the national taxpayer identification number when forming a unique identification number of a foreign trade participant (http://www.eurasiancommission.org/ru/nae/news/Pages/20-01-2020-2.aspx).

Large modern factories-manufacturers of mass export goods quite naturally are located near the nodes of the transport infrastructure-ports, railway stations, and main highways. At the same time, the goods sent for export must pass through customs procedures. In this regard, it is necessary to develop unmanned technologies (without a human driver) for transporting goods for customs procedures to the warehouses of the local customs authority. But it is even more promising and expedient to develop a digital technology "Mobile customs warehouse". Civil servants of the local Customs authority could transport a special set of equipment in an ordinary passenger car, arriving at the manufacturer's factory at the time of loading the goods into standard transport containers. Digital technology, in which customs officers, carrying out customs procedures under continuous video recording, form on-site electronic customs declarations online, will achieve a very significant reduction in economic costs. Since the technology will eliminate the useless movement of export goods from the warehouse of the manufacturer to the customs warehouse, then from this warehouse to the port (to the railway, etc.). This technology also can be used effectively when shipments of imported goods arrive at the transport infrastructure hubs. In this perspective, the digital technology "Mobile Customs Warehouse" can be used similarly when unloading standard transport containers from a ship, car, plane, car, while useless transportation of this product from the place of arrival to customs warehouses is excluded. Both during import and export, the Customs Authority must leave a "digital seal" on each standard shipping container—an inseparable, single-use microchip that stores the digital key to an archived folder containing information about the customs procedure. This block of information contains an electronic customs declaration, a video recording of the participants and officials who participated in the procedure and is stored in the system of the inter-jurisdictional polysubject blockchain (Kalinina et al., 2019).

The *fourth principle* of the development of digital technologies is the development of satellite Internet tracking of the crossing of the borders of jurisdictions by standard transport containers with goods moving along the routes of export–import foreign trade transactions. In the development of these technologies, it will be necessary to create special robotic complexes at the key points of the transport infrastructure—electronic gates (digital technology "electronic state border"). The operation of these electronic gates should be integrated with the technologies of loading (unloading) manipulations with standard transport containers, which are carried out when unloading goods arriving under an import contract, or when loading goods sent for export. At the stage of crossing the "electronic state border" by a particular batch of goods, additional control should be carried out over the proper production of customs procedures provided for during export, as well as during import. The digital technology "electronic state border" should provide for the possibility of adding this additional block of information about the fact of crossing the border, about the participants and officials of this procedure, a file with a video recording of the border crossing, as a new unchangeable link in the blockchain chain for archival storage. In this combination of technologies, the digital mark of the proper crossing of the border by the goods will be combined with the "digital seal" of the customs authority and will be tracked along the route of the transaction, maintaining communication with the satellite. These digital technologies should also provide for information and communication interaction with the interstate Internet portal "Foreign Economic Activity of the Greater Eurasian Partnership online".

The *fifth principle* of digital technology development in cybersecurity. This is a set of measures for the differentiation and storage of continuously changing access codes to technically and legally significant digital technologies of foreign economic activity in the system of inter-jurisdictional polysubject blockchain. This set of measures provides for and ensures the impossibility of relatively fast and cost-effective hacking of technologies in cyber-attacks. The partner states should pay close attention to cybersecurity, because, first, digital technologies of foreign economic activity will continue to develop rapidly and widely (Matytsin, 2021). At the same time, secondly, criminal groups will also intensively try to carry out hacker attacks in order to "intercept the management" of foreign trade commodity flows and assign money proceeds for stolen material values. The prevention of such cybercrimes will be provided on the condition that this technology will be developed in the direction of providing continuous Internet tracking of moving vehicles from space satellites. Within the next 10 years, the quality of the broadcast video image of objects within the size of 1 cm should be achieved, despite the difficult weather conditions and (or) the dark time of the day. In addition, vehicles carrying shipments of goods under foreign trade transactions, for example, sea vessels, railway trains, road trains, should be equipped with drones (for example, copters). These drones must periodically fly over the vehicle's path, while also providing a satellite Internet broadcast of the video image of the moving vehicle online. If necessary and under certain conditions, participants in foreign economic activity should be able to connect to satellite Internet tracking via secure Internet communication channels at any stage of the execution of a foreign trade deal. In our opinion, the most efficient operation of this Internet tracking service should be carried out through the Internet portal "Foreign Economic Activity of the Greater Eurasian Partnership online", where each participant in foreign economic activity can open and operate their account of legal significance.

21.4 Conclusion

The priorities for the development of the space for the use of new-industrial digital technologies 4.0 by foreign trade companies of the EAEU and BRICS member states should be concentrated in two main directions. The first main direction provides for a broad and intensive application of a complex of neo-industrial digital technologies in the space of the Greater Eurasian Partnership (EAEU + BRICS), with China playing a leading role in creating a holistic digital and technological architecture of a single area for foreign economic activity. The second main direction is the creation of new digital technologies used in other economic spheres for foreign trade as end-to-end technologies for transparent is interactive tracking of goods movement from the release from the assembly line of the manufacturer to the point and moment of its receipt by the end consumer.

The principles of the development of the space for the use of new-industrial digital technologies 4.0 by foreign trade companies of the EAEU and BRICS member states represent a set of five basic functional links in the implementation of foreign trade activities, which inherently include digital technologies.

- Digital technologies of continuous online tracking of all stages of the production of goods, combined with high automation and robotization of conveyor production.

- The improvement of online financial settlements between foreign trade counterparties with accelerated integration of smart contract algorithms into remote banking technologies.
- Achieving end-to-end electronic remote control and Internet tracking of goods moving along the route of a foreign trade transaction, while combining digital technology with electronic mobile customs procedures.
- The provision of satellite Internet tracking of goods crossing the borders of jurisdictions moving along the routes of export–import foreign trade transactions.
- Ensuring cybersecurity with the help of an inter-jurisdictional polysubject blockchain.

The space for the use of new-industrial digital technologies 4.0 by foreign trade companies of the EAEU and BRICS member states should be digitized in the Internet portal "Foreign Economic Activity of the Greater Eurasian Partnership online". On this portal, each participant of foreign economic activity can open and operate their account of legal significance. The priorities and principles of the development of the space for the use of new-industrial digital technologies 4.0 by foreign trade companies of the countries participating in the Greater Eurasian Partnership should be fixed in an interstate conventional agreement.

Acknowledgements The reported study was funded by RFBR according to the research project No. 18-29-16132 "Priorities for the legal development of digital technologies of foreign trade activities in the context of international economic integration".

References

BRICS economic partnership strategy until 2025. https://brics-russia2020.ru/images/114/81/1148133.pdf Accessed 22 April 2021.

Glazyev, SYu. (2018). On the use of digital technologies to create the EAEU capital market. Eurasian Integration. *Economics, Law, Politics, 1*, 7–8.

Inshakova, A. O., Goncharov, A. I., Inshakova, E. I., & Stepanova, I. A. (2020a). Prerequisites for the development and popularization of alternative methods of resolving economic conflicts. In A. O. Inshakova, A. V. Bogoviz (Eds.), *Alternative methods of judging economic conflicts in the national positive and soft law*. A volume in the series Popkova. Advances in Research on Russian Business and Management., Information Age Publishing. https://www.infoagepub.com/products/Alternative-Methods-of-Judging-Economic-Conflicts-in-the-National-Positive-and-Soft-Law

Inshakova, A. O., Goncharov, A. I., Salikov, D. A. (2020b). Electronic-digital smart contracts: Modernization of legal tools for foreign economic activity. *Lecture Notes in Networks and Systems, 91*, 3–13.

Kalinina, A. E., Inshakova, A. O., & Goncharov, A. I. (2019). Polysubject jurisdictional blockchain: Electronic registration of facts to reduce economic conflicts. *Studies in Computational Intelligence, 826*, 205–213.

Matytsin, D. E. (2021). Internet-investing as a remote algorithm of retail investment financing. In *Modern global economic system: Evolutional development vs. revolutionary leap*. Lecture Notes in Networks and Systems. Springer Nature.

Moscow Declaration of the XII BRICS Summit. http://kremlin.ru/supplement/5581 Accessed 02 04 2021.

Muller, J., & Voigt, K. (2018). Sustainable industrial value creation in SMEs: A comparison between Industry 4 and Made in China 2025. *International Journal of Precision Engineering and Manufacturing-Green Technology, 5*(5), 659–670.

Statement on the digital agenda of the Eurasian economic union. http://pravo.eaeunion.org/document?id=71553074&byPara=1&sub=1 Accessed 02 March 2021.

Szalavetz, A. (2019). Industry 4.0 and capability development in manufacturing subsidiaries. *Technological Forecasting and Social Change, 145*, 384–395.

Tarakanov, V. V., Inshakova, A. O., & Goncharov, A. I. (2020). Interactive demo online maintenance foreign trade deals of business entities of BRICS jurisdictions. *Lecture Notes in Networks and Systems, 110*, 319–327.

The decision of the supreme Eurasian economic council of 11.10.2017 No. 12. On the main directions of Implementation of the Digital Agenda of the Eurasian Economic Union until 2025. http://pravo.eaeunion.org/document?id=71708158&byPara=1&sub=1 Accessed 12 April 2021

The main directions of the implementation of the digital agenda of the Eurasian Economic Union until 2025 and their Annexes. http://pravo.eaeunion.org/document?id=71708158&byPara=1&sub=1 Accessed 12 April 2021.

The Treaty on the Eurasian Economic Union and its Annexes. http://base.garant.ru/70670880/#ixzz6gzcJ0u7L Accessed 02 February 2021.

Weyer, S., Schmitt, M., Ohmer, M., & Goreck, D. (2015). Towards Industry 4.0–Standardization as the crucial challenge for highly modular, multi-vendor production systems. *IFAC-PapersOnLine, 48*(3), 579–584.

Chapter 22
Supranational Legal Mechanism for the Use of New-Industrial Digital Technologies 4.0 in Foreign Trade by Economic Entities of the EAEU and the BRICS Member States

Alexander I. Goncharov and Agnessa O. Inshakova

Abstract The chapter of the monograph contains the scientific development of an inter-jurisdictional set of legal measures that ensure the effective use of advanced digital technologies by economic entities of the EAEU and BRICS jurisdictions in the implementation of foreign trade activities. The authors substantiate a new scientific and methodological approach to blockchain coordination of regulatory documents intended to regulate the introduction and application of advanced digital technologies for foreign trade. It is proposed to adopt an open-partnership approach as a scientific and methodological basis for the formation of a set of measures for the use of advanced digital technologies. Within this approach, the creation and subsequent implementation of certain regulatory documents on digital technologies are differentiated into certain stages. The results of the implementation of these stages are recorded by the relevant departments of the countries participating in the integration associations in the system of the inter-jurisdictional poly subject blockchain. The authors prove that it is advisable to fix this set of legal measures as part of the provisions of the international convention, to which the countries participating in integration associations will be able to join, thereby ensuring the entry into force of the relevant regulatory provisions in their jurisdictions.

Keywords Foreign trade · New-industrial digital technologies 4.0 · Inter-jurisdictional legal mechanism · Digital platform · Blockchain · EAEU · BRICS

JEL Codes G18 · G24 · K12 · K15 · L14 · L24 · L86 · O16

A. I. Goncharov (✉) · A. O. Inshakova
Volgograd State University, Volgograd, Russian Federation
e-mail: gimchp@volsu.ru

A. O. Inshakova
e-mail: ainshakova@list.ru; gimchp@volsu.ru

A. O. Inshakova and E. E. Frolova (eds.), *Smart Technologies for the Digitisation of Industry: Entrepreneurial Environment*, Smart Innovation, Systems and Technologies 254, https://doi.org/10.1007/978-981-16-4621-8_22

22.1 Introduction

The supranational legal mechanism for the use of non-industrial digital technologies 4.0 in foreign trade by economic entities of the EAEU and BRICS member states has not been considered in scientific developments until recently. Certain regulations that are adopted in the national jurisdictions of these associations purposefully do not address the issues of introduction and application of digital technologies of Industry 4.0 by companies in foreign trade. At the interstate level, there are no documents that are mandatory for the implementation of the EAEU and BRICS members in the field of the introduction and application of neo-industrial digital technologies for conducting foreign economic activity. In the Russian Federation, legislation on expanding the use of digital (computer) technologies based on remote Internet communication is being formed relatively slowly and fragmentally. At the same time, the development of legal regulation in this economic sphere can contribute to a significant increase in the volume and intensity of foreign economic activity in integration associations. In addition, to prevent the emergence of conflicting regulatory decisions that may be taken in national jurisdictions, such legal risks should be initially excluded. For this purpose, it is necessary to apply special procedures for approving the documents being developed that are mandatory for conducting foreign economic activity by economic entities—residents of the EAEU and BRICS member states. Such procedures at the scientific and methodological level are integrated by the authors into an open-partner approach to the formation of a set of measures for the introduction and application of advanced neo-industrial digital technologies for conducting foreign economic activity.

22.2 Materials and Methods

The following documents were examined as part of the regulatory framework:

- Federal Law No. 183-FZ of 18.07.1999 "On Export Control";
- Decree of the President of the Russian Federation of 21.07.2020 No. 474 "On National Development Goals of the Russian Federation for the period up to 2030";
- Passport of the National Project "National Program" Digital Economy of the Russian Federation" of 04.06.2019 No. 7;
- Federal Law of 31.07.2020 No. 258-FZ "On experimental legal regimes in the field of digital innovations in the Russian Federation";
- Incoterms®, 2020-ICC Rules on the Use of Domestic and International Trade Terms;
- Electronic Bills of Landing;
- Uniform Rules for Sea Waybills;
- Rotterdam Rules;
- State Program for the Development of the Digital Economy and Information Society for 2016–2020, Belarus;

- The State program "Digital Kazakhstan" for 2017–2020″;
- Armenia's Digital Transformation Agenda until 2030;
- The program of digital transformation of Kyrgyzstan—"TazaKoom";
- The Treaty on the Eurasian Economic Union and its Annexes. Statement on the Digital Agenda of the Eurasian Economic Union;
- The decision of the Supreme Eurasian Economic Council of 11.10.2017 No. 12 "On the main directions of implementation of the Digital Agenda of the Eurasian Economic Union until 2025";
- The main directions of the implementation of the digital agenda of the Eurasian Economic Union until 2025 and their Annexes;
- Moscow Declaration of the XII BRICS Summit;
- BRICS Economic Partnership Strategy until, 2025;
- Digital Economy Report, 2019.

The study of doctrinal sources covers the scientific works of some Russian scientists, including these authors: Maksyutina, Golovkin, Belikova, Inshakova, Goncharov, Kalinina, Matytsin. Also, scientific works of a number of foreign authors, namely Szalavetz, Muller, Voigt, Kagermann, Gilchrist, Weyer, Schmitt, Ohmer, and Goreck.

The content of this chapter of the monograph is developed on basis of the materialist worldview and the general scientific method of historical materialism. General scientific methods of cognition are applied: dialectical, hypothetical-deductive method, generalization, induction, and deduction, analysis and synthesis, empirical description. The research also uses private scientific methods: dogmatic, comparative-legal, hermeneutic, structural–functional, etc.

22.3 Results

The concept of "digital technologies of Industry 4.0" in the lexicon appeared relatively recently, as a result of which it has not yet been thoroughly studied. Defining this phenomenon, it is necessary to correlate it with such a phenomenon as the Fourth Industrial Revolution. The Fourth Industrial Revolution as the main theme of the World Economic Forum 2016 was characterized primarily as a combination of technologies that ensure the implementation of cyber-physical systems. Introduction into the production of goods, into the movement of goods, and the service of human needs blur the boundaries between the physical, digital, and biological spheres (Maksyutina & Golovkin, 2017).

Entities that are interested in using the above technologies are also foreign trade companies. This subject of legal relations should be interpreted based on the definition of foreign economic activity. According to Federal Law No. 183-FZ of 18.07.1999 (as amended on 8.12.2020) "On Export Control," "foreign economic activity—foreign trade, investment, and other activities, including industrial cooperation, in the field of international exchange of goods, information, works, services, results of intellectual activity (rights to them)." Therefore, a foreign trade company is an organization that carries out activities in the field of cross-border exchange of goods, information, works, services, and the results of intellectual activity (rights to them). The legal regulation of digital technologies of Industry 4.0 in the field of foreign economic activity is a purposeful state social and legal impact. It is carried out on the processes of formation and implementation of cyber-physical systems in the relations of international exchange of goods, information, works, services, results of intellectual activity (rights to them), in the service of human needs, which are implemented by foreign trade companies (Matytsin & Rusakova, 2021).

In the Russian Federation, legislation on expanding the use of digital (computer) technologies based on remote Internet communication in the foreign economic sphere also is being formed slowly and fragmentally. The Decree of the President of the Russian Federation of 21.07.2020 No. 474 "On the National Development Goals of the Russian Federation for the period up to 2030," undoubtedly, has an epochal significance for the country for legal regulation and neo-industrial digital technologies, and also for the digital economy as a whole, because the Decree defines digital transformation as the national goal of Russia's development. The achievement of this goal should be expressed: first, in the "digital maturity" of key sectors of the economy and social sphere, as well as public administration. Secondly, the increase in the share of common socially significant services available in the electronic form to 95%. Third, in the growth of the share of households, those are provided with broadband access to the information and telecommunications network "Internet," up to 97%. Fourth, an increase in investments in domestic solutions in the field of information technology by 4 times compared to the indicator of 2019.

In our opinion, the Decree of the President of the Russian Federation of 21.07.2020 No. 474 is directly related to the regulation of the relations studied in this scientific development since the foreign economic activity of Russian economic entities is certainly one of the key sectors of the domestic economy. As a follow-up to this Decree, a set of 6 federal projects has been developed and is being implemented. In particular, the federal project "Digital Technologies" provides for the creation of "end-to-end" digital technologies mainly based on domestic developments. Also, it provides for the solution of two more tasks. First, the creation of a comprehensive system for financing projects for the development and/or implementation of digital technologies and platform solutions, including venture financing and other development institutions. Secondly, the transformation of priority sectors of the economy, including industry, agriculture, construction, urban development, transport, and energy infrastructure, financial services, through the introduction of digital technologies and platform solutions. In particular, the Federal project "Regulatory Regulation of the Digital Environment" provides for the solution to the problem of creating a system of legal regulation of the digital economy based on a flexible approach in each area. It also provides for the introduction of civil turnover based on digital technologies ("Passport of the national project," National Program "Digital Economy of the Russian Federation" (approved by the Presidium of the Presidential Council for Strategic Development and National Projects, Protocol No. 7 of 04.06.2019).

Following the National Program, Federal Law No. 258-FZ of 31.07.2020 "On Experimental legal regimes in the field of digital Innovations in the Russian Federation" was developed and adopted. Experimental legal regimes in the field of digital innovations can be established following this Federal Law in some areas of development, testing, and implementation of digital innovations, but, unfortunately, the law does not include the direction of foreign economic activity. Therefore, the field of activity is open for Russian legal scholars and legislators to create an appropriate legal regime for the use of neo-industrial digital technologies 4.0 in foreign trade by economic entities of the EAEU and BRICS member states (Inshakova et al., 2020).

At the interstate level, the most authoritative and well-developed source for conducting foreign trade activities is Incoterms®, 2020 ICC Rules for the use of domestic and international trade terms. The Incoterms®, 2020 Rules are an updated version of the Incoterms® 2010 rules and take into account the latest trends and changes in the international trading environment, namely the expansion of free trade zones, the increasing importance of electronic communications in business transactions, changes in transport practices, etc. Incoterms ® 2020 is developed by leading experts of the ICC Commission, working in different parts of the world and representing almost all sectors of trade, which guarantees their universal applicability to business needs. Incoterms®, 2020, provides a clear interpretation of all terms, includes extended instructions and illustrations to help users choose the necessary rule, provides advice on the use of electronic procedures, and describes the possibilities of using Incoterms® terms in domestic trade. (http://iccbooks.ru/catalog/mezhdunarodnyy-biznes/inkoterms-2020-pravila-icc-po-ispolzovaniyu-natsio/). In this series of sources, we should also mention Electronic Bills of Landing, Uniform Rules for Sea Waybills, Rotterdam Rules (https://comite-maritime.org/work). The regulation of new-industrial digital technologies 4.0 related to foreign trade activities is relatively narrow and fragmented in these sources.

Belarus has approved the "State Program for the Development of the Digital Economy and Information Society for 2016–2020" (https://www.mpt.gov.by/ru/gosudarstvennaya-programma-razvitiya-cifrovoy-ekonomiki-i-informacionnogo-obshchestva). In Kazakhstan—the state program "Digital Kazakhstan" for 2017–2020 (https://zerde.gov.kz/images/8B.pdf). In Armenia—"Armenia's Digital Transformation Agenda until 2030" (https://www.gov.am/ru/news/item). In Kyrgyzstan—"TazaKoom" (https://kloop.kg/blog/2018/04/12/). However, there are no documents on neo-industrial digital technologies 4.0 in foreign economic activity at the level of the EAEU and BRICS that are mandatory for all the member states of the integration associations.

The Treaty on the Eurasian Economic Union and its Annexes (http://base.garant.ru/70670880/#ixzz6gzcJ0u7L); Statement on the Digital Agenda of the Eurasian Economic Union (http://pravo.eaeunion.org/documentid=71,553,074&byPara=1&sub=1); Decision of the Supreme Eurasian Economic Council of 11.10.2017 No. 12 "On the main directions of implementation of the Digital Agenda of the Eurasian Economic Union until 2025"(http://pravo.eaeunion.org/document?id=71708158&byPara=1&sub=1); The main directions of the implementation of the digital agenda of the Eurasian Economic Union until 2025 and their Annexes (http://pravo.eae

union.org/document?id=71,708,158&byPara=1&sub=1); Moscow Declaration of the XII BRICS Summit (http://kremlin.ru/supplement/5581); BRICS Economic Partnership Strategy until, 2025 (https://brics-russia2020.ru/images/114/81/1148133.pdf)—it also do not provide a complete legal mechanism for the use of non-industrial digital technologies 4.0 in foreign trade by economic entities of the EAEU and BRICS member states.

Among the official international documents, it is worth mentioning the "Report on the Digital Economy 2019. Value creation and benefits: implications for developing countries. Overview." Its authors are rightly concerned about the quite natural reduction of jobs, cyber threats, over-concentration of profits of monopolies, the growing territorial disparities due to the expansion and intensification of digital technologies in economic activity (https://unctad.org/en/PublicationsLibrary/der2019_overview_ru.pdf).

Foreign colleagues are quite deeply exploring digital technologies of Industry 4.0 (Gilchrist, 2016; Kagermann et al., 2016; Muller & Voigt, 2018; Szalavetz, 2019; Weyer et al., 2015). However, it is quite understandable that the issues of digital support for foreign trade in the EEU and BRICS are not interested at all.

Professor Belikova, referring to the Competition Committee of the OECD Office of Finance and Entrepreneurship, notes that the term "digital economy" is an umbrella term for markets focused on digital technologies, which are usually involved in the trade of information goods or services through electronic commerce. The digital economy operates on a multi-level basis in separate segments connected to separate segments related to data transmission. We cannot fully agree with our esteemed colleague regarding the trade-in information goods or services through electronic commerce because modern foreign trade still represents the movement of huge amounts of material values (things). (Belikova, 2018).

Diachenko O. V., highlighting the process, industry, and technological approaches to the digitalization of industry, as well as system and end-to-end digitalization projects within the framework of the EAEU, differentiates the normative legal acts regulating relations arising in the digital environment. First, these are acts regulating crowdfunding and the turnover of digital financial assets. Second, acts regulating the identification of entities in the digital environment and the requirements for the work of certification centers issuing electronic signatures. Third, acts regulating electronic transactions and electronic archives. Fourth, measures of tax incentives for entities operating in the IT sector. Fifth acts regulating data turnover. We support our colleague's conclusion that the unification of the elements of the digital economy ecosystem and the cooperation of the technological base of the EAEU partner states will be hindered and constrained by the institutional framework. We also support his conclusion that it is necessary to regulate these relations at the supranational level, but there may be procedural difficulties and contradictions in the adoption of a regulatory framework that is mandatory for all members of the EAEU. Let us clarify that in his scientific article, O.V. Dyachenko develops the problems of industry in the economy, without touching on foreign economic activity (Dyachenko, 2020).

As a scientific and methodological basis for a set of measures to form a balanced (taking into account the interests of all participating countries) legal mechanism for the use of neo-industrial digital technologies 4.0 in foreign trade by economic entities of the EAEU and BRICS member states, we propose to apply an open-partnership approach. Within the framework of this approach, the creation and subsequent implementation of each digital technology are differentiated into certain stages, the results of the implementation of these stages are consistently recorded by the relevant departments (ministries) of the countries participating in the integration associations in the mechanism of the inter-jurisdictional poly subject blockchain (Kalinina et al., 2019).

The model for implementing the open-partner approach is as follows. As part of the first stage, on a digital platform independent of any of the member states of the integration associations in the information and telecommunications network "Internet," the specialized commission of the association for foreign trade creates a new root archive directory (folder). The digital platform registers all the archive folders are recorded only if all the subjects of the blockchain agree to this. For example, the root archive folder "Digital marking of goods manufactured for export" is created. As part of the second stage, on the server of each foreign economic agency, in each country participating in the integration association, the same duplicate directory (folder) "Digital labeling of goods manufactured for export" is created. As part of the third stage, the initiator state submits to the digital platform in the root archive folder the primary project text of the document with a description and procedure for using the technology "Digital marking of goods manufactured for export," indicating the IT developer, supplier of this technology, etc. The document in the current text is duplicated in a folder on the server of each foreign economic agency. As part of the fourth stage, each foreign economic agency of the country participating in the integration association finalizes the draft text of the document being created. Initially, the members of integration associations set a limit of five legal iterations, reworking the text to the final content, so that in the end, on the 5th attempt, a consensus was reached and the options received from all participants became the same and turned into a single regulatory document. The inclusion of the texts of the created documents in the root archive folder on the digital platform and in duplicate folders is registered and duplicated at each legal iteration, and thus an open-partner approach is implemented using the inter-jurisdictional poly subject blockchain. The fifth final stage involves the introduction of the final version of the agreed text of the document in the root archive directory (folder) on the digital platform. From the moment of registration of the last consent received by the digital platform from the member country of the association (the subject of the blockchain) with the final version of the text of the document, the specified document becomes legally valid for all member states of the integration association. Initially, members of integration associations set a deadline of five working days for registering consent with the final version of the document text.

In the future, if it is necessary to make changes and/or additions to the current regulatory document on a particular neo-industrial digital technology 4.0 used for foreign trade, the open-partner approach with the help of an inter-jurisdictional poly subject blockchain is repeated.

22.4 Conclusion

The algorithm of the open-partner approach, structured in five stages, which provides for the coordination of versions of the texts of regulatory documents regulating the introduction of neo-industrial digital technologies 4.0 in foreign trade activities in the EAEU and BRICS associations, through several legal iterations, should be applied in the system of inter-jurisdictional poly subject blockchain.

This open-partnership approach, implemented with the help of an inter-jurisdictional poly subject blockchain and used in coordinating the implementation of digital technologies of industry 4.0 in the EAEU and BRICS associations, is proposed to be enshrined in the provisions of the international convention "On Digital Technologies of Foreign Economic Activity used by residents of the Greater Eurasian Partnership jurisdictions." The member states of the EAEU and BRICS integration associations will also be able to join this convention in the system of inter-jurisdictional poly subject blockchain, thereby ensuring the entry into force of the relevant regulatory provisions in their national territories.

Acknowledgements The reported study was funded by RFBR according to the research project No. 18-29-16132 "Priorities for the legal development of digital technologies of foreign trade activities in the context of international economic integration."

References

Armenia's digital transformation Agenda until 2030. https://www.gov.am/ru/news/item. Accessed 2 April 2021.

Belikova, K. M. (2018). Features of the legal regulation of the digital intellectual economy. *Law and Law, 8*, 26–30.

BRICS economic partnership strategy until 2025. https://brics-russia2020.ru/images/114/81/1148133.pdf. Accessed 2 April 2021.

Decree of the President of the Russian Federation of 21 July 2020 No. 474 "On national development goals of the Russian Federation for the period up to 2030". *Rossiyskaya Gazeta* [Russian Newspaper], 159, 22 July 2020.

Digital economy report 2019. Value creation and benefits: implications for developing countries. Review—Geneva: UNCTAD, 2019. 31p. https://unctad.org/en/PublicationsLibrary/der2019_overview_ru.pdf. Accessed 2 April 2021.

Dyachenko, O. V. (2020). Digital interaction of the industry of Russia and the EAEU participating countries. Bulletin of Chelyabinsk State University, 2(436). Economic Sciences, 68, 182–190. https://doi.org/10.24411/1994-2796-2020-10218

Digital transformation program of Kyrgyzstan— "TazaKoom". https://kloop.kg/blog/2018/04/12/. Accessed 2 April 2021.

Electronic Bills of Landing. *Uniform rules for sea waybills; Rotterdam rules*. https://comitemaritime.org/work Accessed: 2 April 2021.

Federal Law No. 183-FZ of 18 July 1999 "*On Export Control*" (as amended on 8 December 2020). RossiyskayaGazeta [Russian Newspaper], 146, 29 July 1999.

Federal Law No. 258-FZ of 31 July 2020 "On Experimental legal regimes in the field of digital innovations in the Russian Federation". *Rossiyskaya Gazeta* [Russian Newspaper], 173, 06 August 2020. Accessed 2 April 2021.

Gilchrist, A. (2016). *Industry 4.0*. Apress.

Incoterms®. (2020). *ICC rules for the use of domestic and international trade terms*. http://iccbooks.ru/catalog/mezhdunarodnyy-biznes/inkoterms-2020-pravila-icc-po-ispolzovaniyu-natsio/ Accessed 2 April 2021.

Inshakova, A. O., Goncharov, A. I., Deryugina, T. V., Stepanova, I. A. (2020). Mediation: legislation and law enforcement activities. In A. O. Inshakova & A. V. Bogoviz (Eds.), *Alternative methods of judging economic Conflicts in the national positive and soft law*. A volume in the series Popkova, E.G. (Ed.) Advances in Research on Russian Business and Management. Information Age Publishing. https://www.infoagepub.com/products/Alternative-Methods-of-Judging-Economic-Conflicts-in-the-National-Positive-and-Soft-Law. Accessed 2 April 2021.

Kagermann, H., Anderl, R., Gausemeier, J., Schuh, G., & Wahlste, W. (2016). *Industrie 4.0 in a global context: Strategies for cooperating with international partners (acatech STUDY)*. Herbert UtzVerlag.

Kalinina, A. E., Inshakova, A. O., & Goncharov, A. I. (2019). Polysubject jurisdictional blockchain: Electronic registration of facts to reduce economic conflicts. *Studies in Computational Intelligence, 826*, 205–213.

Maksyutina, E. V., & Golovkin, A. V. (2017). Neoindustrialization of the Russian economy based on technologies of the fourth Industrial Revolution and human capital development. *Scientific and Technical Bulletin of the St. Petersburg State Polytechnic University. Economic Sciences, 10*(1), 43–52.

Matytsin, D. E. & Rusakova, E. P. (2021). The strategy of quality management in Industry 4.0 and formation of the cognitive economy based on industrial and manufacturing engineering in the Russian Federation and countries of the EU. *International Journal for Quality Research, 15*(4),https://doi.org/10.24874/IJQR15.04-03

Moscow Declaration of the XII BRICS Summit. http://kremlin.ru/supplement/5581. Accessed 2 April 2021.

Muller, J., & Voigt, K. (2018). Sustainable industrial value creation in SMEs: A comparison between Industry 4 and Made in China 2025. *International Journal of Precision Engineering and Manufacturing-Green Technology, 5*(5), 659–670.

Passport of the national project "National Program "Digital Economy of the Russian Federation" (approved by the Presidium of the Presidential Council for Strategic Development and National Projects, Protocol No. 7 of 04.06.2019). URL: https://digital.gov.ru Accessed 2 April 2021.

State program for the development of the digital economy and information society for 2016–2020. Belarus. https://www.mpt.gov.by/ru/gosudarstvennaya-programma-razvitiya-cifrovoy-ekonomiki-i-informacion-nogo-obshchestva. Accessed 2 April 2021.

Statement on the Digital Agenda of the Eurasian Economic Union. http://pravo.eaeunion.org/document?id=71553074&byPara=1&sub=1

Szalavetz, A. (2019). Industry 4.0 and capability development in manufacturing subsidiaries. *Technological Forecasting and Social Change, 145*, 384–395.

The decision of the Supreme Eurasian Economic Council of 11 October 2017 No. 12 "*On the main directions of implementation of the Digital Agenda of the Eurasian Economic Union until 2025*". http://pravo.eaeunion.org/document?id=71708158&byPara=1&sub=1. Accessed 2 April 2021.

The main directions of the implementation of the digital agenda of the Eurasian Economic Union until 2025 and their Annexes. http://pravo.eaeunion.org/document?id=71708158&byPara=1&sub=1. Accessed 2 April 2021.

The state program "Digital Kazakhstan" for 2017–2020. https://zerde.gov.kz/images/8B.pdf. Accessed 2 April 2021.

The Treaty on the Eurasian Economic Union and its Annexes. http://base.garant.ru/70670880/#ixzz6gzcJ0u7L. Accessed 2 April 2021.

Weyer, S., Schmitt, M., Ohmer, M., & Goreck, D. (2015). Towards Industry 4.0—Standardization as the crucial challenge for highly modular, multi-vendor production systems. *IFAC-PapersOnLine, 48*(3), 579–584.

Chapter 23
The Principle of Information Transparency of Rulemaking and Law Enforcement Activities in the Republic of Belarus in the Socio-economic Domain

Grigory A. Vasilevich

Abstract The paternalistic type of relations between the state and the person, the state and society, is replaced by a new type of relationship between these entities. The growth of legal culture, legal consciousness of citizens, is changing the previously established paradigm of relations: the number of people who recognize themselves as full-fledged subjects of the political process is growing. The state's sustainable socio-economic development is directly dependent on how rulemaking and law enforcement effectively affect social processes, to what extent citizens' expectations from public authorities are justified. The development of information technologies significantly changes citizens' role, public organizations in making government decisions, and significantly impacted this process. Public administration is carried out through the adoption of legislative acts and law enforcement. Information transparency of actions in these domains is an essential factor in the legitimacy of government bodies.

Keywords Rulemaking · Law enforcement · Transparency · Principles

JEL Codes K00 · K20 · K40 · O10 · O20

23.1 Introduction

On the role of law in the management of state and public affairs.

Forming an effective state tends to be successful when the economy and social policy are in harmony with constitutional principles and legal values. The management of state and public affairs is successful when the law considers the laws of social development. Public administration is carried out through rulemaking and law enforcement activities. As noted, some features are inherent in public administration, among which it distinguishes (1) continuity and efficiency of activities; (2) functional

G. A. Vasilevich (✉)
Department of Constitutional Law, Belarusian State University, Minsk, Belarus
e-mail: Gregory1@tut.by

A. O. Inshakova and E. E. Frolova (eds.), *Smart Technologies for the Digitisation of Industry: Entrepreneurial Environment*, Smart Innovation, Systems and Technologies 254, https://doi.org/10.1007/978-981-16-4621-8_23

specialization; (3) the presence of legal and functional regimes; (4) application of measures of responsibility; (5) the activities of the management apparatus, referring to its subordination; (6) availability of professional staff; (7) administrative discretion (Tikhomirov, 1998).

Many scientists pay attention to the applied or instrumental nature of positive law (Bondar, 2006; Chernogor & Zaloilo, 2018; Khabrieva & Chirkin, 2007; Tikhomirov, 2007, 2010, 2011). Belarusian scientists also consider this item thoroughly. For example, in 2018, a book was published in Belarus dedicated to forming our state as a legal and social state (Vasilevich & Nikitenko, 2018).

Modern law pays close attention to its development's objective laws, which are reflected in the legal principles (Gadzhiev, 2004). Some scholars emphasize that the law could "normatively anticipate reality" (Tikhomiro, 2007). However, scientists usually emphasize the general importance of law principles (their influence on the entire sphere of regulation, including the development of subsidiary regulation, resolution of conflicts, and others) (Talapina, 2015). They do not always focus on how specific principles, such as the principle of information transparency, affect relations between various law subjects. From the same legal science positions, public administration's information transparency is considered a public law principle. Of course, the position of E. V. Talapina is fair in the sense that there has been a "conceptual revision of the scheme of relationships between subjects in the exercise of the right to information, namely, the transition from the traditional exercise of the citizen's right to receive information to the activity of the state authorities themselves in disclosing information about their activities" (Talapina, 2011).

23.2 Materials

The study was based on the Republic of Belarus's Constitution, the provisions of the Law "On Normative Legal Acts" and other Belarusian legislation acts.

The theoretical basis of the article was the research of such scholars as Tikhomirov (2007, 2010, 2011, 2018), Khabrieva and Chirkin (2007), Bondar (2006), Gadzhiev (2004, 2020), Talapina (2011, 2015), Bachilo (2017), and some others.

Thanks to the doctrinal research of such authors as Matuzov and Malko (1999), Tikhomirov (2007), Kovalenko (2019), Talapina (2011, 2015), Gadzhiev (2020), Inshakova et al. (2020). We reveal the influence of legal principles on the formation of the national legal system, special attention is focused on information transparency of the rulemaking and law enforcement activities of public authorities.

The conclusion is made that two areas are essential for sustainable development: (1) the transition from the traditional realization of the citizen's right to receive information to the activity of the state authorities in disclosing information about their activities; (2) reverse impulses—from citizens toward the authorities. The effectiveness of law in ensuring the individual's development depends on many factors, including citizens' expectations of certain state decisions. That is, we can talk about the effectiveness of the legal regulation of socio-economic relations. After all, the more closely the decisions of state power institutions correlate with these expectations, the more successful their implementation is.

23.3 Methods

The study's methodological foundations were formed by analysis and comparative legal methods, thanks to which it was possible to identify the main problem area in rulemaking and law enforcement to make several proposals for improving legislation and practice.

23.4 Results

It seems to us that the reverse impulses are crucial—from citizens toward the authorities. The effectiveness of law in ensuring the individual's development depends on many factors, including citizens' expectations of certain decisions by the state. That is, we can talk about the effectiveness of the legal regulation of socio-economic relations. After all, the more closely the decisions of state power institutions correlate with these expectations, the more successful their implementation is. The Constitution of the Republic of Belarus enshrines the guarantees of citizens' right to receive, store and disseminate complete, reliable, and timely information on the activities of state bodies, public associations, political, economic, cultural, and international life, and the state of the environment. State bodies, public associations, officials are obliged to provide access to a citizen of the Republic of Belarus with materials affecting his/her rights and legal interests.

The famous Russian scholar N. I. Matuzov distinguishes three models of interaction between state and law: totalitarian-statist, liberal-democratic and pragmatic. The first model states that the state rises above the law and is not bound by it. This model was typical for the Soviet period. According to the second model, the law prevails in the state. The state considers this model as ideal and unattainable at present. As for the pragmatic model, the state creates law and, at the same time, is bound by it, is self-limited for the sake of the common good (Matuzov & Malko, 1999).

In modern conditions, the paternalistic type of relations between the state and the individual, the state, and society is becoming obsolete: a new character of relations between these subjects is being built (Inshakova et al., 2020). The growth of legal culture, legal consciousness of citizens is changing the previously established paradigm of relations; the number of people who recognize themselves as full-fledged subjects of the political process is growing. In our opinion, the state's sustainable socio-economic development is directly dependent on the extent to which rulemaking and law enforcement effectively affect social processes. Therefore, as N. S. Bondar mentions the relevance of legal regulation to the needs of social development, the predictability of the development of law, equality of all before the law and the court, guaranteed justice (the right to an objective, impartial and independent court), the right to qualified legal assistance, to the presumption of innocence and other factors that ensure public consensus become a sensitive issue (Bondar, 2006).

The purpose of state activity is to serve people and create conditions for a decent life. Such criteria can assess the effectiveness of such activities as to whether the life expectancy, quality, and people's satisfaction with the state's services are increasing, whether the interests of society and the individual are considered, the level of protection of rights and freedoms is. In world practice, there exist ratings of state assessments. Among them, one can name not only the traditionally mentioned human development index, the rating of countries by income per capita, and others, but also the rating of countries by the level of happiness. Measurements carried out by sociological services in various countries indicate that happiness does not depend only on material security. It is more important to provide a sense of justice, freedom, equality. The authorities should hear the people, respond to reasonable proposals, provide an opportunity for opponents to legally express their points of view, and convince them of their policy's correctness through arguments. At the same time, according to the outstanding Russian scholar I. A. Ilyin, one of the main requirements is that "each of us must learn to distinguish the matter of justice from the matter of personal interest and not cover up our self-interest with the declamation of justice".

Transparency of public administration is an industry-wide principle of legal regulation. We will pay attention to the definition of the Constitutional Court of the Russian Federation, which in the context of access to information, in the Decision dated December 8, 2011 No. 1624-0-0 on the complaint of T. A. Andreeva, F. V. Morozov and others for violation of their constitutional rights, paragraph 1 of Art. 1 of the Federal Law "On ensuring access to information on the activities of state bodies and local self-government bodies" indicated the following: "Such legislative regulation is aimed at creating conditions (guarantees) that ensure maximum information transparency of state bodies and local self-government bodies for citizens and other civil society actors, and is consistent with the Convention on Access to Official Documents adopted by the Council of Europe on November 27, 2008 (preamble, Art. 3), which emphasizes the particular importance of transparency in the activities of state bodies in a pluralistic and democratic society and which, at the same time, does not exclude restrictions on the right to access official documents, provided that these restrictions are clearly established by law, are necessary and proportionate to the goals of protection generally recognized in a democratic society of values.

The critical constitutional principle reads that state power comes from the people (Art. 3 of the Constitution of the Republic of Belarus). Practice shows that a violation of the principles of democracy and parliamentary (representative) democracy can be the driving motives of protest moods. Therefore, it is imperative to establish such a practice in everyday reality that would indicate that people are the main subjects of relations developing in the state. These goals are served by improving the rulemaking process, involving citizens in it, identifying, and considering the people's will when solving public life's most critical issues.

It is essential to ensure the transparency of the preparation, adoption, and implementation of regulatory legal acts to find the optimal balance of public and private interests, their proper legal regulation. It makes it possible to prevent possible collisions and other defects in rulemaking, carry out public (professional) expertise of projects, and predict possible consequences connected with the introduction of legislative acts. To a greater extent, it ensured the "acceptance" of public authorities' decisions by society, which is no less important than compliance with many legal formalities, albeit quite reasonable.

We agree with those scholars who argue that law is a direct or indirect cause of changes in the socio-economic sphere, and, conversely, the trends in its development are a direct or indirect cause of changes in the law. The fact that the level of a country's economic development affects its legal system became apparent back in the nineteenth century. The fact that the legal system itself, individual legal principles, judicial decisions, and other aspects of law functioning have clear economic explanations has become apparent due to economists and legal scholars' research over the past few decades.

Over the last three decades, information and communication technologies (ICT, IT) are being introduced in everyday life at an accelerated pace. IT's revolutionary impact concerns government structures and institutions of civil society, economic and social domains, science and education, culture, and daily life. The movement along the path of development of today's information society has become a vector of modern life (Mamychev et al., 2020). The desire to anticipate the future is characteristic of the country's leadership pursuing a corresponding course. This is evidenced by decisions taken back in the early 2000s, as well as Decree No. 8, of December 21, 2017, "On the development of the digital economy". Thanks to the adopted in the Republic of Belarus in the early 2000s programs for the development of the information society ("Electronic Belarus") and some legislative acts, state bodies' network infrastructure is actively developing to ensure automated information interaction between them based on the formation of a single national information resource. Law is also one of those areas where they play a very positive role, contributing to a closer and more effective settlement of state and public issues. The use of information technologies makes it possible to de-bureaucratize many relations; IT can radically affect relations between the state and the individual and promote the expansion of rights and freedoms. In such conditions, while maintaining its social orientation, the state is increasingly fulfilling its service role (Bachilo, 2017). One of the results of the introduction of information technologies can be providing comprehensive information openness of the procedure for the preparation and adoption of acts by government bodies, which means great opportunities for citizens to exercise their right to manage state affairs. The development of rulemaking activity's information infrastructure is an essential condition for high quality and timely interaction between government institutions and citizens (Kuznetsov, 2017). The use of ICT should be more actively developed to discuss draft regulations, consider citizens' appeals, and consult them. More than fifteen years ago, the author proposed to use information technologies in the legislative sphere to publish draft normative acts on Web sites for the broad public to be acquainted with them.

Another point was obliging the rulemaking bodies to disseminate the adopted acts via the Internet, which would help ensure the constitutional right to publish or share the content of acts of legislation, as required by Art. 7 of the Constitution (Vasilevich, 2004). Now, this practice has been developed and consolidated in legislation. However, there is still no full feedback from citizens, their public organizations that have expressed their proposals: after the project's discussion, at least a brief information on which proposals have been considered, which have been rejected, and for what reasons—should be posted on the Web sites.

The era of information technology fundamentally changes the content of many previously established relationships. We share the position that the dynamic development of information technologies, extensive access to them by the population has become "a catalyst for the transformation of the established social, economic, legal and managerial models of the twentieth century; the new situation required that the state search for effective means to solve a wide range of managerial tasks and functions" (Zhuravleva, 2002). As the Academician of the Russian Academy of Sciences T. Ya. Khabrieva states that law sources' digitalization and placement in virtual space pose several research tasks (Khabrieva & Chirkin, 2007).

The UN suggests considering e-government as "a new level of digital interaction between public authorities of various branches of government, citizens, organizations and other entities, based on the use and application of information technologies in public administration by public authorities to optimize and integrate processes and procedures for data and information effective management, improving the quality of public service delivery and expanding communication channels to involve people in the political decision-making process".

The international experience shows that law's digitalization ensures a more dynamic development of relations between the state, society, and people. In this regard, ideas are expressed that artificial intelligence will replace lawyers in preparing draft acts, including draft laws. It can be noted that artificial intelligence is extensively penetrating social life; one of its "peak" manifestations used to be playing chess against artificial intelligence. Under the newly created programs, a human cannot defeat artificial intelligence. It suggests that artificial intelligence, created with the help of a team of information specialists, lawyers, sociologists, specialists in the field of social psychology, can provide significant assistance, if not in creating ideal drafts of legal acts, but at least—in their expertise, identifying gaps and inaccuracies, contradictions, other defects that may have an adverse impact. At this stage, information technology can primarily be used for these purposes. Artificial intelligence could be used to prepare a draft version of a future project for legal expertise. However, the final legal assessment at this stage of development should be carried out by legal professionals.

In the legal literature, attention is rightly drawn to the redundancy of legal regulation, the instability of legislation, its constant adjustment, inconsistency, gaps, and other defects in legal norms. These topics are usually the focus of scholars and practitioners' attention who regularly discuss them at conferences, seminars, and round tables. There are indeed reasons for this. If the early 1990s of the XX century lacked contemporary acts corresponding to the new vector of development, now the emphasis should be shifted toward more excellent stability of the established legal norms. The above said is necessary so that citizens, legal entities, courts, and other law enforcement agencies make their plans, including economic ones, at least for the near future, and be sure that they will not undergo unexpected changes. The stability of legal regulation contributes to the strengthening of the rule of law and law and order, as it makes possible the formation of a unified law enforcement practice to ensure the equality of all before the law and the court. So far, it has not been possible to ensure the high quality of legislation, which leads to constant changes and additions. Prominent scholars put forward constructive ideas about new legal technologies that should be used to draft regulatory legal acts.

It has long been the task of providing a package principle for the preparation of regulatory legal acts. In this regard, we believe that when developing a draft law or other legislative act or defining their concept, representatives of the relevant department should be included in the working group. Simultaneously, with preparing a draft legislative act, it is advisable to develop a draft departmental act. Such work could be incredibly intensive after the draft law's approval in the first reading in the House of Representatives of the National Assembly of the Republic of Belarus. In our opinion, it would be helpful if, within the framework of the head commission responsible for preparing the draft law, we discuss the main provisions of the future departmental act when the need for its adoption arises from this draft (the same should be done if the adoption of a government act is expected). The exchange of information between these authorities (Parliament, Government, and Ministry) can also be organized when developing a set of acts using information technology. Information interaction of executive authorities with other rulemaking entities is crucial for improving the rulemaking quality.

Creating a system for collecting, registering, and processing regulatory legal acts in the ministry and state committee can increase the rulemaking process's efficiency through information technology. Within the ministry framework, the state committee would help create a database for monitoring law enforcement, analyzing departmental regulations, citizens' appeals, and electronic versions of documents.

A dossier of significant and vital regulatory legal acts should be formed. It should include information about the initiators and developers of the project, the persons who approved the project, the results of legal, including anti-corruption, expertise, the persons who carried them out, the forecast of regulatory impact, the results of legal monitoring.

The creation of a system for ensuring rulemaking activities, which allows deploying electronic versions of documents, and draft documents' tracking through all stages of the rulemaking process, with the obligatory connection of all rulemaking initiatives' subjects, is a positive experience.

The creation of the Internet systems for posting information on state authorities' activities, a system for recording and monitoring the results of work with citizens' appeals, helps increase the efficiency of public authorities' law enforcement activities. In the future, it is desirable that the heads of departments, directly or using information technologies, communicate with citizens, answer their questions, and complaints. It can intensify the communication with people and solve their problems. In the Russian Federation, communication with citizens through video blogs is becoming a reality.

The National Center for Legal Information of the Republic of Belarus (NCLI) has created a system for forming the National Register of Legal Acts of the Republic of Belarus (AIS NRLA). As the director of the NCLI Kovalenko noted at an international conference, an automated information system is being developed to ensure the rulemaking process (AIS "Rulemaking"). It "will allow the automation of the draft normative legal acts movement at all stages of the rulemaking, including the preparation of a draft text, its approval, appropriate expertise, public discussion, signing of a legal act, and its official release. This system will make it possible to ensure a complete and transparent cycle of preparation of a draft normative legal act and its adoption. It will also allow to drastically reduce correspondence between state bodies, simplify, digitalize and, as a result, speed up the rulemaking process" (Kovalenko, 2019).

The Law of the Republic of Belarus "On Normative Legal Acts" indicates such an essential direction of the legislative process as forecasting the consequences of adopting a normative legal act, assessing its regulatory impact. Of course, when carrying out predicting the consequences of legal regulation of relations, it is necessary to analyze big data. However, it is precisely in the era of information technology that this is possible. Computers are capable of processing them efficiently.

Professor Yu. A. Tikhomirov singles out extensive backbone laws as objects of legal forecasting; institutions of the subsector and branch of legislation, the prospects for their action in the socio-economic context: the system of legislation; tendencies of legal regulation in certain spheres of state and public life and determination of the optimal legal solution option; study and assessment of the development of public law institutions. Simultaneously, Prof. Tikhomirov points out that similar phenomena and processes of development of foreign states' legal systems and international legal features should also be the object of forecasting (Tikhomirov, 2018).

Problems with the forecast of legal regulation are often visible when the rulemaking bodies promptly did not anticipate the events' development and did not establish an optimal transition period for preparing citizens, entrepreneurs to changing, new conditions with the entry into force of the act. In our opinion, to put into effect some critical decisions, such transition periods could be not days or months, but even years: a person, planning his life, activities, must be sure that this area will not be exposed to sudden changes. With this approach, trust in the authorities and the state only increases.

In the legal literature, much attention has been paid to considering possible risks connected with adopting and implementing an act of legislation. Russian authors rightly note that "the inability and unwillingness to objectively assess the situation, identify the factors influencing its development, determine the goals and means of legal influence, haste in the preparation and adoption of acts, the flow of amendments and changes, lack of considerations and analytical information, underestimation of the enforcement mechanism, lack of scientific foresight" are the reasons for the shortcomings of legal regulation of both the previous and modern periods (Tikhomirov, 2010).

In this regard, artificial intelligence can also become a tool that allows predicting all possible risks and offer the best solution. Currently, information technologies can still play a supporting role. In addition to forecasting and examining acts, they create conditions for public discussion of projects to manifest norm-setting initiatives by citizens. Turning to the latter aspect, we note that the Belarusian legislation in this part needs a corresponding amendment.

The rulemaking process's information transparency is an essential condition for the scientific analysis of the draft projects. It can be carried out both under the initiative at the request of individual representatives of science, setting out, for example, their position in the media, or on a planned basis, since following the current Law "On Normative Legal Acts", the scientific approach is considered one of the main principles of such a process. In this regard, it is necessary to formulate a scientific methodology for assessing the state of legislation. The following can be used as relevant indicators: (1) the degree of reflection of constitutional principles (e.g., on the retroactive effect of laws); (2) the presence of basic laws (e.g., software); (3) uniformity of development of branches of legislation; (4) adequacy to the tasks of economic, social and political development, ensuring the rights of citizens; (5) openness and accessibility of legislation; (6) reduction of offenses; (7) Compliance with the principles and norms of international law.

A thorough scientific discussion regarding the concept of essential bills and other acts, their key provisions can draw project developers' attention to those norms that need to be adjusted.

The law "On Normative Legal Acts" names the acts adopted by various bodies. In practice, programs, concepts, strategies, and alike are often adopted. Moreover, some of these acts, for example, programs, can be adopted by the Government and local authorities. The named law does not define these acts. One can often come across a statement that these are acts of the so-called soft law. We can partly agree with this. However, a violation of the requirements (provisions) of these documents (concepts, programs, and others), deviation from the benchmarks (indicators) contained in them can be a legal basis for bringing the perpetrators to justice, primarily in executive power. Discussions of draft programs and concepts should become a common practice both at the republican and local levels. After all, programs are developed to achieve socio-economic development *priorities and goals* (Kazantsev, 2019). Not only individual citizens but also private campaigns can make a significant contribution here. We believe that there should be a dossier of law or another regulatory legal act in the public domain. We mean information about the persons who developed the act's concept, the list of members of the working group that prepared the project, the list of persons who endorsed it. A forecast of the consequences of adopting the act signed by the developers, its regulatory impact on public relations must be attached. We share the opinion expressed in the media and legal science on increasing the preamble of a normative legal act. The preambles of significant acts that significantly affect the legal situation should give a brief analysis of the state of legal regulation, the goals that are set in connection with its adoption, and the forecast of their achievement (In this regard, the directives of the Head of State can be an example).

These essential components can significantly enhance the effectiveness of legislation and the practice of its application, increase responsibility for inadequate preparation of projects and ensure public demand for their possible low quality.

A more comprehensive introduction of information technologies can make preparation, adoption, assessment (expertise) of draft acts more cost-efficient. The transparency of the conclusions regarding the forecast of the consequences of adopting legal acts will serve the correct choice of the extent of their impact on public relations and legal regulation optimization. Behind the purely material consequences that may occur in connection with the adoption of the act, one must see other components of this process, including how the new rules will affect the business activity of citizens, entrepreneurs, their mood, physical and psychological mood, and their sense of justice and rationality of the adopted solutions (Inshakova et al., 2018). The forecast, in our opinion, should first reflect the positive results that can be obtained following the adoption of the new act. Any act always entails social consequences, so their forecast should always be. The idea of forecasting is to ensure reasonable legal regulation, and even if the population is not happy with the adopted but necessary decision, one must see the possible social consequences. This notion is especially true for significant acts or those that fundamentally (substantially) change legal rules.

Resolution of the Council of Ministers of the Republic of Belarus of 25.01.2019, No. 54 "On predicting the consequences of the adoption (promulgation) of regulatory legal acts" approved the Instruction for predicting the consequences of the adoption (promulgation) of regulatory legal acts. According to the Instruction, forecasting the consequences of the adoption (promulgation) of a normative legal act includes forecasting:

- financial and economic consequences of the adoption (promulgation) of a regulatory legal act to assess the impact of a regulatory legal act on the amount of income, expenses, and (or) sources of financing the deficit (guidelines for using the surplus) of the republican budget, local budgets, budgets of state extra-budgetary funds for the current financial year and medium-term perspective, as well as a general assessment of the consequences of the adoption (promulgation) of a normative legal act for the income and expenses of citizens and legal entities;
- the regulatory impact of the draft normative legal act on the conditions for carrying out the entrepreneurial activity to identify provisions establishing excessive obligations, prohibitions and restrictions, unreasonable costs for legal entities and individual entrepreneurs;
- social consequences of the adoption (promulgation) of a normative legal act to assess the impact of a normative legal act on the legal status of citizens, the process of their interaction with state bodies (organizations), the level of income, and the quality of life of citizens;
- the environmental consequences of the adoption (promulgation) of a regulatory legal act to identify those provisions in the draft of the regulatory legal act, the

implementation of which may harm the environment and (or) is associated with the irrational use of natural resources;
- other consequences that will arise due to the adoption (promulgation) of a regulatory legal act.

It is stipulated that forecasting the consequences of the adoption (promulgation) of a normative legal act is carried out at the stages of planning a normative legal act and preparing a draft normative legal act.

An accurate assessment of the regulatory impact allows government agencies to see the consequences of their decisions and, as correctly noted by Belarusian scientists, "introduce effective regulatory elements" (Shablinskaya, 2020).

One of the crucial scientific guidelines should be the development of the theory of economic analysis of law. Without touching broadly on the problem of goal setting in law, we note the importance of such an approach, according to which the assessment of positive law is carried out in terms of its economic efficiency (Matytsin & Rusakova, 2021). This referral's significance is predetermined by the positive results of a theoretical and applied nature that can be obtained by providing a symbiosis of legal thought and economic science. Specific measures should be taken, starting with individual acts and branches of law. Specialists in information technology, sociologists, specialists in psychology, including social psychology, should be involved in this task. It is necessary to develop an algorithm for conducting such an analysis, a methodology for modeling the consequences of the adoption and performance of acts, to reveal the possible impact on society's political, economic, social, and spiritual sphere.

Analysis of legal literature shows that the study of efficiency is mainly carried out within the procedural law framework: more often, it is about the principle of procedural efficiency when considering criminal cases (Nepomnyashchaya, 2010) and civil cases (Sorokin, 2009). The same applies to legal liability. The principle of efficiency of criminal law repression is also applicable here. It is rightly believed that the principle of repression efficiency presupposes compliance with lawmaking and law enforcement measures. In civil law, attention is drawn to the fact that the implementation of civil rights and obligations is subordinated to the principle of efficiency (Yagofarova, 2004).

Legal regulation's efficiency is manifested in establishing certain boundaries within which one can act but which cannot be violated.

Currently, the problem of interpretation (clarification) of normative legal acts has become urgent again. The need to streamline this work, the solution of several controversial and ambiguously implemented in practice issues, is actively raised by representatives of business entities, specialists, and citizens who have had to face legislative acts in their lives. They draw attention to the fact that quite often representatives of state bodies (officials) give answers to their requests, legal entities and individuals act according to the clarifications received, and then a higher organization or regulatory (supervisory) bodies, disagreeing with this clarification, hold liable a legal or natural person for failure to comply with an act of legislation. Questions also arise about where it is possible to challenge the clarification given to the subject, with which he disagrees, in what form the interpretation (clarification) should be given, what are the consequences in connection with the new "turn" of application practice because of this interpretation (clarifications), and alike. These issues are essential for practice since defects in their solution give rise to many disputes, introduce instability elements, undermine confidence in the rulemaking body, law enforcement agencies, entail property, and other liability. The way out is the creation of a system of administrative justice and the introduction of the institution of a constitutional complaint.

If you get acquainted with the acts of legislation posted on the National Legal Internet Portal of the Republic of Belarus or in the Consultant Plus database, you can see a very diverse picture, which, in our opinion, is undesirable. Thus, some state bodies, giving explanations and posting them on their Web sites or in the Consultant Plus database, do not indicate which of the officials explains, the name of the act (document), its other details (e.g., the date of adoption). There are examples when "clarification" on important issues is given on behalf of government bodies' press centers. Of course, in this case, a certain amount of assistance is provided to those who raised the question and other potential performers of this act. However, as noted above, a higher level of the legal order is inherent in the rule of law. It is especially worrisome for subjects engaged in entrepreneurial activity, which often, contrary to this explanation (response) from a state body, with which a higher authority or a supervisory body does not agree, brings them to legal responsibility.

When, because of the interpretation of an act, rights and freedoms are limited, and new obligations arise, then the act of interpretation must be put into effect at least from the day of promulgation. It is necessary to streamline the clarifications of acts, determine the persons authorized within the department to implement them, and exclude citizens and business entities' responsibility if they followed this explanation of the department representative earlier (Vasilevich, 2019). One of the options for improving the situation could be adopting a resolution of the Government of the Republic of Belarus aimed at streamlining the practice of giving clarifications of regulatory legal acts by the executive authorities subordinate to the government. It could enshrine several fundamental rules that should be followed.

Information technologies make it possible to convey information about the adopted regulatory legal acts to citizens and business entities more quickly. The author proposed "on the entry of normative acts after their official publication in electronic form" (Vasilevich, 2005). At the same time, it is advisable to publish the essential departmental acts concerning citizens' rights and freedoms in print media. Here it is necessary to exclude the so-called digital discrimination when some citizens do not have a daily opportunity to use Internet services.

Transparency is essential not only in the rulemaking process but also in law enforcement. Information openness can be ensured by previously established standards (publication of acts, informing representative bodies about their activities at relevant meetings, answering inquiries and appeals of deputies, and others) and expanding the information field through information technology.

The Constitutional Court of the Republic of Belarus was among those state bodies that have long appreciated IT's potential. From the point of view of direct contact of new technologies and precedents of the Constitutional Court of the Republic of Belarus, it is interesting its Conclusion of October 25, 2004, "On the Compliance with the Constitution and legislative acts of the Republic of Belarus, international legal acts ratified by the Republic of Belarus, decisions of the State Tax Committee of the Republic Belarus dated May 8, 2001, No. 62 "On the procedure for paying tax on income of foreign legal entities receiving income from operations with securities from sources in the Republic of Belarus", and dated May 25, 2001, No. 72 "On approval of the Instruction on taxation of "other income" of foreign legal entities that do not operate in the Republic of Belarus through a permanent establishment." For the first time, there was a mention of the placement on its official Web site of inaccurate information by a party to the case (public authority), which led to the defendant's onset of adverse consequences. Having studied the practice, the Constitutional Court noted that the situation with the application of specific amounts of tax rates was significantly aggravated by the fact that the official Web sites of some state bodies, responsible for conducting economic and tax policy, and in the official printed edition of the Bulletin of the Ministry of Taxes and Duties of the Republic Belarus, aimed at taxpayers, contained information not only on the tax rates on the income of banks, non-bank credit, and financial institutions, insurance, and 23 reinsurance organizations but also a position on the payment of income tax of a specific subject—a foreign legal entity through a permanent representation in a smaller size than it was established. For the first time in its practice, the Constitutional Court, in its reasoning part, referred to an information resource located in the global network and considered the posting of information on it significant, therefore, important for making a final decision. Another interesting example of the "intersection" of information technologies with the practice of the Constitutional Court was the Conclusion of September 27, 2002, "On the Compliance of the Constitution of the Republic of Belarus" with Part Two of Art. 6 of the Law "On the Procedure for Exit from the Republic of Belarus and Entry into the Republic of Belarus of Citizens of the Republic of Belarus", Clause 13 of the Rules for Processing Documents for Traveling Abroad for Citizens of the Republic of Belarus, Subclause 25.1 of Clause 25 of the Instructions on the "Procedure for Issuing a Passport of a Citizen of the Republic of Belarus for Traveling Abroad in terms of establishing five years of validity of a mandatory mark in the passport of a citizen of the Republic of Belarus, temporarily leaving Belarus." The Constitutional Court recognizes that the general civil passport valid in the Republic of Belarus complies with the established requirements and is, according to the law, valid for leaving the Republic of Belarus to all countries of the world. At the same time, it indicated that the Republic of Belarus did not have a proper, technically complete computer database of persons with restriction of their right to free travel (i.e., convicts, committed in crimes, wanted persons, and alike), which must be present at all border crossings of the republic, and suggested that the relevant authorities carry out work within a reasonable time to create this database. In this regard, the global trend of tightening the requirements for identity documents is also interesting. The terrorist acts of September 11, 2001, in the USA, the growth of acts of terrorism radically changed the approaches to ensuring security in most foreign states. The result was the newly increased requirements for passports, including the availability of new information (fingerprinting, retinal scanning, and others).

The potential of information technology for practical application is immense. Nowadays, according to the experience of other countries, they could be used to prevent crimes. Of course, it is still difficult to imagine that appropriate chips can be used to monitor children's whereabouts, but the chips can be used to find stolen vehicles. In 2005, the author suggested that in the future, certain administrative cases, concerning, for example, traffic violations, would be considered based on the developed computer program. In this case, there would be less abuse. Certain elements have already been introduced when fixing speed violations and bringing administrative responsibility. At present, we believe it is necessary to take measures for possible consideration using artificial intelligence of cases of customs and tax offenses. It is in this area that legal relations should be the most "formalized:" rights, obligations, responsibility (circumstances mitigating and aggravating responsibility) should be clearly "described" in the law. These measures will reduce corruption in these areas, including when resolving conflict situations. Of course, a citizen who disagrees with the "cold" calculation of the machine should have the right to file a complaint with a judge in its traditional sense. When resolving cases related to customs or tax offenses, "improvisation" is hardly needed. Business entities complain more about the unacceptable "improvisation" of the regulatory authorities. For the latter, a program can also be created, the data of which could indicate the need to conduct an audit or, conversely, to refuse to conduct it. Legality will also be ensured by the transparency of judicial practice and recognizing the legal (judicial and administrative) precedent as a source of law.

Artificial intelligence assistance is already possible when advising on legal issues; going to courts using IT is already gaining ground. For example, for typical questions, there may be databanks of typical answers. The possibility of utilizing artificial intelligence when considering legal disputes is evidenced by the fact that artificial intelligence can create works of art that are indistinguishable from human works and sometimes even surpass them, i.e., it is capable not only of learning but also of improvisation. Artificial intelligence can help a citizen (client) predict a court case's consequences: determine the percentage of success in his/her case. It may facilitate the adoption of a decision to apply to a court or other body competent to consider a legal conflict or refuse such an appeal.

23.5 Conclusion

In the context of the expansion of information transparency of state institutions and IT, the law's organizing role increases significantly. The life of society is influenced by many factors, both internal and external. This fact necessitates their systematic analysis. The transparency of social processes makes it possible to have accurate baseline data for such an analysis.

We share the view of the judge of the Constitutional Court of the Russian Federation, Professor G.A. Gadzhiev, who emphasizes the importance of new technologies for ensuring freedom and equality, reliable guarantees for the protection of human rights. The creation of digital technologies presupposes state institutions' functioning aimed at a new type of social organization of society, which is characterized by the collection, processing of data, and management based on the analysis of a vast database (Gadzhiev, 2020).

From our point of view, two guidelines are essential for the sustainable development of the state and society: (1) the transition from the traditional implementation of the citizen's right to receive information to the active steps of the state authorities in disclosing information about their activities; (2) reverse impulses—from citizens toward the authorities.

The effectiveness of law in ensuring the individual's development depends on many factors, including citizens' expectations of certain decisions on the part of the state. That is, we can talk about the effectiveness of the legal regulation of socio-economic relations. After all, the more closely the decisions of state power institutions correlate with these expectations, the more successful their implementation is.

References

Bachilo, I. L. (2017). *Information law: Textbook for academic bachelor's degree* (5th ed.) Rev and am (p. 419). Yurayt. (in Rus.)

Bondar, N. S. (2006). *Constitutionalization of the socio-economic development of the Russian statehood (in the context of the decisions of the Constitutional Court of the Russian Federation)* (p. 224). LLC "Vikor-Media. (in Rus.)

Chernogor, N. N., & Zaloilo, M. V. (2018). *Acute problems of lawmaking* (p. 144). Institute of Legislation and Comparative Law under the Russian Federation's Government: LLC "Pravaya Firma Contract". (in Rus.)

Gadzhiev, G. A. (2004). *Objective nature of legal principles and their consideration in economic legislation. Legal reform, judicial reform, and constitutional economics*. Compiled by P. D. Barenboim (p. 208). (in Rus.)

Gadzhiev, G. A. (2020). The doctrine of reasonable expectations in the digital era. In the book: Law and laws in a programmed society: Proceedings of scientific articles.

Ilyin, I. A., *On the education of the Russian people for justice*. http://www.odinblago.ru/nashi_zadachi_1/78. Accessed March 10, 2021.

Inshakova, A., Goncharov, A., & Kazachenok, O. (2018). The imperatives of financial policy in the sphere of the digital economy: impacts on increasing investment activity and tax potential of Russian regions. In *Proceedings of the International Scientific Conference "Competitive, Sustainable and Secure Development of the Regional Economy: Response to Global Challenges" (CSSDRE 2018)* (pp. 337–342). Atlantis Press.

Inshakova, A. O., Frolova, E. E., Galkina, M. V., & Rusakova, E. P. (2020). Development of social market economy under the influence of noneconomic factors: Modeling and regulation. *International Journal of Sociology and Social Policy*, 1–11.

Kazantseva, N. (2019). *Economic incidents of legislation* (p. 280). Prospect. (in Rus.)

Khabrieva, T. Ya., & Chirkin, V. E. (2007). *The theory of modern constitution* (p. 320). Norma. (in Rus.)

Korobeeva, A. I. (2008). Complete course of criminal law: In 5 volumes. *Vol. 1: Crime and Punishment* (pp. 88–90). (in Rus.)

Kovalenko, E. I. (2017). In some areas of further enhancing the culture of publication and enactment of normative legal acts. In the book: Topical issues of the development of legal information in designing the information society. Proceeding's materials. National Center for Legal Information (p. 464). National Center of Legal Information of the Republic of Belarus. (in Rus.)

Kovalenko, E. I. (2019). Electronic government: Current state and development prospects. Constitutional foundations of the Republic of Belarus's development as a social state in modern conditions. In *Proceedings of the international scientific conference*, Minsk, October 3–4, 2019 (p. 556). Belorussian State University, BSU. (in Rus.)

Kuznetsova, P. U. (2017). *Information technology in legal practice: A textbook for academic bachelor's degree* (3rd ed.) rev and am (p. 325). Yurayt. (in Rus.)

Mamychev, A. Yu., Kim, A. A., & Frolova, E. E. (2020). "The future" as an attractor of modern political-legal and socio-economic transformations: An overview of the main problems and approaches. *Advances in Law Studies, 8*(S5), 3–17. (In Rus.).

Matuzov, N. I., & Malko, A. V. (1999). *Political science for lawyers* (pp. 149–150). (in Rus.)

Matytsin, D. E., & Rusakova, E. P. (2021). The strategy of quality management in Industry 4.0 and formation of the cognitive economy based on industrial and manufacturing engineering in the Russian Federation and countries of the EU. *International Journal for Quality Research, 15*(4). https://doi.org/10.24874/IJQR15.04-03

Nepomnyashchaya, T. V. (2010). *Criminal punishment measures: Problems of theory and practice.* Doctor in Law Thesis, 12.00.08 (p. 46) (in Rus.)

Shablinskaya, D. V. (2020). Assessment of the regulatory impact of legislation in economic activity. In V. C. Ladutko & M. N. Satolina (Eds.) (p. 278). Belaruskaya Nauka. (in Rus.)

Sorokin, V. V. (2009). *Procedural form of legal activity in modern Russia: Theory and practice.* Doctor in Law Thesis, 12.00.01. (p. 24). (In Rus.)

Talapina, E. V. (2011). Legal problems of public administration in the information society. *Proceedings of the Institute of State and Law of the Russian Academy of Sciences* (Vol. 4, pp. 175–194). (in Rus.)

Talapina, E. V. (2015). The principle of transparency in public law. *State and Law, 5*, 40–47. (in Rus.).

Tikhomirov, Yu. A. (1998). *Administrative law and process* (p. 798). Yurinformtsentr. (in Rus.)

Tikhomirov, Yu. A. (2007). *Law-based management* (p. 485). Formula Prava. (in Rus.)

Tikhomirov, Yu. A. (2010a). *Legal regulation: Theory and practice* (p. 400). Formula Prava. (in Rus.)

Tikhomirov, Yu. A. (2010b). *The effectiveness of legislation in the economic sphere: Research and practical research* (p. 384). Walters Kluver. (in Rus.)

Tikhomirov, Yu. A. (2011). *State: Continuity and novelty* (p. 80). IL "Jurisprudence". (in Rus.)

Tikhomirov, Yu. A. (2018). *Legal forecasting: Scientific and practical guide* (p. 168). Institute of Legislation and Comparative Law under the Government of the Russian Federation. (in Rus.)

Tikhomirov, Yu. A., Tolmacheva, N. N., & Bogolyubov, S. A. (2019). *How to ensure the effectiveness of the law.* Proceeding material (p. 208). Moscow. (in Rus.)

United Nations E-Government Survey 2014: E-Government for the Future We Want. United Nations. https://publicadministration.un.org/egovkb/Portals/egovkb/Documents/un/2014-Survey/E-Gov_Complete_Survey-2014.pdf. Accessed January 11, 2021.

Vasilevich, G. A. (2004). Constitutional guarantees of the right of citizens to information. In G.A. Vasilevich (Ed.), *Proceedings of the II International Research and Practical Conference "Legal Informatization"*, October 28–29 (pp. 26–27). (in Rus.)

Vasilevich, G. A. (2005). Information technologies in the Republic of Belarus. *The Book: Problems of Legal Informatization, 2*, 6–12. (in Rus.).

Vasilevich, G. A., Nikitenko, P. Q. (2018). Constitutional foundations for the development of a legal, social state in the Republic of Belarus: textbook. In G. A. Vasilevich, & P. G. Nikitenko (Eds.). *Vysheyshaya shkola* [Higher School] (p. 319). (in Rus.)

Vasilevich, G. A. (2019). The role of legal science in eliminating gaps, contradictions, and other defects in legislation. *Pravo.by, 6*, 33–40. (in Rus.)

Vasilevich, S. G. (2014). On public administration: Concept, goals, performance criteria. *Management Problems, 4*(53), 147–151. (in Rus.).

Vasilevich, S. G. (2016). The role and significance of the preamble of a normative legal act in rulemaking and law enforcement. *Pravo.by*, *2*, 21–26. (in Rus.)

Yagofarova, I. D. (2004). *Law as a measure of restriction of freedom.* Doctor of Law thesis, 12.00.01. (p. 28) (in Rus.)

Zhuravleva, L. V. (2002). Information technologies in the modern world (overview). *Law and informatization of society: Proceedings of articles.* RAS INION, Center for Social Scientific-Information Research (pp. 75–88). (in Rus.)

Chapter 24
Sustainable Development and the Legal Regulation of Forced Migration in Russia

Dmitry V. Ivanov and Nikita Yu. Molchakov

Abstract The contemporary status of the sustainable development concept is closely related to the adoption of the Sustainable Development Goals by the UN General Assembly in 2015. The following SDGs are directly related to the legal regulation of forced migration: the elimination of poverty and hunger, good health and well-being, gender equality, decent work and economic growth, peace, justice, and strong institutions. The modern States, including the Russian Federation, carry out activities to implement Sustainable Development Goals in their national legislation. A striking proof of this fact was the development and adoption of the Concept of the State Migration Policy of the Russian Federation for 2019–2025, which is partly aimed at achieving the above-mentioned Sustainable Development Goals. Currently, the subject of discussion in the state bodies of Russia and the academic community is the necessity of systematizing the legislation on forced migration in the form of a Migration Code. We believe that the adoption of such an act will contribute to the progressive development of legal regulation in this area in light of Sustainable Development Goals.

Keywords Sustainable development · State migration policy · Forced migration · "Mass influx" of refugees · Migration legislation of the Russian Federation · Concept of the State Migration Policy of the Russian Federation for 2019–2025

JEL Code K37

The original version of this chapter was revised: The misspelt co-author name "Nikita Yu. Molchkov" has been changed to "Nikita Yu. Molchakov". The correction to this chapter is available at https://doi.org/10.1007/978-981-16-4621-8_31

D. V. Ivanov (✉) · N. Yu. Molchakov
Moscow State Institute of International Relations (MGIMO University), Moscow, Russia

A. O. Inshakova and E. E. Frolova (eds.), *Smart Technologies for the Digitisation of Industry: Entrepreneurial Environment*, Smart Innovation, Systems and Technologies 254, https://doi.org/10.1007/978-981-16-4621-8_24

24.1 Introduction

As rightly pointed out in the doctrine, "international migration has become a commonplace of everyday life and, at the same time, it has generated many real or imaginary problems that deeply affect the areas of economics, politics and, in fact, all spheres of social life" in modern Russia (Zayonchkovskaya et al., 2009). The relevant problems are certainly observed in the field of legal regulation of both forced and labor migration. However, we deliberately limit the subject of this very research only to the sphere of forced migration, since it is precisely in this issue that the national legislation of the Russian Federation needs radical and systemic reform.

We are convinced that, on the one hand, the Russian legislation should be amended, bearing in mind the concept of a durable solution to the refugee problem (Ivanov, 2019), and, on the other hand, the amendments should pursue the achievement of Sustainable Development Goals (Abashidze et al., 2016; Kukushkina, 2017). In particular, the following SDG are worthy of attention, regarding the forced migration: the elimination of poverty and hunger, good health and well-being, gender equality, decent work and economic growth, peace, justice, and strong institutions. Furthermore, in this context, we also cannot fail to mention the Section VI of the 2000 UN Millennium Declaration ("Protecting the vulnerable"), which refers to the determination of States "to help all refugees and displaced persons to return voluntarily to their homes, in safety and dignity and to be smoothly reintegrated into their societies" (United Nations Millennium Declaration, 2000).

The Concept of the State Migration Policy of the Russian Federation for 2019–2025 (hereinafter referred to as "the 2018 Concept"), approved by the President of Russia in 2018, should become the basis for the appropriate amendments to the Russian legislation (Decree of the President of the Russian Federation No. 622). In this document, the major and most pressing issues and deficiencies in the field of legal regulation of forced migration were formulated. It particularly reflects, for the first time in Russia, the need for a comprehensive approach to solving problems in the field of population movement, based on the concept of regulated and controlled migration. This fact is confirmed by the signing by the Russian Federation, in December 2018, of the Global Compact for Safe, Orderly, and Regular Migration in Marrakech, to which more than 130 States are Parties.

24.2 Methodology

Any State defines the legal foundations of its migration policy in special documents that contain a general understanding of the migration situation, directions of development of migration legislation, approaches to resolving issues related to the forced migration both in the short and long terms. The Russian Federation is also no exception to this rule. A significant number of legal acts have been adopted throughout the history of Russian migration legislation.

One of the first legal instruments that determined the foundations of the State migration policy was the Decree of the Government of the Russian Federation dated May 18, 1992, No. 327 "On the Republican Long-Term Migration Program." It particularly noted the necessity to "minimize uncontrolled migration flows," to "promote the socio-economic adaptation and integration of migrants in the Russian Federation by establishing legal and human conditions for their reception and accommodation," and to "cooperate with international organizations in the field of migration" (List of instructions on the implementation of the Concept of State Migration Policy for 2019–2025). Two years later, by the Decree of the President of the Russian Federation dated August 9, 1994, No. 1668 "On the Federal Migration Program," the priority areas of activity of federal executive bodies in the field of regulating migration processes in the Russian Federation, protecting the rights of refugees and forced migrants and addressing the adverse consequences of irregular migration were determined (Decree of the President of the Russian Federation No. 1668).

The next stage in the process of developing the legal framework of the State migration policy was the "Concept for the Regulation of Migration Processes in the Russian Federation," approved by the Order of the Government of the Russian Federation dated March 1, 2003, No. 256-r (Order of the Government of the Russian Federation No. 256-r). However, as the doctrine rightly points out, "the legal status of this document was so low that it did not become a guideline for the subsequent development of laws and mechanisms for regulating migration processes, and the development of federal target programs" (Vorobieva et al., 2016).

In 2012, the President of the Russian Federation approved the "Concept of the State Migration Policy of the Russian Federation for the period up to 2025" (hereinafter referred to as "the 2012 Concept") (2012). A distinctive feature of this legal act was the lack of a comprehensive approach to the problem of migration while it focused generally on issues of labor migration. As for forced migrants, this legal instrument mainly contains provisions on the fulfillment of humanitarian obligations concerning them. It is also important to mention that the 2012 Concept does not contain provisions regulating the legal status of foreigners (including stateless persons) who have received temporary asylum in the territory of the Russian Federation. In general, the provisions of this document reflected the situation with migration that existed at that time, characterized by the increasing number of labor migrants, but not forced migrants. Consequently, the 2012 Concept did not contain provisions aimed at improving the Russian migration legislation in terms of regulating situations of "mass influx" of forced migrants. It was the latter circumstance precisely that became one of the major arguments in favor of developing a new Concept of State Migration Policy, which was approved in 2018 (Inshakova et al., 2017).

24.3 Results

In 2015–2018, the migration situation in the Russian Federation has changed significantly, which is explained by the events in the south-east of Ukraine. The number of persons granted temporary asylum in the territory of the Russian Federation has significantly increased in the first place. If in 2014, according to official statistics, the number of persons who received temporary asylum in the territory of the Russian Federation reached 2882 persons, of which not a single person was a citizen of Ukraine, then in 2015, this number of migrants reached 237,780 persons, of which 234,360 persons were citizens of Ukraine. In 2016, the number of migrants amounted to 313,707 persons, of which 311,134 were citizens of Ukraine, and in 2017, out of 228,392 migrants, 226,044 were citizens of Ukraine. And in 2018, out of 125,442 persons who moved to Russia, 123,434 of them were citizens of Ukraine (Federal Statistics Service of the Russian Federation, 2019).

Problems arising from the 2015–2018 migration situation in the Russian Federation could not be resolved within the framework of those areas of migration policy that were defined in the 2012 Concept. Consequently, it was necessary to adopt a new document that pays much more attention to the forced migration of the population. In October 2018, the President of the Russian Federation signed Decree No. 622 "On the Concept of the State Migration Policy of the Russian Federation for 2019–2025" (hereinafter referred to as "the 2018 Concept"). It should be noted that this document especially emphasized, "In recent years, migration activity has significantly increased near the external borders of the Russian Federation and in the area of its interests."

In general, the expert community had a positive view of this new document. Thus, I.V. Ivakhnyuk rightly notes the following: "The 2018 Concept is a logical document, concise and explicit in its language and meaning. This is fundamentally important since only an extremely clear and convincing document that does not allow for ambiguities and double interpretations can serve as a guide for public authorities involved in the development of legislation and the implementation of specific measures in the area that the Concept is devoted to, as well as local self-government, business, civil society institutions and, finally, the media" (Ivakhnyuk, 2018). The new Concept is true, in contrast to previous documents of this kind, an instrument that approaches the definition of the main directions of State migration policy both in the field of forced and labor migration as comprehensively as possible. At the same time, this document itself emphasizes the principle of the complexity of solving the challenges of migration policy.

In the 2018 Concept, the goal of the State migration policy is formulated in more detail in "creating a migration situation that contributes to solving problems in the field of socio-economic, spatial and demographic development of the country, improving the quality of life of its population, ensuring State security, protecting the national labor market, maintaining inter-ethnic and inter-religious peace and harmony in Russian society, as well as in the field of protection and preservation of Russian culture, the Russian language and the historical and cultural heritage of the peoples of Russia, which form the grounds of its cultural (civilizational) code" (Decree of the President of the Russian Federation No. 622). It should be noted that this list has been supplemented with humanitarian issues in comparison with the 2012 Concept.

The 2018 Concept provides for several measures to achieve this goal, particularly improving the migration legislation of the Russian Federation. At the same time, as it is noted in the document, this process involves "the establishment of rules that are simple, transparent and enforceable for citizens, which meet the goals, principles, and objectives of migration policy, free from administrative barriers and their costs" (Decree of the President of the Russian Federation No. 622). The very same State migration policy should be carried out in accordance with the principles of a democratic State of law established by the Constitution of the Russian Federation, and universally recognized principles and rules of international law, as well as considering the principles defined in such documents as the National Security Strategy of the Russian Federation, the Strategy for countering extremism in the Russian Federation until 2025, the Strategy of the State National Policy of the Russian Federation for the Period until 2025, the Strategy for Scientific and Technological Development of the Russian Federation (Matytsin & Rusakova, 2021), the Strategy of State Cultural Policy of the Russian Federation for the Period up to 2030," the Strategy for Socio-Economic Development of the Far East and the Baikal Region for the Period until 2025," the Concept of Public Safety in the Russian Federation, the Concept of Demographic Policy of the Russian Federation for the period up to 2025, the Concept for the Long-term Socio-Economic Development of the Russian Federation and the period up to 2020, the Concept of the Demographic Policy of the Far East for the period up to 2025, and the Forecast of the Long-term Socio-Economic Development of the Russian Federation for the period up to 2030.

The 2018 Concept provides for the need to improve the efficiency of administrative procedures in the field of migration, including the introduction of electronic forms of interaction between recipients of public services and the bodies and organizations providing them, the use of the multifunctional centers as well as other organizational and technical solutions. Furthermore, the improvement of administrative procedures should contribute to the prevention, detection, and suppression of violations of the migration legislation of the Russian Federation. It is also obvious that increasing the efficiency of administrative procedures is closely related to the informatization of migration management processes. This issue is given special attention in the List of instructions on the implementation of the 2018 Concept, approved by the President of the Russian Federation in March 2020 (Section IV "Informatization of the Sphere of Migration Management") (List of instructions on the implementation of the Concept of State Migration Policy for 2019–2025).

The 2018 Concept provides for "assist foreign citizens seeking protection in the territory of the Russian Federation following universally recognized principles and rules of international law and the legislation of the Russian Federation" as one of the priority tasks of the migration policy. In this regard, paragraph 27 of the said document deserves special attention, which defines the main directions of migration policy in the field of assisting foreign citizens seeking protection in the territory of the Russian Federation (Inshakova et al., 2018). In particular, the Russian Federation aims at the following: (1) to maintain high standards and further development of mechanisms for providing assistance to foreign citizens seeking protection in the territory of the Russian Federation, in accordance with the international legal obligations of the Russian Federation and with due regard for the interests of Russian citizens; (2) to ensure the readiness of the concerned federal executive bodies in cooperation with executive bodies of the constituent entities of the Russian Federation to receive and accommodate foreign citizens in case of their urgent mass arrivals in the Russian Federation; (3) to provide refugees, persons who have received temporary asylum, and applicants for obtaining the appropriate status of assistance in social and cultural adaptation, taking into account their vulnerable situation; (4) to the establishment of additional instruments for regulating migration flows, taking into account the socio-political and socio-economic situation in the countries from which foreign citizens arrive in the Russian Federation. We should mention that the 2012 Concept did not contain similar or analogous provisions. It is also important to note that for the first time the term "mass arrival" of foreign citizens was introduced into Russian legislation, which can be considered as an analog of the term "mass influx" of refugees, enshrined in the conclusions of the Executive Committee of the United Nations High Commissioner for Refugees (UNHCR, 1981, 2004).

The provisions on international cooperation of the Russian Federation in the field of migration have also an important place in the 2018 Concept, which includes: (1) the implementation of international treaties in the field of migration; (2) the conclusion, if necessary, of new international treaties, taking into account the goals, principles, objectives, and major directions of migration policy; (3) the ensuring the mutual interests of the Member States of the Eurasian Economic Union; (4) the participation of the Russian Federation in the activities of international organizations, cooperation with foreign partners, and international institutions in order to improve the quality of regulation of international migration flows; (5) ensuring the meaningful involvement of the Russian Federation in interaction with foreign States, international organizations, and associations in the establishment of a positive, balanced and unifying international agenda and in the development of decisions on migration issues; (6) the exchange of experience in the implementation of migration policy and management of migration processes; (7) increasing the efficiency of mechanisms for countering illegal migration, including expanding agreements on readmission; (8) improving the mechanisms for the exchange of information on citizenship and other information related to migrants, as well as improving the level of security of such information during its processing and cross-border transfer; (9) monitoring of the socio-political and socio-economic situation in foreign States, whose citizens are provided with protection in the Russian Federation. This list differs significantly from similar provisions of the 2012 Concept, not limiting the international cooperation of the Russian Federation in the field of migration only by the implementation of international treaties in force and the conclusion of new international treaties, as well as cooperation with other States within the framework of international organizations.

We must note the fact that in the 2018 Concept, special attention is paid to the issues of combating illegal (unlawful) migration. This document particularly defines and substantiates the need to implement measures aimed at identifying and suppressing the organization of illegal migration, human trafficking, the use of forced labor, including together with the competent authorities of foreign States.

The 2018 Concept as one of the priority directions of the State migration policy refers to "the improvement of the mechanism of migration control to solve the problems of ensuring the national security of the Russian Federation, including countering criminal, terrorist, and extremist organizations." This perspective is also given special attention in the List of instructions of the President of the Russian Federation on the implementation of the 2018 Concept (Section III "Improving the Mechanisms for Ensuring Security and Law and Order in the Migration Sphere") (List of instructions on the implementation of the Concept of State Migration Policy for 2019–2025). In particular, it is supposed, on the one hand, "to introduce flexible and convenient migration rules for the temporary stay in the Russian Federation of law-abiding foreign citizens," and on the other hand, "to improve the institute of undesirability of stay (residence) of a foreign citizen in the territory of the Russian Federation," which implies the establishment of the "loyalty agreements" mechanism. The meaning of the latter comes to the adoption by the migrant of obligations, "the violation of which may entail the announcement of the undesirability of stay (residence) within the territory of the Russian Federation and the loss of the right to stay in the Russian Federation" (List of instructions on the implementation of the Concept of State Migration Policy for 2019–2025). It is about strict compliance with the legislation, non-participation in political activities and activities of organizations prohibited in the Russian Federation, etc. However, we shall agree with those authors who argue that "the institution of "loyalty" requires additional consideration" (Paukova, 2021) due to the lack of a clear legal framework based on the principle of legal certainty. At the same time, as we see it, the above approach to the development of migration legislation in the context of the necessity to ensure national security is very balanced.

24.4 Conclusions

An analysis of the provisions of the 2018 Concept shows that achieving the goals specified in it requires a deep reform of the current migration legislation of the Russian Federation, as well as the system of public administration and control in the field of migration. We suppose, particularly, that the achievement of these goals implies the solution of the following tasks:

1. Creation of the unified mechanism for statistical and information support of the migration policy of the Russian Federation, since at present this function is carried out by several federal executive bodies.
2. Establishment of the separate and independent federal executive body, whose competence will include not only control and supervision in the field of migration but also legal regulation of migration relations, as well as determining the directions of State migration policy.
3. Development of normative legal acts in the field of forced migration, which provides, in particular: (a) amendments and additions to the current legislation, providing for the explicit identification of categories of forced migrants and determination of their general and special legal status on the basis of the forms of territorial asylum that have developed in Russia (granting political asylum, refugee status and temporary asylum); (b) development and adoption of

a separate Federal Law on temporary asylum, taking into account the increasing number of this category of forced migrants in the Russian Federation and the negative experience of the European system of asylum and the provision of humanitarian protection; (c) amendments and additions to the existing normative legal acts governing the procedure for granting and the status of persons who have received temporary asylum within the territory of Russia, as well as the mechanism for granting this category of foreigners the opportunity to obtain the Russian citizenship; (d) amendments and additions to the current legislation providing for a special procedure for granting asylum within the territory of the Russian Federation in situations of "mass arrival" ("mass influx") of foreign citizens.

4. Comprehensive and objective assessment of the current migration legislation in terms of Russia's international legal obligations in the field of forced migration at the universal, regional, and bilateral levels, especially within the framework of the Commonwealth of Independent States (CIS) and the Eurasian Economic Union (EEU).

The above proposals can serve as a basis for the establishment of an effective model of legal regulation of forced migration of the population since they also consider the current migration situation as well as trends in the legal regulation of forced migration by national and international law.

References

Abashidze, AKh., et al. (2016). Achievement of sustainable development goals (2016–2030): International legal dimension. *RUDN Journal of Law, 1*, 65–78.

Concept of the State Migration Policy of the Russian Federation for the period up to 2025. Approved by the President of the Russian Federation on June 13, 2012. https://www.garant.ru/products/ipo/prime/doc/70088244/. Accessed April 24, 2021.

Decree of the President of the Russian Federation dated August 9, 1994 No. 1668, On the Federal Migration Program. http://www.pravo.gov.ru/proxy/ips/?docbody=&prevDoc=102041177&backlink=1&&nd=102031729. Accessed April 24, 2021.

Decree of the President of the Russian Federation dated October 31, 2018, No. 622, On the Concept of the State Migration Policy of the Russian Federation for 2019–2025. http://www.consultant.ru/document/cons_doc_LAW_310139/. Accessed April 24, 2021.

Federal Statistics Service of the Russian Federation. (2019). *The number of internally displaced persons, refugees, and persons who received temporary asylum as of June 4, 2019*. http://www.gks.ru/free_doc/new_site/population/demo/tab-migr4.htm Accessed April 24, 2021.

Ivakhnyuk, I. V. (2018). *New message to migration authorities, society and migrants*. https://russiancouncil.ru/analytics-and-comments/analytics/novyy-posyl-migratsionnym-organam-obshchestvu-i-migrantam/. Accessed April 24, 2021.

Ivanov, D. V. (2019). Interstate cooperation for durable solution of the refugee problem: International legal aspects. *Migration Law, 4*, 19–23.

Kukushkina, A. V. (2017). The concept of sustainable development (international legal aspects). *Tomsk State University Journal of Law, 23*, 29–39.

List of instructions on the implementation of the Concept of State Migration Policy for 2019–2025. http://kremlin.ru/acts/assignments/orders/62960. Accessed April 24, 2021.

Lokosova, V. V., & Rybakovsky, L. L. (2014). *Migration processes in Russia*. Econ-Inform.

Order of the Government of the Russian Federation dated March 1, 2003, No. 256-r, On the Concept of Regulation of Migration Processes in the Russian Federation. http://www.consultant.ru/document/cons_doc_LAW_113093/ae34219c5777239ace76672ef4455433e0a5b907. Accessed April 24, 2021.

Paukova, Yu. V. (2021). On the necessity to reform the institution of the undesirability of stay (residence) of foreign citizens in Russia in the context of digitalization. *Law and Politics, 1*, 23–33.

UNHCR . (1981). *Protection of Asylum-Seekers in Situations of Large-Scale Influx, No. 22 (XXXII)*. https://www.unhcr.org/excom/exconc/3ae68c6e10/protection-asylum-seekers-situations-large-scale-influx.html. Accessed April 24, 2021..

UNHCR . (2004). *Conclusion on International Cooperation and Burden and Responsibility Sharing in Mass Influx Situations No. 100 (LV)*. https://www.unhcr.org/en-us/578371524.pdf. Accessed April 24, 2021.

United Nations Millennium Declaration. General Assembly resolution 55/2 of 8 September 2000. https://www.ohchr.org/EN/ProfessionalInterest/Pages/Millennium.aspx. Accessed April 24, 2021.

Vorobieva, O. D., Rybakovsky, L. L., & Rybakovsky, O. L. (2016). *Migration policy of Russia: History and modernity*. Econ-Inform.

Zayonchkovskaya, Zh., Mkrtchyan, N., & Tyuryukanova, E. (2009). *Russia facing the challenges of immigration. Post-soviet transformations: Reflection in migrations*. Adamant.

Part IV
Environmental Entrepreneurship Technologies in the Context of Industry 4.0 in the Russian Federation, the EAEU, and BRICS Countries

Chapter 25
The Concept and Types of Environmental Entrepreneurship

Aleksey P. Anisimov and Denis E. Matytsin

Abstract The chapter defines important concepts for the legal regulation of environmental entrepreneurship, which are important for the correct interpretation of the content and conclusions of the monographic study, and also for the development of environmental law as a branch of science and legislation. These include the following concepts: "ecological entrepreneurship", "ecological tourism", "ecological insurance", "ecological contract", "ecological audit" and a number of others. The authors argue that environmental entrepreneurship is a type of entrepreneurial activity carried out by legal entities and individual entrepreneurs at their own risk and under their responsibility to make a profit. This activity is aimed at the production of environmental protection products, conducting research and credit and financial activities, performing environmentally significant works, and providing services. The most developed type of environmental entrepreneurship in Russia is related to the provision of environmental services. Among the latter, in turn, it is necessary to distinguish between environmental audit, environmental insurance, eco-tourism, and waste management of production and consumption. The chapter discusses the specifics of the relevant agreements. It is suggested that over time, new types of contractual relations will also develop, in particular, in the field of voluntary environmental certification, for example, in relation to organic agricultural products. Currently, two types of environmental contracts have been mainly developed—the provision of services for the management of solid municipal waste, as well as for the provision of eco-tourism services. The remaining contracts are either not in demand

A. P. Anisimov (✉)
Volgograd State University, Volgograd, Russia

D. E. Matytsin
Department of Business Law, Arbitration and Civil Procedure, Institute of Law, Volgograd State University, Volgograd, Russia

Department of Civil Law and Procedure, Volga Branch of the International Law Institute, Volgograd, Russia

D. E. Matytsin
e-mail: Dmatytsin@ya.ru; ppagp@volsu.ru

A. O. Inshakova and E. E. Frolova (eds.), *Smart Technologies for the Digitisation of Industry: Entrepreneurial Environment*, Smart Innovation, Systems and Technologies 254, https://doi.org/10.1007/978-981-16-4621-8_25

at all, or their number in the Russian Federation is still measured by an insignificant amount.

Keywords Environmental entrepreneurship · Industry 4.0 · Environmental agreements · Environmentally friendly technologies · Environmental audit

JEL Codes K32 · L26 · O44 · P28 · P48 · Q52 · Q56 · F18

25.1 Introduction

At present, neither the scientific literature nor the regulatory legal acts clearly define the category of "environmental entrepreneurship". However, in most European countries, its characterization focuses on the criterion of the production of goods and the provision of services aimed at reducing or eliminating the consequences of environmental harm. To reduce emissions and discharges of harmful substances, reduce or recycle the amount of waste, as well as the use of various environmentally friendly technologies. At the moment, the United States, Japan, and Western European countries are the leaders in the implementation of environmental technologies. As noted in the scientific literature, Germany alone accounts for 43% of environmental patents for products and goods in demand in all countries. The volume of products produced in the environmental market, according to some experts, is estimated at 600 billion to 2 trillion US dollars, and the growth rate is from 5.5 to 7% per year. The domestic market of the United States for the production of environmental products and services is $37 billion, Japan—$30 billion, Germany—$20 billion, France—$10 billion. The market for environmental goods and services in Eastern Europe is estimated at $20 billion, and in the Czech Republic, Hungary, and Bulgaria it exceeds $600 million (Urazova, 2007).

In Russia, the formation of environmental entrepreneurship is at the very beginning. According to Article 2 of the Civil Code of the Russian Federation, entrepreneurial activity is an independent activity carried out at its own risk, aimed at systematic profit-making. Obtaining profit from the use of property, the sale of goods, the performance of works or the provision of services by persons registered in this capacity in accordance with the procedure established by law. It follows that environmental entrepreneurship is the activity of citizens-entrepreneurs and legal entities aimed at the production of goods, performance of works, and provision of services that allow for the prevention, limitation, or elimination of negative environmental impacts (Ryzhenkov, 2018).

According to E. V. Varennikova, environmental entrepreneurship should include proactive economic activities carried out taking into account environmental requirements and restrictions, which are aimed at avoiding (reducing) the negative impact on the environment, as well as improving environmental performance to maximize profits. The criteria for classifying entrepreneurs as environmentally oriented are the introduction of environmental management systems into their economic activities.

As well as the use of a significant share of environmental innovations, the significant role of environmental utility in the overall utility of the results of their activities (Varennikova, 2011).

In turn, S. V. Zlobin believes that environmental entrepreneurship is a socially significant, independent, initiative activity of individuals, which is associated with "conscious acceptance of the potential risk of unfavorable consequences in the production of environmental products. And which consists of conducting research, credit and financial activities, performing environmentally significant works, and providing services. Moreover, the activity is aimed at making a profit (income)" (Zlobin, 2011). Sharing the latter approach, we will continue to adhere to it further.

25.2 Materials and Methods

The legislative base of the study was made up of the following documents. The Civil Code of the Russian Federation and the Federal Laws "On Specially Protected Natural Territories", "On the Basis of Tourist activity in the Russian Federation", "On the Safety of Hydraulic Structures", "On the Industrial Safety of Hazardous Production Facilities", "On Production and Consumption Waste", the Code of Merchant Shipping (the document introduces some terms and procedures analyzed in the work). These laws contain the main parameters of environmental agreements, including the form of the agreement, the essential conditions, the specifics of accounting for the state of the environment or individual natural objects, etc.

Environmental and civil legislation of the Russian Federation as a set of normative legal acts regulating business activities in the field of environmental protection has become the subject of research in the works of some authors. Among them: Bekisheva (2011), Varennikova (2011), Ruden (2015), Urazova (2007), and some other authors.

The definition and characteristics of individual environmental agreements were investigated: Inshakova et al. (2018) (environmental insurance), Korostelev (2008) and Ryzhenkov (2014) (eco-tourism), Serov (2000) (environmental audit), and problems of environmental education and education—Selivanova (2019).

In the process of research, general scientific methods are used, such as formal-logical, dialectical, system-structural, critical cognition. Methods of synthesis, classification, and generalization were used to interpret the results of the study. The paper also uses private scientific methods: formal-legal, the principle of assessing legal processes, the method of comparative analysis, etc.

25.3 Results

25.3.1 Main Directions of Environmental Entrepreneurship

At the moment, in the Russian Federation, we can distinguish three areas of development of environmental entrepreneurship:

1. environmental entrepreneurship can consist both in the performance of environmental works (construction of environmentally safe houses) and in the provision of environmental services. Among the latter, there is no doubt about environmental audit; environmental insurance; eco-tourism; waste management of production and consumption. Over time, contractual relations in the field of voluntary environmental certification will also develop (for example, concerning organic agricultural products) (Inshakova et al., 2018);
2. in addition to activities related to the direct performance of works or the provision of services, environmental entrepreneurship may also include research (development) activities. Which is aimed at the creation of environmentally friendly building materials, the development of biogas technologies, the production of new generations of solar panels, wind generators, and other renewable energy sources;
3. ensuring environmental interests in the field of entrepreneurship can also be carried out by taking into account environmental requirements when privatizing state-owned enterprises. This issue was addressed by the Order of the Ministry of Natural Resources of the Russian Federation No. 469 of 21.11.1995 "On accounting for the environmental factor in the privatization of state and municipal enterprises and organizations," but it has lost its force, and no new legal acts have been adopted.

The above-mentioned types of business activities in the field of environmental protection are presented in civil law contracts with varying degrees of detail. At the same time, the term "environmental contract" is not used in Russian civil law but is quite widespread in the foreign science of civil and environmental law (Ruden, 2015).

Environmental contracts are a separate type of civil law contract aimed at performing environmental works or providing environmental services that are not related to the use of natural resources.

In our opinion, contracts related to the transfer of land plots, water bodies, or other natural resources to lease or on other property or obligation rights should be called "natural resource contracts" (land lease agreement, water use agreement, etc.).

25.3.2 Main Types of Environmental Agreements

Let us consider the content of the main types of environmental agreements.

1. contract for the provision of services for the management of solid municipal waste (MSW). According to Article 24.7 of Federal Law No. 89-FZ of 24.06.1998 "On Production and Consumption Waste", regional operators conclude contracts for the provision of services for the management of MSW with the owners of such waste. Moreover, the contract for the provision of services for the management of these wastes is public for the regional operator. This means that the regional operator does not have the right to refuse to conclude this contract to the owner of MSW, which are formed and the places of accumulation of which are located in the area of its activity. Under the contract for the provision of services for the treatment of MSW, the regional operator undertakes to accept MSW in the amount and in the places (on the sites) of accumulation, which is defined in this contract. He undertakes to ensure their transportation, processing, neutralization, burial in accordance with the legislation. At the same time, the owner of solid municipal waste undertakes to pay for the services of the regional operator at a price determined within the limits of the unified tariff for the service of the regional operator approved in accordance with the established procedure.

The specificity of this contract is that its regulation is carried out not only by the norms of civil law but also by the norms of environmental law, which specify the requirements for the parties to the contract, its essential conditions (including the requirements for the calculation of fees under the contract).

2. an environmental audit contract. An environmental audit is an independent, comprehensive, documented assessment of compliance by a legal entity or individual entrepreneur with the requirements. This includes regulations and regulatory documents, federal norms and rules in the field of environmental protection, and the requirements of international standards, as well as the preparation of recommendations for improving such activities (Serov, 2000).

Currently, there is no mandatory environmental audit in the Russian Federation, only a voluntary environmental audit is applied. A voluntary audit is carried out on the initiative of the audited entity. Its implementation is mainly related to the requirements of foreign investors and banks when providing investments or credit funds for the development of production or promotion of Russian products in foreign markets, which must be certified according to international standards (Matytsin & Rusakova, 2021). The subject of an environmental audit contract is usually the environmental documentation of a business entity or other activity. However, there are reasons to talk about the need to include in the environmental audit the physical condition of treatment facilities and other environmental-saving facilities, as well as the need to comply with environmental requirements in the implementation of production processes.

A separate discussion requires the need to further expand the subject of the environmental audit contract, which is due to new challenges to humanity in the era of globalization, including a significant expansion of activities related to greenhouse gas emissions into the atmosphere. The essential terms of the environmental audit contract are not defined by law, but, in addition to the subject of the contract, they should include the price and the term. It would be advisable to regulate these features by a special federal law "On Environmental Audit", which has not yet been adopted in Russia.

3. agreement on the sale of a tourist product. The contract in question is a type of contract for the sale of a tourist product, which, in turn, is a type of civil contract for the provision of paid services. The main requirements for the provision of tourist services are formulated in Federal Law No. 132-FZ of 24.11.1996 "On the basics of Tourist activity in the Russian Federation". The environmental specifics of the agreement on the sale of a tourist product can be discussed both in relation to territories with a general (ordinary) ecological and legal regime and in relation to the conduct of tourist tours in specially protected natural territories (SPNT) that fall within the scope of Federal Law No. 33-FZ of March 14, 1995 "On Specially Protected Natural Territories."

At the moment, within the borders of all categories of SRNT, including state nature reserves, it is allowed to organize educational eco-tourism, but within the borders of SPNT, additional prohibitions and restrictions are imposed on tourists and tour organizers. For example, they are related to the prohibition of free movement within the boundaries of SPNT outside of specially designated ecological trails, the time of stay, the rules of behavior of tourists, etc., which are not present on other tourist routes. Eco-tourism is today one of the most promising areas of state support for environmental entrepreneurship, accounting for about 10–20% of the total world tourism market and is the most dynamically developing industry. The World Wildlife Fund defines eco-tourism as tourism involving travel to places with relatively not spoilt nature. Travel to get an idea of the natural and cultural-ethnographic features of the area, which does not violate the integrity of ecosystems and creates such economic conditions in which the protection of nature and natural resources becomes profitable for the local population.

In our opinion, the specificity of eco-tourism is that it perfectly meets the goals and objectives of sustainable development of the country, combining the solution of social, economic, and environmental problems of Russia.

How does this manifest itself? First, the economic task is to make a profit for legal entities and individual entrepreneurs involved in the provision of eco-tourism services and to pay taxes to the budget. Scientists have calculated that tourism (including environmental tourism) directly or indirectly affects the development of 32 sectors of the economy, and also has a great impact on the development of the country's infrastructure (construction of new roads and railways, airports, etc.). Eco-tourism is less resource-intensive than other types of tourism. There is no need for an extensive and developed tourist infrastructure (hotels, restaurants, entertainment venues). Therefore, due to the smaller amount of necessary tourist infrastructure for each ruble invested, the profit will be higher than in other types of tourism. Accordingly, eco-tourism is economically profitable (Inshakova et al., 2018). In South Africa, eco-tourism ranks second in terms of profitability after diamond mining and in Alaska-second after oil production. Kenya's annual income from the use of national parks reaches $450 million, while Ecuador receives more than $180 million from eco-tourism in the Galapagos Islands (Bekisheva, 2011).

However, in the Russian Federation, this figure does not yet reach $12 million. For comparison, the Valley of Geysers in Yellowstone National Park (USA) is visited by two million people annually, and the Valley of Geysers in Kamchatka (Russia) is visited by only 3000 people a year (Bukharova, 2010).

Second, eco-tourism as a kind of ecological entrepreneurship performs a number of important social functions. Compared to other types of tourism, its pronounced social orientation is manifested in the creation of new jobs for the local population (especially in rural areas). Also in promoting traditional forms of environmental management, the production of environmentally friendly food. Also in increasing investment in both infrastructure and services, as well as in nature conservation. Also in the growth of the well-being of the local population and the development of special education aimed at the acquisition of tourist and environmental professions. Also in the development of crafts; the development of local self-government, etc. (Ecotourism, 2002). The importance attached to the development of eco-tourism in Russia is evidenced by the fact that out of the first seven special economic zones of the tourist and recreational type, six were supposed to develop eco-tourism. It is necessary to pay attention to the fact that the main development of ecological tourism in the Russian Federation will be carried out in territories with a special ecological and legal regime (dendro-parks, botanical gardens, national and natural parks).

Third, the importance of eco-tourism is very important for environmental protection. The fact is that the most effective protection of land is not through the establishment of prohibitions and penalties, but by creating effective economic guarantees. Which are aimed at ensuring the sustainable development of not individual land plots, but large territories. In this case, not only environmental problems are solved (residents do not arrange garbage dumps on tourist routes, do not drain pollutants into rivers, etc.), but also economic ones. The latter is important because many environmental offenses stem from poverty. Citizens do not buy garbage containers and do not pay for garbage collection, not because they do not feel sorry for nature, but because of a lack of money. Therefore, very often ravines in rural areas are one big dump. Since in the areas of mass eco-tourism, most of the residents are involved in trade, food, hotel business, etc., nature becomes profitable for everyone. Then, the pollution of land and other natural objects will be less. It won't be profitable. Otherwise, ecological tourists will not come here (Ryzhenkov, 2014).

Fourth, the contribution of eco-tourism to environmental education and education is important. Getting to know nature in reality, tourists realize the need to treat it carefully. Thus, the formation of ecological culture and ecological legal awareness takes place (Selivanova, 2019).

To date, the Government of the Russian Federation has set a goal to increase the number of visitors to SPNT to 16 million people by 2035 and to involve at least half of the national parks in eco-tourism programs. That is why in the federal and regional programs of the four types of contracts for environmental services, only eco-tourism is constantly mentioned as a priority, while environmental insurance and audit, unfortunately, are not.

With all the advantages of eco-tourism, it, as a complex legal phenomenon, has negative consequences:

(a) the mass appearance of tourists in the territories of fragile ecological systems leads to a deterioration in their quality. In Russia, these problems have not yet been thoroughly investigated, but the negative consequences of eco-tourism have already been established. In Kamchatka, visits to the valley of geysers led to a violation of the regime of thermal springs, and some geysers stopped pouring out. In the Arkhangelsk region, a visit to the Pinezh caves disturbed the heat and sound balance and led to the mass death of bats. On the island of Valaam in the 80s of the XX century, the degradation of landscapes occurred due to the excess recreational load (Korostelev, 2008). However, the development of eco-tourism in the Arctic poses a particular threat, given the high level of fragility and irreplaceability of many northern ecological systems.
(b) the low level of skills of the local population does not always allow them to be involved in the eco-tourism industry. In cases where the local population refuses to follow the rules of travel companies and improve their skills, this can lead to conflicts between environmental entrepreneurs and residents.
(c) environmental damage from tourism and its compensation. Increased construction of tourist sites increases the pressure on scenic landscapes and can lead to land degradation in addition to deforestation. In addition, tourism infrastructure can reduce the esthetic appeal of the area and have direct or indirect impacts on wildlife (for example, by limiting the migration routes of animals). Exceeding the number of visitors over the ability of nature to withstand this anthropogenic load can put enormous pressure on nature and lead to harmful consequences. This can be soil erosion, increased pollution, discharges into water bodies, and deterioration of the animal habitat. This could be increasing pressure on endangered species and increasing the vulnerability of forests to wildfires. Add to this the increased pressure on water use. This can lead to a lack of drinking water for the local population. And we will also add an increase in the noise level, illegal placement of solid municipal waste.
(d) minimization of negative consequences for the environment in the development of eco-tourism largely depends on the level of ecological and legal culture, both of the local population and the tourists themselves. In addition to throwing household garbage by tourists (which can be eaten by animals, and then die), they also often frighten animals, try to feed them harmful products (for example, sweets). They also trample down glades and pick flowers, and carve their names and other inscriptions on stone and wood. They buy poaching products derived from endangered animal species, etc.

4. environmental insurance contract. Environmental insurance in the Russian Federation refers to one of the types of civil liability insurance. This is the responsibility of citizens and legal entities whose activities are associated with increased danger to others, in particular, as a result of negative environmental impacts (Zelenova, 2018). However, there is no mechanism to implement the provisions of Article 18 of the Federal Law "On Environmental Protection", which provides for environmental insurance, yet.

An experiment conducted in the mid-1990s on environmental insurance in nineteen constituent entities of the Russian Federation showed the insufficiency of legal regulation of this type of insurance. But even in the absence of the relevant federal law, experimental calculations have shown the following. The environmental risk insurance system allows you to compensate for up to 40% of the losses caused, while maintaining a sufficiently high-financial stability of insurance operations. In addition, the insurer's interest in obtaining maximum profit is better than, any controller for monitoring the state of environmental protection equipment at the policyholder. From the point of view of the current legislation on environmental insurance, it can be mandatory and voluntary.

At the moment, there is no mandatory environmental insurance in the Russian Federation as such, but a similar type of insurance is widely used in three related areas of public relations. First, compulsory insurance of civil liability for damage caused as a result of an accident of a hydraulic structure is provided for in Article 15 of Federal Law No. 117-FZ of July 21, 1997 "On the Safety of Hydraulic Structures". Secondly, compulsory insurance of civil liability for damage caused as a result of an accident or incident at a hazardous production facility is provided for in Article 15 of Federal Law No. 116-FZ of 21.07.1997 "On Industrial Safety of Hazardous Production Facilities". Third, Article 249 of the Commercial Shipping Code of the Russian Federation No. 81-FZ of April 30, 1999, provides for marine insurance, which, among other things, is applied in the case of transportation of dangerous goods by sea.

Currently, voluntary environmental insurance is not very common in the Russian Federation (although there are some examples of such contracts). This was due to the lack of a proper regulatory framework, and due to the lack of free funds for economic entities, which could be accumulated in the form of insurance deductions in case of unforeseen events.

25.4 Conclusion

Environmental entrepreneurship is a type of entrepreneurial activity carried out by legal entities and individual entrepreneurs at their own risk and under their responsibility in order to make a profit. This activity is aimed at the production of environmental protection products, conducting research and credit and financial activities, performing environmentally significant works, and providing services. The most developed in Russia is environmental entrepreneurship, which is associated with the provision of environmental services. Among the latter, environmental audit, environmental insurance, eco-tourism, and waste management of production and consumption have been developed to a greater or lesser extent.

Over time, contractual relations in the field of voluntary environmental certification (for example, in relation to organic agricultural products) may also develop. This trend suggests the emergence of a new type of intersectoral agreements-environmental agreements. At the same time, contractual relations in the field of environmental services are currently developing unevenly. Today, two types of environmental contracts are most in demand in Russia—the provision of services for the management of solid municipal waste, as well as for the provision of eco-tourism services. Contracts for environmental insurance, environmental audit, and voluntary environmental certification are currently not properly applied and distributed in the Russian Federation.

Acknowledgements The work was supported by Russian Science Foundation (project No. 20-18-00314).

References

Federal Law. (1995). *On Specially Protected Natural Territories.* Adopted by the State Duma on February 15, 1995, No. 33-FZ.

Federal Law. (1996). *On the Basics of Tourist Activity in the Russian Federation.* Adopted by the State Duma on October 4, 1996, No. 132-FZ.

Federal Law. (1997a). *On the Safety of Hydraulic Structures.* Adopted by the State Duma on June 23, 1997, No. 117-FZ.

Federal Law. (1997b). *On Industrial Safety of Hazardous Production Facilities.* Adopted by the State Duma on June 20, 1997, No. 116-FZ.

Federal Law. (1998). *On Production and Consumption Waste.* Adopted by the State Duma on May 22, 1998, No. 89-FZ.

Federal Law. (1999). *Code of Merchant Shipping.* Adopted by the State Duma on March 31, 1999, No. 81-FZ.

Bekisheva, S.D. (2011). Development of ecological entrepreneurship as a direction of ecological policy of the Republic of Kazakhstan. *Environmental Law, 5*, 27–32.

Bukharova, O. (2010). According to the Laws of Nature. *Rossiyskaya Gazeta* [Russian Newspaper], November 26, 5347.

Eco-tourism on the way to Russia. Principles, recommendations, Russian and foreign experience (2002). Grif and K.

Inshakova, A., Goncharov, A., & Deryugina, T. (2018a). Crop insurance with state support: Ensuring the sustainability and competitiveness of the agricultural sector of the Volgograd region. In *Proceedings of the international scientific conference "competitive, sustainable and secure development of the regional economy: Response to global challenges" (CSSDRE 2018)* (pp. 654–658). Atlantis Press.

Inshakova, A., Goncharov, A., & Kazachenok, O. (2018b). The imperatives of financial policy in the sphere of the digital economy: Impacts on increasing investment activity and tax potential of

Russian regions. In *Proceedings of the international scientific conference "competitive, sustainable and secure development of the regional economy: Response to global challenges" (CSSDRE 2018)* (pp. 337–342). Atlantis Press.

Korostelev, E. M. (2008). *Practicum on ecological tourism. Educational and methodical manual.* Saint Petersburg State University.

Matytsin, D. E., & Rusakova, E. P. (2021). The strategy of quality management in industry 4.0 and formation of the cognitive economy based on industrial and manufacturing engineering in the Russian Federation and countries of the EU. *International Journal for Quality Research, 15*(4). https://doi.org/10.24874/IJQR15.04-03

Order of the Ministry of Natural Resources of the Russian Federation of 1995, On the Consideration of the Environmental Factor in the Privatization of State and Municipal Enterprises and Organizations (Adopted by the Ministry of Natural Resources of the Russian Federation on November 21, 1995, No. 469).

Ruden, E. V. (2015). *Legal regulation of environmental contractual relations in Ukraine: Abstract of the dissertation of the Candidate of Legal Sciences.* Taras Shevchenko National University of Kyiv.

Ryzhenkov, A. Y. (2014). Legal regulation of ecotourism development in Russia: Problems and prospects. Business. Education. Right. *Bulletin of the Volgograd Institute of Business, 4*, 259–264.

Ryzhenkov, A. Y. (2018). Legal problems of the development of environmental entrepreneurship in the Russian Federation. *Laws of Russia: Experience, Analysis, Practice, 1*, 66–70.

Selivanova, S. I. (2019). The importance of environmental education in the formation of ecological culture. *Industry and Agriculture, 12*, 96–98.

Serov, G. P. (2000). *Environmental audit. Conceptual and organizational frameworks.* Exam.

Urazova, L. P. (2007). *Management of the development of environmental entrepreneurship: Abstract of the dissertation of the Candidate of Economic Sciences.* St. Petersburg Academy of Management and Economics.

Varennikova, E. V. (2011). Environmental policy as a form of organization of environmentally oriented entrepreneurship. *Bulletin of the Leningrad State University Named after A.s Pushkin, 6–1*, 60–67.

Zelenova, D. A. (2018). Environmental insurance as a way to ensure environmental safety. *Education and Science in Russia and Abroad, 6*, 125–129.

Zlobin, S. V. (2011). *Legal regulation of environmental entrepreneurship in the Russian Federation: Abstract of the dissertation of the candidate of legal sciences.* Volgograd State University.

Chapter 26
Legal Incentives as a Means of Mediating the Development of Environmental Entrepreneurship

Agnessa O. Inshakova, Tatiana V. Deryugina, and Albert V. Tumakov

Abstract The chapter raises and solves the problem of using effective mechanisms of legal regulation aimed at the development of environmental entrepreneurship. Based on the analysis of Russian, foreign, and international legislation, scientific doctrine, and the experience of some foreign countries, conclusions are drawn about the need to use positive means of legal regulation (legal incentives) and negative ones (legal restrictions, prohibitions, and obligations). It is proved that when regulating rights in private sectors, it is necessary to use incentives to a greater extent, and restrictions in public sectors. The nature of relations related to environmental entrepreneurship is investigated. The key categories that encourage business entities to link their activities with the protection of the natural environment are needs and interests. The authors point out the complex structure of interests concerning the protection of the natural environment. Legal mechanisms for observing the interests of all participants in legal relations are proposed. The analysis of the institute of the best available technologies in the field of environmental protection is carried out. Conclusions are drawn about the prospects of its use in business activities and further development. The nature of the concept of ecological entrepreneurship is investigated. Conclusions are drawn about the too narrow approach outlined in the normative legal acts. It is pointed out that it is necessary to consider any business activity in the process of which technologies related to the protection of the natural environment are used as environmental entrepreneurship. The necessity of changing the system of legal regulation by increasing the number and quality of legal incentives is justified. A system of legal incentives is proposed to optimize the processes of environmental protection in the course of business activities.

Keywords Environmental entrepreneurship · Interest · Legal incentives · Legal restrictions · Environmental protection · Mechanism of legal regulation

---

A. O. Inshakova (✉)
Volgograd State University, Volgograd, Russia
e-mail: ainshakova@list.ru; gimchp@volsu.ru

T. V. Deryugina · A. V. Tumakov
Moscow State University of the Ministry of Internal Affairs of Russia Named After V.Ya. Kikot, Moscow, Russia

A. O. Inshakova and E. E. Frolova (eds.), *Smart Technologies for the Digitisation of Industry: Entrepreneurial Environment*, Smart Innovation, Systems and Technologies 254, https://doi.org/10.1007/978-981-16-4621-8_26

JEL Codes K32 · L26 · P28 · O44 · P18 · Q52 · Q56

26.1 Introduction

The establishment and improvement of the institute of environmental entrepreneurship are a key factor in the development of modern society. At the same time, it is important not only to focus the state policy on the introduction of the best available technologies in the field of environmental protection but also to make businesses aware of the need to use modern means of environmental production. Only the coincidence of these aspirations can lead to the desired result.

The legal regulation of relations related to the protection of the natural environment has a complex nature, which is primarily due to the need to respect public and private interests (Inshakova et al., 2018a). One of the legal means to solve this problem is legal incentives. Optimization of legal incentives is one of the most effective mechanisms for regulating relations, including those related to environmental entrepreneurship.

The establishment of legal incentives is aimed at synchronizing public and private interests, where each of the parties can get its benefits: the state and society—a favorable environment, and the business entity-profit from conducting eco-friendly production. Legal incentives are designed to motivate entrepreneurs to invest in environmental protection.

Incentives are an objective necessity for the development of entrepreneurial activity in any sphere, especially in the environmental and social spheres. The effectiveness of the development of any sector of the economy depends on how the system of legal incentives is formulated. Any legal system should contain both certain incentives and restrictions, which together will create an effective mechanism for protecting the natural environment. At the same time, it is important to remember that the effectiveness of legal regulation largely depends on the consistency and consideration of the interests of all participants in legal relations (the state, society, and the business entity).

The basis of environmental and legal incentives is the method of positive motivation, aimed primarily at the formation of the business entity's economic interest in environmental protection measures. An important role is played by the formed environmental needs of the society and the economic assessment of measures to protect the environment.

When improving the legal regulation, it is necessary to investigate the nature of the legal relations that arise, as well as the main motivations that influence the formation of interest in the business entities in carrying out measures to protect the natural environment. At the same time, it is necessary to take into account both the specifics of entrepreneurial activity, the main purpose of which is to make a profit, and the specifics of environmental activity, and first of all, its public and state importance, but at the same time, the economic cost with minimal financial efficiency. All these factors should be taken into account when building a system of legal regulation with the use of mechanisms of legal incentives and restrictions.

26.2 Materials and Methods

Russian and international sources of legal regulation were used in the preparation of the research. Among the Russian legal acts are the following. Federal Law No. 7-FZ "On Environmental Protection" (State Duma of the Russian Federation, 2002). Civil Code of the Russian Federation (State Duma of the Russian Federation, 1994). Federal project "Introduction of the best available technologies", developed within the framework of the national project "Ecology" (Presidium of the Presidential Council for Strategic Development and National Projects, 2018). Information and technical Guide to the best available technologies (ITS) 46-2017 "Reduction of emissions of pollutants, discharges of pollutants during the storage and warehousing of goods" Rosstandarat (2017a). ITS-2017 'Improving energy efficiency in the implementation of economic or other activities' (Rosstandarat, 2017b).

Among the international acts, the following can be distinguished: Directive 2008/1/EEC of the European Parliament and of the Council of 15 January 2008 on Integrated Pollution Prevention and Control (European Parliament and Council of the European Union, 2008); Council Directive 85/337/EEC of 27 June 1985 on the environmental impact assessment of certain projects implemented by the public and private sector (European Parliament and Council of the European Union, 1985). Much attention is paid to the analysis of the Model Law of the Union of Independent States (CIS) (2000) "On the Basis of Environmental Entrepreneurship" (Inter-Parliamentary Assembly of the Member States of the Union of Independent States, 2000).

The scientific basis of the study was the works of Russian lawyers Barkov and Grishina (2018); Vafin and Zolotova (2019), Kalinichenko (2001), who paid great attention to environmental entrepreneurship in their works. The works of Gukasyan (1970), Vaneeva (1988), Pershina (2002), Gribanov (2001), Malko (1995), (Inshakova et al., 2018a), Matytsin (2021), and others are devoted to the problems of interest, its correlation with needs, incentives and restrictions, and mechanisms of legal regulation.

The topic identified in the study is important not only for individual national systems but also for the entire international community. Because of this, the scientific literature of foreign countries is represented by numerous works in the field of environmental entrepreneurship. When conducting the study, the following works were used: He et al. (2020), Sun et al. (2020), Omri (2018), Ranjan (2019), Schlange (2009), Gurău and Dana (2018), also used the works of other authors.

The study used a dialectical method, which allowed us to consider the mechanism of legal regulation of environmental entrepreneurship as a single system, to investigate contradictions, to carry out systematization. Using the method of analysis, individual elements of the mechanism of legal regulation of environmental entrepreneurship are studied, a detailed analysis of Russian, foreign, and international legislation is made, and scientific doctrines are studied. The used method of legal analogy made it possible, based on the study of the experience of foreign countries, international experience, to formulate development prospects and recommendations for the further development of the current legislation in the field of environmental entrepreneurship. Using the dogmatic method, formal and logical connections between the presence of legal incentives and the development of environmental entrepreneurship is established. The method of the systematic approach allowed us to consider all the elements of the mechanism of legal incentives for environmental business entities.

26.3 Research

The problem of ecological entrepreneurship and environmental protection is relevant both for individual countries (Barkov & Grishina, 2018; Vafin & Zolotova, 2019; He et al., 2020), continents (Sun et al., 2020), and the entire international community (Omri, 2018). The law of various States provides for a variety of legal means aimed at minimizing the consequences of conducting business. Among the latter, there are various incentives and restrictions imposed on business entities to interest them in using various methods aimed at preventing harm to the environment.

Every action is based on needs and interests. Without going into the controversy about the relationship of these categories (Gukasyan, 1970), we take as a basis the statement that the need is a prerequisite for the formation of interest (Pershina, 2002). In turn, interest can be considered as the goal of the implementation of subjective civil law (Gribanov, 2001) and its limit. Civil legislation (Article 1 of the Civil Code of the Russian Federation (hereinafter referred to as the Civil Code of the Russian Federation) (the State Duma of the Russian Federation, 1994) proclaimed the possibility of acquiring and exercising subjective rights by one's own will and in one's interest. However, the interests of one subject are always opposed by the interests of other subjects (in absolute legal relations—this is an unlimited and indefinite circle of persons, in relative legal relations-specific subjects and their interests), which means that it is possible to consider the interest as the limit of the exercise of a subjective right when the interests of one subject are limited to the interests of other persons. In the case of environmental entrepreneurship, the interests of both the state, the entrepreneur and individuals must be aimed at achieving one goal—the protection of the natural environment. Only with this approach can you achieve tangible results.

However, the entrepreneur's interest is primarily in making a profit (paragraph 1 of Article 2 of the Civil Code of the Russian Federation), otherwise, such activities will not qualify as entrepreneurial. In contrast to this interest, the citizens' interest in a favorable environmental environment is opposed. And here the task of the state, which carries out the legal regulation of both business relations and relations aimed at protecting the natural environment, is to harmoniously use the set of legal means that allow the entrepreneur to realize his interest in making a profit and take measures aimed at preventing environmental consequences from doing business.

Returning to our original statement that any action is based on interest, we state that in order for a business entity to need to implement the best available technologies for environmental protection, it must have an interest. Ensuring this interest is in the field of the possibility of the state, which, through the legal regulation of relations, can establish various incentives that contribute to the emergence of interest in the subject of entrepreneurial activity. At the same time, it should be noted that in addition to incentives, legal restrictions are also an effective legal means of solving problems of environmental protection. Only a harmonious combination of incentives and restrictions will solve the problem of introducing the best available technologies for environmental protection by business entities.

An analysis of environmental legislation shows that most of it are restrictive and punitive. The restrictive nature is expressed in the establishment of legal prohibitions, and the punitive one is aimed at applying legal protection measures in the commission of environmental offenses (State Duma of the Russian Federation, 2002). The main message on which the system of legal norms is based is the presumption of the environmental danger of any business activity (Article 3 of the Federal Law "On Environmental Protection"). Very little attention has been paid to promoting eco-friendly business.

As an economic regulation, the legislation provides for the collection of fees for negative impact on the environment (Article 16 of the Federal Law "On Environmental Protection"); environmental insurance; incentives for the termination of non-ecological production; a system of state support measures, including the development of investments; the introduction of new technologies; educational and information support; the use of renewable energy; new methods of controlling environmental pollution. Among the economic instruments are the provision of tax benefits; funds from the federal and regional budgets; reduction of the level of payment for negative impact on the environment, which in general should be assessed positively. However, this provision is offset by the norm laid down in paragraph 4 of Article 16 of the Federal Law "On Environmental Protection", which formulates the conditions for obtaining state support.

First, government support can be obtained by implementing the best available technologies, and we will discuss this issue below. Secondly, state support can be obtained under the condition of the design, construction, reconstruction, and installation of environmental systems listed in the federal law in the closed list. This, of course, prevents the introduction of the latest technologies that are unknown at the time of preparation of the law and are not included in the closed list for this reason. Thus, the use of a closed list in a mandatory norm, in our opinion, significantly reduces the effectiveness of this legal provision.

It should be noted that in general, Russian legislation in the field of environmental protection is similar to foreign legislation, where, as a rule, the most popular legal incentives are state subsidies for environmental protection measures. Interesting in this regard is the experience of Australia, where, unlike most countries, subsidies are paid not for a specific environmental action (for example, the construction of wastewater treatment plants, etc.), but activities in the field of environmental protection—ecosystem services (Ranjan, 2019).

As a means aimed at protecting the natural environment, the Russian legislator uses the institutions of licensing, environmental certification of economic activities, obtaining a comprehensive environmental permit, declaring the impact on the environment, zoning, establishing special legal regimes, planning, monitoring, methods of control and supervision, state accounting, and others. At the same time, the bulk of the listed legal means are aimed at prohibiting, restricting, or burdening the business entity.

The implementation of the best available technologies is based on the federal project "Implementation of the best available technologies", developed within the framework of the national project "Ecology" (Presidium of the Presidential Council for Strategic Development and National Projects, 2018). One of the components of this project is a system of information and technical reference books on the best available technologies. An analysis of these reference books, in particular, the information and technical reference book ITS 46-2017 "Reducing emissions of pollutants, discharges of pollutants during the storage of goods' (Rosstandarat, 2017a), ITS-2017 "Improving energy efficiency in the implementation of economic or other activities" (Rosstandarat, 2017b) shows that they are standardized rules for the implementation of a particular activity, which, in general, is not bad. However, it cannot be called a breakthrough technology aimed at a comprehensive solution to the problem of negative impact on the environment.

Among the economic incentives, the federal project names only the possibility of obtaining subsidies to reimburse part of the cost of obtaining coupon income on bonds issued to implement investments in the introduction of the best protective technologies (Matytsin, 2021).

It should be noted that the Russian federal project "Introduction of the best available technologies" corresponds to the main trends in the implementation of such projects in the countries of the world. Following the Organization for Economic Cooperation and Development's best available technology policy, various countries (Germany, Sweden, India, China, the United States, and others) have developed their own best available technology concepts (OECD, 2019). At the same time, if we consider the legal means laid down in such concepts, they are primarily associated not with the development of the best technologies themselves, but with careful control over the compliance of business activities with the standards developed in the best available technologies directories. This is not to say that this approach is bad enough. However, the national legislation of some countries shows the same trend as the Russian legislation. Maximum attention is paid to restrictions and prohibitions, and very few legal and economic incentives are prescribed.

The review of the economic instruments laid down in the environmental legislation shows that there are only four economic incentives (tax incentives, the possibility of receiving subsidies from the federal and regional budgets, reduced fees for negative environmental impacts, and reimbursement of part of the cost of obtaining coupon income), as opposed to numerous restrictions. We believe that this approach is not entirely justified. The legislator should use a legal technique based on a harmonious combination of incentives and restrictions. The establishment of numerous prohibitions reduces the effectiveness of tools aimed at introducing the latest environmental technologies.

The analysis of international acts in the field of environmental entrepreneurship shows the focus of the international community on solving the problem of negative impact on the environment as a result of entrepreneurial activity. For about fifty years, the European Community has been shaping environmental policy (Kalinichenko, 2001). Since 1973, the European Community's environmental action programs have been adopted, which form the main areas of activity in the field of environmental protection and recommend that the participating countries make appropriate changes to their national legislation. The work of the European Committee on Environment and Health is also important. The Organization for Economic Cooperation and Development, which unites the countries of North and South America, Europe, and the Asia-Pacific region in two hundred committees, conduct developments on environmental protection (OECD, 2019).

Directive 2008/1/EEC of the European Parliament and of the Council of the European Union of 15 January, 2008 on Integrated Pollution Prevention and Control (European Parliament and Council of the European Union, 2008) can be considered as one of the main documents that form the policy in the field of environmental entrepreneurship. Council Directive 85/337/EEC of 27 June, 1985 On the Environmental Impact Assessment of Certain Projects Implemented by the public and private sector (European Parliament and Council of the European Union, 1985) can also be considered. Both documents are aimed at establishing a set of measures to control and prevent pollution. Among the proposed tools, the directives provide for preventive and control measures. Preventive measures include issuing permits, licensing new facilities, setting standards, using the best available technologies, and public participation in project discussions. Control measures include state, cross-border, and public control. Among the incentives, we can single out only the encouragement of public associations that take part in the discussion and examination of environmental projects.

The member states of the Union of Independent States (CIS) have developed a model law (2000) "On the Basics of Environmental Entrepreneurship " (Inter-Parliamentary Assembly of the Member States of the Union of the Independent States, 2000), aimed at regulating the activities of performing works and providing environmental services. The analysis of this law allows us to identify a number of negative trends. The first is the proclamation of the principle of redistributing the costs of environmental protection to entrepreneurs, which, together with the presence of a fee for a negative impact on the environment, can be considered as placing a double burden on business. The second is an extremely narrow approach, indicated in the formulation of the concept of environmental entrepreneurship, which is understood as activities related to the production of goods, the performance of works and the provision of services in order to ensure the preservation, protection and restoration of the natural environment (Inshakova et al., 2018b). Thus, this definition allows us to say that there can be environmental entrepreneurship and other things that do not have the goal of preserving the natural environment. We believe that this is incorrect. Any business, in any field of activity, should be eco-friendly (to a greater or lesser extent), it should always have the goal (along to make a profit) of protecting the natural environment. This can be expressed in the use of economical and energy-saving technologies, the development and implementation of protective equipment and equipment, control of wastewater, and so on. But in any activity-this approach should be considered as a priority.

The presented model law proceeds from the fact that the subjects of environmental entrepreneurship are persons whose production of eco-friendly products is the main activity and the share of these products is at least 75% of the annual volume of production (Article 4 of the model law "On the Basics of Environmental Entrepreneurship"). At the same time, the volume of incentives is spelled out as general directions. Thus, incentives also play an insignificant role in the basis of legal regulation at the CIS level, which, of course, does not form an entrepreneurial interest for activities in this area.

At the heart of any business, activity is a certain interest and due to the specifics of business activity, this interest is associated with making a profit. On the other hand, in order to preserve, restore and protect the natural environment, the task of any activity should be the preservation and enhancement of environmental values. Thus, the target of any business, not only environmental, can now be defined as "making money by solving environmental problems" (Schlange, 2009).

Considering environmental entrepreneurship from the point of view of the interest of the business entity in the protection of the natural environment, we can note the experience of a number of countries related to the construction of an interdependent system. In these countries, the relationship is built between the local community and business in relation to the natural environment of a particular region. This model is usually associated with community entrepreneurship (Gurău & Dana, 2018). The positive aspects of such a system are the presence of an interconnected and mutually conditioned interest on the part of all the subjects of relations. The local community, as well as local businesses, has a direct interest in preserving the habitat in which they live.

The review of the Russian, foreign and international legislation in the field of environmental protection shows a significant preponderance of negative regulators (restrictions, bans, binding, including the establishment of fees for use, regulatory compliance, licensing, etc.) over positive regulators (benefits, subsidies, incentives). Although positive regulators contribute to the development of the entrepreneur's interest in the introduction of eco-friendly technologies and equipment in order to protect the natural environment.

Of course, the first place should go to encourage business entities to modernize production, reduce the negative impact on the environment, reduce resource consumption, and introduce waste-free technologies, and so on. At the same time, it is necessary to correctly combine incentives and restrictions. The general trend in the use of these legal means should be as follows: incentives should be used more for private industries, and restrictions should be used more for public ones.

The legal theory relates legal incentives to subjective rights, interests, benefits, and incentives (Malko, 1995). In light of the above, it is necessary to improve the legal regulation of relations related to environmental entrepreneurship. Such improvement should be carried out in the following areas: to ensure the interest of the business entity in participating in activities aimed at protecting the natural environment, by establishing additional subjective rights, benefits, and incentives.

26.4 Conclusions

The conducted research allows us to draw a number of conclusions aimed at forming the interest of the business entity in the implementation of environmental activities through establishing legal incentives. Any business (not only environmental business), in any field of activity, should be eco-friendly, it should always have the goal (along with the goal of making a profit) to protect the environment. This approach should be considered a priority.

An analysis of the current Russian foreign and international legislation in the field of environmental protection has shown a significant preponderance of negative regulators (restrictions, prohibitions, bindings) over positive regulators (benefits, subsidies, incentives). These regulators contribute to the development of the entrepreneur's interest in the introduction of eco-friendly technologies and equipment in order to protect the natural environment. The necessity of introducing a number of legal and economic incentives is proved.

It is proved that for a business entity to need to implement the best available technologies for environmental protection, and it must have an interest. Ensuring this interest is in the field of the possibility of the state, which, through the legal regulation of relations, can establish various incentives that contribute to the emergence of interest in the subject of entrepreneurial activity.

A system of incentives is proposed that allows the business entity to form an interest in environmental protection measures from the following elements. Financial and credit mechanisms (provision of concessional loans or interest-free loans, subsidies, provision of subventions); tax [tax benefits or tax exemption (temporary or permanent)]; price (regulation of prices for primary resources, energy carriers); formation of a state electronic information system to promote best environmental practices; introduction of accelerated depreciation of fixed assets with environmental purposes.

The article substantiates the need to use the institution of payment for ecosystem services as a legal incentive.

Acknowledgements The work was supported by Russian Science Foundation (project No. 20-18-00314).

References

Barkov, A. V., & Grishina,Ya. S. (2018). Methodological foundations of constructing the Russian model of legal support for environmental entrepreneurship. *Civil Law, 4*, 4–8.

Gribanov, V. P. (2001). *Interest in Civil Law. Implementation and protection of civil rights*. Statute.

Gukasyan, R. (1970). *The problem of interest in the Soviet civil procedure law*.

Gurău, C., & Dana, L.-P. (2018). Environmentally-driven community entrepreneurship: Mapping the link between natural environment, local community and entrepreneurship. *Technological Forecasting and Social Change, 129*, 221–231.

He, J., Nazari, M., Zhang, Y., & Cai, N. (2020). Opportunity-based entrepreneurship and environmental quality of sustainable development: A resource and institutional perspective. *Journal of Cleaner Production, 256*, 120390.

Inshakova, A., Goncharov, A., & Deryugina, T. (2018a). Crop insurance with state support: Ensuring the sustainability and competitiveness of the agricultural sector of the Volgograd region. *Proceedings of the international scientific conference "competitive, sustainable and secure development of the regional economy: Response to global challenges" (CSSDRE 2018)* (pp. 654–658). Atlantis Press.

Inshakova, A., Goncharov, A., & Kazachenok, O. (2018b). The imperatives of financial policy in the sphere of the digital economy: Impacts on increasing investment activity and tax potential of Russian regions. *Proceedings of the international scientific conference "competitive, sustainable and secure development of the regional economy: Response to global challenges" (CSSDRE 2018)* (pp. 337–342). Atlantis Press.

Inter-Parliamentary Assembly of the Member States of the Union of Independent States. (2000). *Model Law "On the Fundamentals of Environmental Entrepreneurship"* (June 13, 2000, No. 15–6).

Kalinichenko, P. A. (2001). *Environmental protection in the activities of the European Community (international legal aspect)*. Ph.D. in Law thesis.

Malko, A. V. (1995). *Incentives and restrictions in law: Theoretical and informational aspect: Dissertation of the Doctor of Law*. State Academy of Law.

Matytsin, D. E. (2021). Internet-investing as a remote algorithm of retail investment financing. Modern Global Economic System: Evolutional Development vs. Revolutionary Leap. *Springer book series "Lecture Notes in Networks and Systems"*. Springer Nature.

OECD (2019). Best Available Techniques for Preventing and Controlling Industrial Pollution, Activity 3: Measuring the Effectiveness of BAT Policies, Environment, Health and Safety, Environment Directorate, OECD/OECD, 2 rue André-Pascal, 75775 Paris Cedex 16, France. https://www.oecd.org/chemicalsafety/risk-management/measuring-the-effectiveness-of-best-available-techniques-policies.pdf

Omri, A. (2018). Entrepreneurship, sectoral outputs, and environmental improvement: International evidence. *Technological Forecasting and Social Change, 128*, 46–55.

Pershina, I. V. (2002). *Interest in law*. Ph.D. in Law thesis.

Presidium of the Presidential Council for Strategic Development and National Projects. (2018). *Passport of the national project "Ecology"* (December 24, 2018, No. 16).

Ranjan, R. (2019). Deriving double dividends through linking payments for ecosystem services to environmental entrepreneurship: The case of the invasive weed Lantana camara. *Ecological Economics, 164*, Article 106380.

Rosstandarat. (2017a). *Information and technical reference ITS 46-2017 "Reduction of emissions of pollutants, discharges of pollutants during storage and warehousing of goods"* (October 10, 2017, No. 2141).

Rosstandarat. (2017b). *Information and Technical Reference Book ITS-2017 "Improving energy efficiency in the implementation of economic or other activities"* (September 29, 2017, No. 2060).

Schlange, L. E. (2009). stakeholder identification in sustainability entrepreneurship. *Greener Management International, 55*, 13–32.

State Duma of the Russian Federation. (1994). *Civil Code of the Russian Federation (Part One)* (November 30, 1994, No. 51-FZ; revised July 31, 2020).

State Duma of the Russian Federation. (2002). *Federal Law "On Environmental Protection"* (January 10, 2002 No. 7-FZ, revised July 31, 2020).

Sun, H., Pofoura, A. K., Mensah, I. A., Li, L., & Mohsin, M. (2020). The role of environmental entrepreneurship for sustainable development: Evidence from 35 countries in Sub-Saharan Africa. *Science of Total Environment, 741*, 140132.

The European Parliament and the Council of the European Union. (1985, June 27). *Council of the European Union Directive 85/337/EEC on the environmental impact Assessment of certain Public and Private Sector Projects.*

The European Parliament and the Council of the European Union. (2008, January 15). *Directive 2008/1/EEC of the European Parliament and of the Council on Integrated Pollution Prevention and Control.*

Vafin, V. V., & Zolotova, O. A. (2019). Theoretical and legal aspects of the development of environmental entrepreneurship in the region. *Civil Law, 2*, 16–20.

Vaneeva, L. A. (1988). Implementation of the constitutional right of citizens of the USSR to judicial protection in civil proceedings.

Chapter 27
Main Trends and Prospects for the Development of Legislation on Environmental Entrepreneurship

Agnessa O. Inshakova and Tatiana V. Deryugina

Abstract The chapter of the monograph raises and solves the problems of the development of legislation on environmental entrepreneurship. The problems related to both the legal regulation of environmental entrepreneurship relations and the problems related to the development of the institute are analyzed. Based on the study of the main concepts of the development of legal regulation of environmental business relations, formulated in the works of both domestic and foreign civilists, conclusions are drawn about the main trends and prospects for the development of legislation. The article evaluates the theories of legal regulation of environmental entrepreneurship that exist in the doctrine, as well as the legal means and tools used to regulate legal relations. Approaches to defining the principles of environmental entrepreneurship are analyzed. It is pointed out that it is necessary to combine the principles of private and public law, as well as to develop its principles of the institute of environmental entrepreneurship. It is proved that the principle of any business activity should be the preservation and multiplication of natural resources, the creation of a favorable natural environment. The analysis of normative sources of law concerning environmental entrepreneurship is carried out. In the analysis of Russian normative legal acts and foreign and international acts, the same trends of legislation are noted in relation to the presumption of causing environmental harm as a result of business activities. The negative aspects of this approach are pointed out. A system of legal means aimed at improving legislation concerning environmental entrepreneurship is formulated. The author points out the need to allocate legal means and legal instruments. Their classification is carried out. Preventive and follow-up, regulatory, and protective measures are allocated. Separate types of the specified means are formulated.

A. O. Inshakova (✉)
Volgograd State University, Volgograd, Russia
e-mail: ainshakova@list.ru; gimchp@volsu.ru

T. V. Deryugina
Moscow State University of the Ministry of Internal Affairs of Russia Named After V. Ya. Kikot, Moscow, Russia

A. O. Inshakova and E. E. Frolova (eds.), *Smart Technologies for the Digitisation of Industry: Entrepreneurial Environment*, Smart Innovation, Systems and Technologies 254, https://doi.org/10.1007/978-981-16-4621-8_27

Keywords Environmental entrepreneurship · Mechanism of legal regulation · Legal means · Legal incentives · Natural environment · Legal instruments of legal regulation

JEL Codes K32 · L26 · Q56 · Q52 · O44 · P28 · P48

27.1 Introduction

Recently, the study of problems related to ecology and the conduct of environmental entrepreneurship occupies one of the leading positions among representatives of various branches of science. The environmental component of the development of society, along with the economic and social, requires more and more attention. More and more in science, at the state and international level, there is a talk about the need for socialization and greening of entrepreneurship. This implies the resolution of environmental and social problems using entrepreneurial activity-entrepreneurial tools (Batalina et al., 2008). Many scientists point to the inseparability of the concepts of ecology and entrepreneurship (Barkov & Grishina, 2018).

Among the key issues that require quick and high-quality resolution, the following are highlighted. Financial (problems related to the financing of projects); tax-economic (issues aimed at stimulating environmental activities); educational (development of environmental education and awareness and formation of environmental culture); informational (informing citizens and government agencies on environmental issues, committing environmental offenses, etc.). One of the tools for solving these problems is the creation of a regulatory framework that adequately regulates the existing legal relations in this area.

The complexity of creating a legal regulation of the sphere of environmental entrepreneurship is due to the multidimensional nature of the existing problems. This is the presence of many gaps and contradictions; the need to use numerous means and tools of the legal regulation mechanism. This is the intersectoral nature of environmental and business legal relations, which makes it necessary to use legal means of both private and public branches of law. This is the lack of a unified concept of building a system of legal regulation. This is the debatable nature of many approaches and the lack of a unified position outlined in the doctrine regarding the problems associated with environmental entrepreneurship and ways to solve them. Complicating the problem is the lack of a comprehensive legal act regulating environmental entrepreneurship.

We should also note a number of factors that negatively affect the development of environmental entrepreneurship itself, which also affect the construction of a unified system of legal regulation. These include low resource potential of environmental entrepreneurship; lack of professional personnel; low level of environmental literacy.

Taking into account the above, the task of this study is to develop the main provisions of the concept of legal regulation of environmental entrepreneurship, aimed at solving a complex of problems related to ecology and entrepreneurship, where entrepreneurship is the main tool for solving environmental issues.

27.2 Materials and Methods

The theoretical basis of the study was the works of Russian and foreign authors. When analyzing the concepts of building a mechanism for the legal regulation of environmental entrepreneurship, we used the works of domestic legal scholars, Inshakova et al. (2017, 2018), Matytsin and Rusakova (2021), Barkov and Emelyanova (2020), Naumov (2006), Bobkova (2013), Mityakina and Turanin (2019), Volkov (2018), Mokhov (2018), Novoselov (2012), Poiseev and Zamorshchikova (2013), Chechunov (2000), Khamdamov et al. (2018), and Vafin and Zolotova (2019). Among foreign authors, we can highlight Dean and McMullen (2007), Gregori and Holzmann (2020), Gurău and Dana (2018), and Omri (2018). In the science of international law, environmental entrepreneurship is considered by He et al. (2020), Sun et al. (2020), and others.

As a kind of social entrepreneurship, they analyze environmental entrepreneurship in their works (Mokhov, 2018; Barkov & Grishina, 2018; Meek et al., 2010).

Among the international and domestic legal acts that served as the basis for this study, we can single out the Model Law (2000) "On the Basics of Environmental Entrepreneurship" (Interparliamentary Assembly of the Member States of the Union of Independent States, 2000). We can highlight the Council of the European Union Directive 85/337/EEC of June 27, 1985; On the environmental impact assessment of certain public and private sector projects (European Parliament and Council of the European Union, 1985); Directive 2008/1/EEC of the European Parliament and of the Council of the European Union of January 15, 2008, can be highlighted; On Integrated Pollution Prevention and Control (European Parliament and Council of the European Union, 2008). We can distinguish the Federal Law "On Environmental Protection" (the State Duma of the Russian Federation, 2002); the Federal Law "On the Development of Small and Medium-sized Businesses in the Russian Federation" (the State Duma of the Russian Federation, 2007).

The present research is based on general research methods (dialectical and systematic approaches). General scientific methods of analysis and synthesis, system-functional method, induction, and deduction were used in the analysis of scientific doctrines in relation to the legal regulation of environmental business relations. In the study of environmental entrepreneurship as a type of social entrepreneurship, the method of analogy was used. Conclusions regarding the mechanism of legal regulation of environmental entrepreneurship are made using the modeling method.

The use of special legal methods (dogmatic, legal forecasting, comparative legal, and legal analogy) made it possible to conduct a comparative analysis of normative legal acts both domestic and international. To make forecasts regarding trends and prospects of development of legislation in relation to environmental entrepreneurship; to identify formal and logical links between social and environmental entrepreneurship; to conduct a comparative analysis.

27.3 Results

The scientific literature presents several key concepts of legislative reform in relation to environmental entrepreneurship.

The first theory is based on the model of social entrepreneurship, the purpose of which is the socio-ecological mission and service to the interests of society (Barkov & Emelyanova, 2020), the ecological utility of business (Naumov, 2006). With this approach, environmental entrepreneurship is viewed through the prism of social entrepreneurship, which is generally quite reasonable. The legal regulation of environmental entrepreneurship is in the process of formation, while the norms of the institute of social entrepreneurship are reflected in the legislation (State Duma of the Russian Federation, 2007). This makes it possible to use the legal tools and means of the institute of social entrepreneurship in regulating the relations of environmental entrepreneurship. Legal and entrepreneurial tools for implementing such a model are the creation of socially useful corporations. The key issue in the implementation of this concept in practice is the assessment of the public utility of the corporation in terms of its impact on the environment and social processes. In foreign legislation, this problem is solved by creating evaluation standards (Barkov & Serova, 2019). The authors of the theory propose to introduce both a mechanism of state assessment and a mechanism of non-state assessment by means of certification of socially oriented corporations (Inshakova et al., 2018).

In order to formulate a special legal regime for environmental entrepreneurship, the idea of the need for doctrinal development with subsequent implementation in the legislation of the following concepts is justified in science. These are environmental business entities; goods, works, and services classified as environmental; a system of state support measures and a set of restrictions (Mokhov, 2018).

Representatives of another concept suggest using a set of public legal restrictions based on the prohibitions of environmental and environmental legislation when building a system of legal regulation (Bobkova, 2013). In the mechanism of legal regulation, the authors propose to use the tools of private and public branches of law (business, civil, environmental, energy, and others). The special legal regime of environmental entrepreneurship should consist of special financing, taxation, lending, and other components. In general, representatives of this position are based on the intersectoral nature of environmental entrepreneurship, which, accordingly, entails the need for legal regulation to also use an intersectoral approach and legal means of various branches of law (Mityakina & Turanin, 2019). The authors also express a position on the need to create an intersectoral codified regulatory act of the Environmental Code (Volkov, 2018).

There is another approach, the authors of which proceed from the fact that environmental entrepreneurship should include only those that specialize in the production of equipment for the protection of the environment, eco-friendly goods (Chechunova, 2000), the provision of environmental services (Novoselov, 2012), and the development of energy-saving technologies (Poiseev & Zamorshchikova, 2013). In light of this, the legal regulation of environmental entrepreneurship includes only normative legal acts that regulate business activities and establish standards in the environment. This approach can be described as narrow. However, it is reflected in modern Russian legislation. In particular, within the framework of the national project "ecology" (the Presidium of the Presidential Council for Strategic Development and National Projects, 2018), the federal project "introduction of the best available technologies" was developed, the main concept of which is the use of environmentally friendly and resource-saving technologies.

The introduction of environmental audit (Bryansk, Irkutsk, Kaliningrad, Kirov, and other regions) (Vafin & Zolotova, 2019) and environmental insurance (Kaliningrad, Kirov, Orenburg regions, the Republic of Ingushetia, Sakha (Yakutia), Tyva, and others) (Vafin & Zolotova, 2019) also can be mentioned as legal means used in Russian federal and regional legislation. Such measures can be classified as preventive measures aimed at preventing environmental violations and creating the possibility of compensation in the event of their occurrence.

On the contrary, the broad approach is denoted by J. Khamdamov, E. Gafforova, and T. Yershova, which combines the first and third positions described above (Khamdamov et al., 2018). Thus, the authors refer to environmental entrepreneurship as all types of business activities related to the production of environmental products. This is the production of eco-friendly goods, works, and services; the solution of environmental problems in the interests of society, both in the process of carrying out business activities and in the process of carrying out special activities aimed at solving environmental problems (Matytsin & Rusakova, 2021). Thus, the authors combine the idea of creating a model of environmental entrepreneurship, acting in the interests of society, and a narrow approach to regulating relations related to the production of environmentally friendly goods, works, and services.

Perhaps, the broadest approach to the problem is reflected in the works of the authors, which substantiate the thesis on the establishment of the principle of priority of environmental economic activity for all business entities (Pisarev, 2019). Environmental entrepreneurship in the designated position is not distinguished as an independent category, since all entrepreneurship should initially be aimed at preserving a favorable natural environment.

In order to develop the most optimal means of legal regulation of environmental business relations, it is necessary to refer foreign and international experience. Environmental entrepreneurship in foreign countries is usually based on the experience of the establishment and development of the institute of social entrepreneurship (Meek et al., 2010). Foreign authors also emphasize the use of entrepreneurship as a means of recovery or positive impact on the natural environment (Dean & McMullen, 2007). Among the means of regulating the development of environmental entrepreneurship, it is proposed to use digital technologies (Gregori & Holzmann, 2020). Attention is also drawn to the need to develop community-based entrepreneurship, which makes it possible to establish a direct link between the local population, entrepreneurship, and the natural environment (Gurău & Dana, 2018). All these mechanisms fit into the general concept of considering environmental entrepreneurship as social entrepreneurship.

In a number of foreign countries (China, the United States, Sweden, Germany, India, and others), as in Russia, there are standards for the best available technologies (OECD, 2019). In general, this is a positive experience. The negative characteristics of this approach include the initial presumption in the legislation of causing harm to the natural environment, and, accordingly, the construction of legislation on the principle of coercion and responsibility. We believe that along with the legal means used, it is necessary to use incentive mechanisms to a greater extent. When an entrepreneur has an interest in preserving and increasing the natural environment, then this process will develop at a good pace.

In the science of international law, environmental entrepreneurship is considered from the perspective of a resource and institutional approach (He et al., 2020), as a means of sustainable development (Sun et al., 2020) and as a tool for improving the state of the natural environment (Omri, 2018). Among the international legal acts, it is necessary to pay attention to the Model Law (2000) "On the Basics of Environmental Entrepreneurship" (Interparliamentary Assembly of the Member States of the Union of the Independent States, 2000). This is one of the few legal acts that deals with the regulation of environmental entrepreneurship. However, when analyzing it, some negative trends can be noted. First, the law demonstrates a narrow approach, referring to environmental entrepreneurship only entities that produce products, works, and services for environmental purposes. Secondly, the law mainly uses prohibitions and restrictions as legal means. Third, the authors of the law, following global trends, proceed from the presumption of harm caused by business activities.

Among the international acts, the following can also be distinguished: Council Directive 85/337/EEC of June 27, 1985; On the environmental impact assessment of certain public and private sector projects (European Parliament and Council of the European Union, 1985); Directive 2008/1/EEC of the European Parliament and of the Council of the European Union of January 15, 2008; On Integrated Pollution Prevention and Control (European Parliament and Council of the European Union, 2008). Regulatory acts also do not pay the necessary attention to measures of economic incentives for business entities. The mechanism of legal regulation in these acts is based on permissive and control means aimed at preventing the commission of an offense and bringing it to justice in the event of its commission.

An important element of building a system of legal regulation of environmental entrepreneurship is the development of a system of principles as fundamental principles, guidelines for law-making, and law enforcement. At the same time, business law refers to private law, and environmental and environmental law refers to public law. This determines the existence of different legal principles due to different methods of legal regulation and legal means used in the construction of legal institutions (Inshakova et al., 2017). It is necessary to develop common principles of a single institution of environmental entrepreneurship. This approach should be recognized as correct from the point of view of methodology.

Considering the fixed principles of environmental protection (Article 3 of the Federal Law "On Environmental Protection") (respect for the rights to a favorable environment; a combination of environmental, economic, and social interests of a person; protection and reproduction; rational use of resources; and others), it should be noted the following. Their content is not disclosed, with rare exceptions (e.g., much attention is paid to the payment of natural resources and the assessment of the impact on the environment).

The analysis of normative legal acts also allows us to conclude that both Russian and international legislation proceeds from the presumption of causing harm to the environment by conducting business activities. This approach needs to change. We believe that initially, any business entity should have not only the goal of making a profit but also the goal of preserving and multiplying natural resources, creating a favorable natural environment. We believe that it is necessary to develop a system of measures aimed at the development of environmental entrepreneurship. These measures should be divided into preventive and subsequent and regulatory and protective.

Considering the main vectors of creating legal regulation of environmental entrepreneurship, it is necessary to proceed from the need to create (1) conditions for the growth of interest of business entities in solving environmental problems. At the same time, the legislator must take into account both the economic interests of entrepreneurs and the environmental interests of society. Creating a reasonable balance of such interests will solve the problem of business involvement in solving environmental problems. The solution to this problem is the most difficult, since at the intersection of the problem we have the interests of three main groups of subjects. This is a state that needs to develop products to solve socio-economic problems, which currently causes direct damage to the natural environment. These are entrepreneurs whose main goal is to make a profit. This is a society that has an interest in the existence of a favourable natural environment; (2) encourages the conduct of environmentally sound business activities.

27.4 Conclusions

The analysis of normative legal acts and theoretical concepts in relation to the legal regulation of environmental entrepreneurship has shown the following. For the most part, both legal sources and authors suggest using prohibitions and restrictions as a mechanism for the legal regulation of relations and also proceed from the presumption of harm caused by entrepreneurial activity. The mechanism of legal regulation in these acts is based on permissive and control means aimed at preventing the commission of an offense and bringing to justice in the event of its commission. We believe that along with the legal means used, it is necessary to use incentive mechanisms to a greater extent. When an entrepreneur has an interest in preserving and increasing the natural environment, then this process will develop at a good pace.

From the point of view of methodology, it should be recognized as necessary to develop common principles of the institute of environmental entrepreneurship. Initially, any business entity should have not only the goal of making a profit but also the goal of preserving and multiplying natural resources, creating a favorable natural environment. Thus, the principle of preserving and multiplying natural resources, creating a favorable natural environment should become the leading principle of business activity.

The legal measures used for the development of the institute of environmental entrepreneurship should be divided into preventive and subsequent ones. Preventive measures are aimed at preventing the commission of environmental offenses and creating a reserve for eliminating the consequences in the event of violations. This includes environmental audits; environmental insurance; monitoring; environmental certification; financing of environmental projects; provision of tax benefits and benefits for payment (up to exemption from payment) for negative environmental impacts; reimbursement of part of the cost of obtaining coupon income. Follow-up measures are aimed at bringing justice for causing harm to natural resources and compensation for harm.

It is necessary to fix in the current legislation the following legal instruments aimed at the development of environmental entrepreneurship. These are information tools (aimed at widespread dissemination and implementation of the best available technologies. This is informing business entities about preventive and protective measures aimed at preventing environmental offenses and legal consequences for their commission). These are educational tools (aimed at the formation of professional personnel) and educational tools (aimed at the general knowledge of the population). This should include environmental education, advanced training and professional retraining of personnel, improving the level of environmental literacy of the population, and economic tools.

All these instruments should be divided into regulatory and protective ones. Regulatory measures include economic and legal means. Protective measures consist of legal means.

Regulatory economic means include financing (allocation of funds from the federal and regional budgets); tax and economic incentives (tax incentives and incentives for paying for negative environmental impacts); economic and environmental impact assessment. Legal means include: fixing the requirements in the field of environmental protection; establishing the legal regime of business activity. Protective instruments consist of preventive and subsequent measures of legal responsibility.

Acknowledgements The work was supported by Russian Science Foundation (project no. 20-18-00314).

References

Barkov, A. V., & Emelyanova, V. D. (2020). Legal support for assessing the degree of socio-ecological utility of mission-oriented corporations. *Civil Law, 2*, 8–11.

Barkov, A. V., & Grishina, Ya. S. (2018). Methodological foundations of constructing the Russian model of legal support for environmental entrepreneurship. *Civil Law, 4*, 4–8.

Barkov, A. V., & Serova, O. A. (2019). International ISO standards and "B Corporation" certification in the legal model of ensuring environmental entrepreneurship. *Legal Research, 7*, 26–31.

Batalina, M., Moskovskaya, A., & Taradina, L. (2008). *Review of the experience and concept of social entrepreneurship, taking into account the possibility of its application in modern Russia* (p. 233). Higher School of Economics.

Bobkova, A. G. (2013). Trends in the development of legislation on environmental entrepreneurship. Business law. *Business and Law in Russia and Abroad, 3*, 5–8.

Chechunova, E. V. (2000). The effectiveness of environmental entrepreneurship in the conditions of implementation of environmental programs (on the example of city N. Tagil), Ph.D. in Economics thesis, No. 22.

Dean, T., & McMullen, J. (2007). Toward a theory of sustainable entrepreneurship: Reducing environmental degradation through entrepreneurial action. *Journal of Business Venturing, 22*, 50–76.

Gregori, P., & Holzmann, P. (2020). Digital sustainable entrepreneurship: A business model perspective on embedding digital technologies for social and environmental value creation. *Journal of Cleaner Production, 272*, 122817.

Gurău, C., & Dana, L.-P. (2018). Environmentally-driven community entrepreneurship: Mapping the link between natural environment, local community and entrepreneurship. *Technological Forecasting and Social Change, 129*, 221–231.

He, J., Nazari, M., Zhang, Y., & Cai, N. (2020). Opportunity-based entrepreneurship and environmental quality of sustainable development: A resource and institutional perspective. *Journal of Cleaner Production, 256*, 120390.

Inshakova, A., Goncharov, A., & Kazachenok, O. (2018). The imperatives of financial policy in the sphere of the digital economy: impacts on increasing investment activity and tax potential of Russian regions. *Proceedings of the international scientific conference "competitive, sustainable and secure development of the regional economy: Response to global challenges" (CSSDRE 2018)* (pp. 337–342). Atlantis Press.

Inshakova, A. O., Goncharov, A. I., & Sevostyanov, M. V. (2017). Institutional ambiguity of regulation of possessory relations in modern Russia. *Proceedings of the international scientific conference "overcoming uncertainty of institutional environment as a tool of global crisis management", Contributions to Economics* (pp. 207–212).

Inter-Parliamentary Assembly of the Member States of the Union of Independent States. (2000). *Model law "on the fundamentals of environmental entrepreneurship"* (June 13, 2000, No. 15–6).

Khamdamov, ZhKh., Gafforova, E. B., & Yershova, T. V. (2018). Ecological entrepreneurship in the sphere of recycling: A case of firms of PrimorskyKrai. *Bulletin of the TSEU, 1*(85), 6–10.

Matytsin, D. E., & Rusakova, E. P. (2021). Strategy of quality management in industry 4.0 and formation of cognitive economy based on industrial and manufacturing engineering in the Russian Federation and countries of the EU. *International Journal for Quality Research, 15*(4). https://doi.org/10.24874/IJQR15.04-03

Meek, W., Pacheco, D., & York, J. (2010). The impact of social norms on entrepreneurial action: Evidence from the environmental entrepreneurship context. *Journal of Business Venturing, 25*, 493–509.

Mityakina, N. M., & Turanin, VYu. (2019). Subjects of agricultural activity as addressees of legal requirements of rational nature management in agriculture. *Environmental Law, 4*, 17–20.

Mokhov, A. A. (2018). Special legal regime of ecological entrepreneurship. *The Lawyer, 12*, 32–38.

Naumov, A. A. (2006). *Ecological utility in the system of interrelation of ecological and economic needs and ecological entrepreneurship*. Studied in Russia. http://zhumal.ape.relarn.ru/articles/2006/214.pdf

Novoselov, S. N. (2012). Environmental services: Foreign experience and problems of the formation of the domestic market. *Statistics and Economics, 3*(2).

OECD. (2019). Best available techniques for preventing and controlling industrial pollution, activity 3: Measuring the effectiveness of BAT policies, environment, health and safety, environment directorate, OECD/OECD, 2 rue André-Pascal, 75775 Paris Cedex 16, France. https://www.oecd.org/chemicalsafety/risk-management/measuring-the-effectiveness-of-best-available-techniques-policies.pdf

Omri, A. (2018). Entrepreneurship, sectoral outputs and environmental improvement: International evidence. *Technological Forecasting and Social Change, 128*, 46–55.

Pisarev, G. A. (2019). Principles of business law as a basis for the development of eco-oriented business. *Law and Economics, 5*, 33–38.

Poiseev, I. I., & Zamorshchikova, A. A. (2013). The market of environmental services. *International Journal of Applied and Fundamental Research, 8*, 74.

Presidium of the Presidential Council for Strategic Development and National Projects. (2018). *Passport of the national project "Ecology"* (December 24, 2018, No. 16).

State Duma of the Russian Federation. (2002). *Federal law "on environmental protection"* (January 10, 2002, No. 7-FZ, revised July 31, 2020).

State Duma of the Russian Federation. (2007). *Federal law "on the development of small and medium-sized businesses in the Russian Federation"* (July 24, 2007, No. 209-FZ, revised June 08, 2020).

The European Parliament and the Council of the European Union. (1985, June 27). *Council of the European Union Directive 85/337/EEC on the environmental impact assessment of certain public and private sector projects.*

The European Parliament and the Council of the European Union. (2008, January 15). *Directive 2008/1/EEC of the European Parliament and of the council on integrated pollution prevention and control.*

Sun, H., Pofoura, A. K., Mensah, I. A., Li, L., & Mohsin, M. (2020). The role of environmental entrepreneurship for sustainable development: Evidence from 35 countries in Sub-Saharan Africa. *Science of Total Environment, 741*, 140132.

Vafin, V. V., & Zolotova, O. A. (2019). Theoretical and legal aspects of the development of environmental entrepreneurship in the region. *Civil Law, 2*, 16–20.

Volkov, G. (2018). On the methodology of natural resource branches of law. *Environmental Law, 1*, 11–12.

Chapter 28
Ensuring Food Security as a Legal and Technological Problem

Aleksey P. Anisimov and Denis E. Matytsin

Abstract At the moment, there are several definitions of food security in Russian agricultural and legal science. After analyzing their content, the authors argue that the country's food security is an integral part of national security. It describes the degree of realization of the human right to food in the country and the achievement of the Sustainable Development Goals. It guarantees the physical and economic availability to citizens of high-quality and safe food in the quantities necessary for a healthy diet. The advantages of this definition are to take into account international recommendations, as well as to avoid a narrow economic perception of the goals and objectives of ensuring food security in the Russian Federation or another country. Ensuring food security is a strategic socio-economic and environmental goal for any State. When solving this problem in Russia, it is necessary to clearly distinguish between "food security" and "food independence". If the first category characterizes the degree of provision of the population with quantitative and qualitative food, its physical and economic accessibility and security. In the second category, the emphasis is on "import substitution", that is, the percentage of products produced in the country and imported. In the Russian Federation, the main focus of public authorities is on ensuring food independence to the detriment of the goals and objectives of ensuring food security. Food security is one of the guarantees of sustainable agricultural development and is located at the intersection of three types of national security: economic, social, and environmental. Since Russia is a federal state, the Constitution of the Russian Federation provides not only for the subjects of the jurisdiction of

A. P. Anisimov (✉)
Volgograd State University, Volgograd, Russia
e-mail: ppagp@volsu.ru

D. E. Matytsin
Department of Business Law, Arbitration and Civil Procedure, Institute of Law, Volgograd State University, Volgograd, Russia

Department of Civil Law and Procedure, Volga Branch of the International Law Institute, Volgograd, Russia

D. E. Matytsin
e-mail: Dmatytsin@ya.ru; ppagp@volsu.ru

A. O. Inshakova and E. E. Frolova (eds.), *Smart Technologies for the Digitisation of Industry: Entrepreneurial Environment*, Smart Innovation, Systems and Technologies 254, https://doi.org/10.1007/978-981-16-4621-8_28

the Russian Federation but also for the institution of joint jurisdiction of the Russian Federation and its subjects. This is a number of important issues of state-building, within which the subjects of the Russian Federation can adopt their legislative acts. One of such areas of legislative regulation in the subjects of the Russian Federation is the issue of ensuring food security. Having studied the relevant laws of the subjects of the Russian Federation, we concluded that there can be no different levels of food security in the state since the Russian Federation is a set of its subjects. It follows from this that the subject of the Russian Federation cannot set food security indicators lower than it is provided for at the federal level. However, the subject has the right to introduce additional indicators (or indicators higher than the federal ones), which are achieved by the forces and means of such a subject of the Russian Federation. In addition, during the analysis of the legislative acts of the subjects of the Russian Federation on food security, we found that there is massive duplication of the text. The text is from the Food Security Doctrine of the Russian Federation in 2020, which is hardly appropriate.

Keywords Ensuring food security · Sustainable agricultural development · Food independence · Legal regulation · Federal legislation

JEL Codes K32 · L26 · P28 · P48 · Q57 · Q56 · Q52

28.1 Introduction

The country's food security is an integral part of national security, which characterizes the degree of realization of the human right to food in the country and the achievement of the Sustainable Development Goals. As well as ensuring the physical and economic availability of high-quality and safe food for citizens in the quantities necessary for a healthy diet. The problem of ensuring food security has always existed in Russia, becoming more acute during the period of collectivization and the Great Patriotic War.

At the moment, ensuring food security is an element of the strategy for sustainable agricultural development. In the recommendations of the UN and other international organizations, it is periodically noted that the development of agriculture can be considered sustainable only with the stable implementation of economic activity on agricultural land. The development will ensure the growth of agricultural production, provided that the social rights of the inhabitants of rural settlements are properly guaranteed. Environmental protection measures are also required on agricultural land (including measures to preserve the fertility of agricultural land) and in rural settlements.

In this regard, the question arises about the dynamics of the ratio of the legal categories "right to food" and "food security". The human right to food is the necessary "minimum" with which one can normally exist in society, engage in productive work, and lead a healthy and active lifestyle. The human right to food is one of the existing international standards of human life (article 25, paragraph 1, of the Universal Declaration of Human Rights, Article 11 of the International Covenant on Economic, Social and Cultural Rights of 1966). If we try to consider international guarantees of the right to food, we can find that no specific measures of domestic food policy have been established for state parties to international agreements. Moreover, at the international level, the criteria for "adequate nutrition" are not disclosed, which makes it much more difficult to assess the state of food security of the population, both within individual countries and on a global scale as a whole. It follows from this that ensuring the right to food is complex and multidimensional, since, along with humanitarian issues, it also affects the production of food, its distribution, storage, development of domestic agriculture, and the food market. The globalization of economic processes, the increase in the rate of food exports and imports is currently in conflict with the interests of national producers.

The right to food, the provision of which is the goal of State food policy, due to its complexity, requires reflection in almost all areas of modern public administration and the context of global climate change and other environmental threats, new tasks are assigned to the public authorities of any country (Reid, 2002; Peters, 2010).

Agriculture, as one of the main sources of anthropogenic greenhouse gas emissions, contributes to climate change (Parker-Flynn, 2014). Climate change threatens global food production by increasing the frequency and severity of droughts, floods, and hurricanes, reducing crop yields, and putting additional strain on limited water resources (Gonzalez, 2012). Today, agricultural food production remains an under-appreciated and poorly understood aspect of global climate change. Although it is noted that from the work of tractors, trucks, the use of fertilizers, the production of electricity, agriculture forms up to 9% of the total contribution to climate change in terms of CO_2 emissions. Although as a percentage of all global greenhouse gas emissions, direct CO_2 emissions from direct agricultural activities are only about 1.4% (Birdsong, 2013).

Noting the complex nature of the problem of ensuring food security it should also be noted that there is a close relationship between the categories of "food" and "national" security. As noted in the scientific literature there are objectively some differences between, first, nations that can feed their people by any means. Secondly, nations can feed their people at the expense of their resources. A country that can feed its people, under any external influence, will have sufficient stability to act with maximum independence (Fromherz, 2012; Sternick, 2012). Other scholars emphasize the need to develop international trade as the main means of guaranteeing the right to food in one's own country (McCabe & Burke, 2013). In the Russian Federation, the authorities adhere to the first strategy, and therefore, ensuring a stable level of food security is traditionally given great attention.

28.2 Materials and Methods

The study analyzed the provisions of the federal laws “On the general principles of the organization of Legislative (Representative) and Executive Bodies of State Power of the Subjects of the Russian Federation”, “On the Development of Agriculture”, “On the quality and safety of food products”, as well as some other federal legal acts. Special attention in this chapter was paid to the study of the provisions of the laws of the subjects of the Russian Federation on food security, to identify their main advantages and disadvantages, and to assess the level of their legal technique.

As a result of the analysis of scientific works of representatives of world science, it was found that the works of leading foreign scientists reflect the problems of the relationship between the categories “food security” and “social justice”, Inshakova et al. (2017, 2018). Also, the consequences of the loss of food security for the country (Frommherz, 2012; Sternick, 2012). The necessity to give priority to the national producer in ensuring food security is also justified (McCabe & Burke, 2013). At the same time, they pay special attention to the relationship between food and environmental security, the importance of environmental protection measures in the course of agricultural production (Reid, 2002; Peters, 2010), which is most important in the context of global climate change (Gonzalez, 2012; Birdsong, 2013; Parker-Flynn, 2014). Meanwhile, these foreign studies do not reflect the specifics of the policy of ensuring food security in federal states, including the issues of the division of competence between the federal center and the regions.

As applied to Russia, this gap was partially filled in the works of Eregin, Zolotukhina, Lapaeva, Lichichan, and some other authors. However, the problems of harmonization of federal and regional legislation were considered by them outside the context of the country’s agricultural problems. In addition, the legal aspects of the import substitution policy implemented in the Russian Federation seem to be insufficiently studied both in Russian and in foreign science, as well as the peculiarities of the development of the legislation of the subjects of the Russian Federation in terms of ensuring regional food security.

The study used the methods of scientific cognition generally recognized in Russian legal science, including the dialectical method, the method of system analysis, comparative-legal, logical, concrete historical, and others. The method of system analysis in combination with the comparative-legal method allowed us to consider the problems of ensuring food security in the region in the context of the concept of sustainable development, the problems of realizing the human right to food, and the policy of import substitution. The logical method made it possible to analyze the problem under consideration in all the variety of its connections and to take into account not only the legal but also the economic and organizational aspects of the problem under study. In combination with the concrete historical method, this allowed us to identify the main modern problems of ensuring food security in the regions, as well as possible ways to overcome them. The empirical basis of the study was the results of the activities of bodies and organizations involved in ensuring food security in the Russian Federation, materials of law enforcement practice. The authors also analyzed the laws of the constituent entities of the Russian Federation that regulate the issues of ensuring food security in the respective regions.

28.3 Results

28.3.1 Regulatory Regulation of Food Security at the Federal Level

At the federal level, there is no separate law dedicated to ensuring food security. The following laws are aimed at solving this problem to varying degrees: Federal Law No. 184-FZ of October 6, 1999 "On the General Principles of the Organization of Legislative (Representative) and Executive Bodies of State Power of the Subjects of the Russian Federation"; Federal Law No. 264-FZ of December 29, 2006 "On the Development of Agriculture"; Federal Law No. 29-FZ of January 2, 2000 "On the quality and safety of food products"; Federal Law of 2 December 1994 No. 53-FZ "On Procurement and supply of agricultural products, raw materials and food for State needs", as well as some other federal legal acts.

To ensure food security in Russia, the Food Security Doctrine of the Russian Federation has been adopted (approved by Presidential Decree No. 20 of January 21, 2020). This Doctrine is a strategic planning document that reflects the official views on the goals, objectives, and main directions of the state socio-economic policy in the field of ensuring food security of the Russian Federation. The Doctrine distinguishes between two important categories—"food security" and "food independence". Food security refers to the state of socio-economic development of the country, which ensures the food independence of the Russian Federation, guarantees the physical and economic availability for every citizen of the country of food products that meet the mandatory requirements, in amounts not less than the rational norms of food consumption necessary for an active and healthy lifestyle. Accordingly, the food independence of the Russian Federation is the self-sufficiency of the country with the main types of domestic agricultural products, raw materials, and food. Food security is one of the main directions of ensuring the national security of the country in the long term, a factor in preserving its statehood and sovereignty, and an important component of socio-economic policy. As well as a necessary condition for the implementation of the strategic national priority-improving the quality of life of Russian citizens by guaranteeing high standards of life support.

To assess the provision of food security, the achievement of threshold values of several indicators is used as the main indicators. Namely: food independence, economic and physical accessibility of food, and compliance of food products with the requirements of the legislation of the Eurasian Economic Union on technical regulation. Food independence is defined as the level of self-sufficiency as a percentage, calculated as the ratio of the volume of domestic production of agricultural products, raw materials, and food to the volume of their domestic consumption and having threshold values. For example, concerning grain-not less than 95%; sugar-not less than 90%; vegetable oil-not less than 90%; meat and meat products (in terms of meat)—not less than 85%. For milk and dairy products (in terms of milk)—not less than 90%; for fish and fish products (in live weight—raw weight)—not less than 85%; potatoes—not less than 95%; vegetables and melons-not less than 90%; fruits and berries-not less than 60%. For seeds of the main crops of domestic selection—not less than 75%; food salt-not less than 85%. The economic availability of food is defined as the ratio of the actual consumption of basic food products per capita to the rational norms of its consumption that meet the requirements of a healthy diet and have a threshold value of 100%. The physical availability of food is defined as the percentage of the actual provision of the population with various types of retail facilities for the sale of food products and objects for the sale of public catering products to the standards established by the Government of the Russian Federation. Compliance of food products with the requirements of the legislation of the Eurasian Economic Union on technical regulation is defined as the specific weight of all samples of food products. Samples that do not meet the mandatory requirements in the total volume of food samples examined within the framework of state control (supervision) and monitoring of the quality and safety of food products.

Ensuring food security involves risks and threats that can significantly reduce it. Such risks and threats fall into the following categories. First, these are economic risks caused by the possibility of deterioration in the domestic and foreign economic environment, a decline in the growth rates of the global and national economies, high inflation, and a crisis in the banking system. As well as reducing the investment attractiveness of domestic agriculture and fisheries, as well as reducing the competitiveness of domestic products (Matytsin & Rusakova, 2021). Secondly, these are technological risks caused by lagging behind developed countries in the level of technological development of the production base, as well as unauthorized use of medicines for veterinary use in the process of agricultural production. Third, these are climatic and agroecological threats caused by adverse climate changes and abnormal natural phenomena of a natural nature. As well as an increase in the share of degraded land, a decrease in the fertility of agricultural land due to their irrational use in agriculture, the consequences of natural and man-made emergencies. Fourth, these are foreign policy risks that may limit the potential for the development of domestic agriculture and fisheries, caused by fluctuations in market conditions, as well as the use by foreign countries of measures of state support for agriculture that distort international trade. Fifth, these are veterinary and phytosanitary risks associated with the emergence and spread of mass infectious animal diseases that have not been previously registered on the territory of the Russian Federation, as well as with the spread of plant diseases and pests. Sixth, these are sanitary and epidemiological threats associated with the emergence and spread of infectious. And non-communicable diseases of the population due to violations of mandatory requirements for ensuring the safety and quality of products at all stages of its turnover in the consumer market. Finally, seventh, these are social threats caused by the decline in the attractiveness of the rural lifestyle.

Mechanisms for ensuring food security are established in the relevant regulatory legal acts that determine the conditions for the functioning of the country's economy and its sectors and are provided with financial resources from the federal budget and the budgets of the constituent entities of the Russian Federation.

28.3.2 Theory of Food Security Levels and Discussions on the Capabilities of the Russian Federation's Constituent Entities to Implement "Advanced Rule-Making"

The Russian Federation occupies a huge area, which determines the economic, cultural, natural, climatic, and other diversity of its territory, which cannot be taken into account otherwise than through the creation of a federal state. That is why the Constitution of the Russian Federation assumes not only the subjects of the Russian Federation's jurisdiction but also the institution of joint jurisdiction on some important issues of state construction, within which the subjects of the Russian Federation can adopt their legislative acts.

The state of affairs in many spheres of public relations depends on the extent to which the subjects of competence and powers are adequately differentiated between different levels of public authority, as well as on the activity of subjects in the Russian Federation in the implementation of their assigned powers. One of the areas of legislative regulation in the subjects of the Russian Federation is the issues of ensuring food security. When discussing the concept of food security, the question of its "levels" often arises. Some Russian authors consider food security as a hierarchical system, which is divided by subjects into global, national, regional, and household food security. At the same time, scientists argue that the improvement of food policy should be aimed at strengthening the economic activity of the subjects of the Russian Federation with the transfer of the center of gravity in solving food problems to their level (Belkharoev, 2003).

L. N. Deineka identifies a different list of levels of food security: global (global), mega-regional, national, regional, and at the level of an individual, family, or social group (Deineka, 2004). V. A. Dadalko distinguishes among the levels of food security individual (individuals); local (households); local (cities, districts, municipalities, free economic zones, biosphere territories); territorial (regions); regional (two or more subjects of state education); national (states) (Dadalko, 2013).

While fully agreeing that in a federal state, the subjects of the federation should take a much more active (than now) part in solving the problems of food security, the following should be noted. Based on the international recommendations of the United Nations and the established national practice, the allocation of the "grassroots level' of food security (human, family, social group) is impractical. Since the food supply of individual citizens is only a consequence of the food policy of the state and the international community. The allocation of the local level is also controversial due to the limited capacity of Russian municipalities to influence food policy. And the purpose of allocating a "biosphere territory", where wildlife objects live, is not at all clear. It would be more appropriate to limit ourselves to three levels: international, national, and regional. Accordingly, at the international level, food security issues are regulated by international acts, including treaties and acts of "soft law" (declarations and resolutions). At the federal level by federal laws and by-laws and at the regional level by-laws of the subjects of the Russian Federation.

From the point of view of the sectoral nature, the laws adopted by the subjects of the Russian Federation, by and large, are complex, since they affect public relations regulated by the norms of land, civil, environmental, constitutional, administrative, and financial and other branches of law (Inshakova et al., 2017). That is why in Article 72 of the Constitution of the Russian Federation "agriculture" is mentioned along with other subjects of joint management (nature management, environmental protection, etc.). Thus by adopting laws on ensuring food security, the subjects of the Russian Federation realize the opportunity provided for by the Constitution of the Russian Federation to carry out "advanced rule-making". Filling in legislative gaps at the federal level and expanding the mechanism for guaranteeing the implementation of the provisions of the Doctrine by the forces and means of the regions. This issue cannot be attributed to the subjects of exclusive jurisdiction of the subjects of the Russian Federation, since some federal legal acts are devoted to it. At the moment, there is no consensus in science on the feasibility of implementing the right of "advanced rule-making" by the subjects of the Russian Federation.

Thus, the first group of authors considers dangerous and unacceptable the practice of advanced rule-making of the subjects of the Russian Federation, which is allowed by the current Russian legislation and supported by some scientists (Zolotukhina, 2004).

V. V. Lapaeva writes, "We need such a concept of interpretation of the relevant constitutional and legal provisions, which would completely exclude the possibility of advanced rule-making of subjects in the sphere of joint jurisdiction and would limit their legislative powers only by specifying the adopted federal laws. Only such a solution will make it possible to remove the main contradictions of the current legal model of the division of powers between the center and the regions in the sphere of joint management" (Lapaeva, 2002). V. M. Platonov, in turn, notes that "regional legislation can play a leading role in legislative regulation in those subjects of joint jurisdiction of the Russian Federation and its subjects, on which the processes of making legislative decisions at the federal level are delayed for various reasons, usually of a political nature" (Platonov, 2009). Representatives of the third (compromise) concept write that the Russian Federation should establish "common principles, i.e., the most important, main, most essential provisions, without detailed instructions on the procedure for implementing the legislative powers of the bodies of the subjects of the Russian Federation. Which have the right to carry out legal regulation, at their discretion, detailing the federal legal principles, filling them with specific content" (Plotnikova, 2003). Ideally, for the subjects of joint jurisdiction, it is necessary to adopt the basis of federal legislation that establishes general principles for the separation of powers of the Federation and its subjects. And already the subjects detail them, which "will allow us to determine the boundaries of federal participation in the framework of joint subjects of competence of the Federation and its subjects" (Platonov, 2010).

Fully sharing the latter position, which implies the expansion of the scope of powers and responsibility of the public authorities of the subjects of the Russian Federation to the population of the region (which, accordingly, solves the problem of "advanced rule-making"), it should be noted that at the moment we are still far from this ideal. Therefore, the question of the validity of the "advanced rule-making" of the regions remains relevant. It seems that today the concept of V. M. Platonov deserves support. Since the federal center does not always have time to quickly solve the economic, social, environmental, and other problems that arise on the ground and the role of the legislation of the subjects of the Russian Federation in these conditions increases.

A typical example of this problem is the issue of ensuring food security, both in the Russian Federation as a whole and in all its subjects. To date, certain aspects of food security are set out in several federal laws. However, its general concept is formulated in the Doctrine of Food Security, which is not a normative legal act. The State Duma of the Russian Federation has repeatedly raised the issue of the need to adopt a special federal law on food security, but such a law has not yet been adopted. In this situation, to implement the provisions of the Doctrine, the subjects of the Russian Federation were forced to adopt their regulatory legal acts on ensuring food security. In which they tried to create additional guarantees for its provision at the regional level.

28.3.3 Analysis of the Provisions of Regional Legislation on Ensuring Food Security

At the moment, only a few regions of the Russian Federation have adopted regional laws on food security. The comparative analysis of their provisions allows us to formulate the following conclusions.

1. The laws of the subjects of the Russian Federation expand the list of terms used in the Doctrine. So, in the Saratov region, the concepts of "wholesale food complex of the region" and "authorized body in the field of food security" were introduced. The law of St. Petersburg uses the terms "basic types of food", "agricultural production", "wholesale and distribution center" and a number of others. In other laws (the Bryansk region), there is a duplication of terms from the Doctrine, for example, "food security", which are adapted for the subject of the Russian Federation.

Note that the list of such practically verbatim reproducible concepts can be quite wide, and include indicators and criteria for food safety, rational norms of food consumption, physical and economic accessibility of food products (Sverdlovsk Region). A more interesting list of terms is found in regional laws adopted more than a decade ago. Thus, in the Ryazan region, the terms "food crisis", "the need for agricultural products and food", "the necessary level of food" is used; in the Primorsky Territory—"vital food products", "the nutritional value of a food product", "the safety of food products", etc.

Accordingly, in the "young" laws of the subjects of the Russian Federation (2010–2020), terms unknown to the federal Doctrine are rarely used (although there are a number of exceptions to this rule, for example, in the law of St. Petersburg).

2. Among the mandatory articles in all laws of the subjects of the Russian Federation are articles of a general nature on the subject of the law, the goals, and objectives of ensuring regional food security, as well as the directions of such provision. Among the latter, it is necessary to mention support for the stability of the food market in the region (including by monitoring the state of the food market), also, the implementation of state support for the agro-industrial complex of the subject of the Russian Federation and the regulation of the market of agricultural products, priority support for citizens in need, etc. Among the latter, it is necessary to mention support for the stability of the food market in the region (including by monitoring the state of the food market), the implementation of state support for the agro-industrial complex of the subject of the Russian Federation, and the regulation of the market of agricultural products. Also priority support for citizens in need, etc. At the same time, some such measures may be specified in a separate article (for example, on state support for individuals and legal entities engaged in the production or processing of agricultural products). Separately, we should highlight the fact that the legislator of the subject of the Russian Federation often connects the issues of food security and the transition to sustainable development, which we practically do not see, for example, in relation to environmental legislation. The category of "sustainable development" is mentioned in the laws of the constituent entities of the Russian Federation, mainly in the context of food and raw materials production and rural development.

Some laws of the subjects of the Russian Federation contain original articles on monitoring the state of food security (Kursk Region), information in the field of food security (Trans-Baikal Territory, Kostroma Region), scientific support of food security (Ryazan region, Primorye Territory), and to ensure the quality and safety of food products (Kabardino-Balkarian Republic, Kursk Region), etc. One of the tasks of ensuring food security set in the Kursk and Kostroma regions, as well as in the Primorye Territory, looks original. The goal is to achieve and maintain the "food independence" of these regions, although the Federal Doctrine does not set such a task for individual subjects of the Russian Federation. An element of the mechanism for ensuring food security in a number of subjects of the Russian Federation is to promote practical activities in the field of research on the impact on human health of food products obtained from genetically modified sources (Kostroma Region). The latter measure, however, is rarely found in the laws of the subjects of the Russian Federation. The creation of a special "Food Security Council" operating in the Nizhny

Novgorod Region, which has no analogs in other regions of the Russian Federation, also deserves mention.

3. The laws of the constituent entities of the Russian Federation pay special attention to the powers of the representative and executive bodies of the constituent entities of the Russian Federation in the field of food security. As noted in the scientific literature, "in the most general form, the subjects of competence are the spheres of public relations assigned by the Constitution of the Russian Federation to the competence of a particular level of state power. And the powers, in turn, should be considered as specific rights and obligations of state authorities to regulate the subjects of competence, or, in other words, specific areas of influence" (Eregin, 2008).

The powers of the legislative body of the subject of the Russian Federation and the governor, as a rule, are set out formally and are reduced to the adoption of laws and the protection of human rights and freedoms in the field of food security (although there are some exceptions. For example, holding parliamentary hearings on food security issues in the Sverdlovsk Region). The articles of laws dealing with the powers of governments and specially authorized bodies are more informative. It is the Government that develops and approves state programs in the field of food security. It also determines the Authorized Body, monitors the state of food security of the subject of the Russian Federation, and interacts with the federal executive authorities in this area. The powers of the Authorized Body that assesses the state of food security of the subject of the Russian Federation are often specified in a separate article.

4. It should be noted that some laws of the subjects of the Russian Federation contain criteria for assessing food security, which may either correspond to those set out in the Doctrine or differ from them.

For example, in the Ryazan region, food security is considered to be achieved if "the provision of food of own production is not less than 65% of the population of the Ryazan region in food in accordance with physiological norms" (Article 4).

A similar indicator in the Trans-Baikal Territory, the Republic of Bashkortostan, and the Kostroma Region is at least 80% of the volume of food consumption by the population of the region in accordance with the established rational norms of food consumption. In the Kursk region, it is 75%, and in the Primorsky Territory—60%. How can such discrepancies be assessed? On the one hand, the legislation of the subject of the Russian Federation exists to take into account regional peculiarities, including the economic plan. On the other hand, the state cannot have different levels of food security, since the Russian Federation is a set of its subjects. It follows from this that the subject of the Russian Federation cannot set food security indicators lower than it is provided for at the federal level (Inshakova et al., 2018). However, the subject has the right to introduce additional indicators (or indicators higher than the federal ones), which are achieved by the forces and means of such a subject of the Russian Federation. In the scientific literature on this issue, there is a very common opinion that "forming a single whole with the federal legislation", the legislation of the subject of the Federation develops it, reproducing, concretizing and supplementing it. At the same time, the reproduction (duplication) of the norms of federal legislation in the legislation of the subject of the Russian Federation performs a dual role: on the one hand, duplication unnecessarily increases the federal legislation and complicates its application. On the other hand, the duplication of federal norms in the legislation of the constituent entities of the Russian Federation ensures the uniform application of federal norms throughout the state. Therefore, it would be wrong to categorically refuse or unconditionally use this technique. In addition, as a rule, norms-concepts, norms-goals, norms-principles are duplicated, without which it is often impossible to build an independent normative act of a subject of the Russian Federation (Safina, 2000).

E. A. Bondareva notes that it is necessary to eliminate or minimize the duplication of the laws of the subjects of the Russian Federation of the provisions contained in the acts of the federal level. At the same time, in some cases, it is impossible to completely exclude the duplication of the norms of federal laws in the legislation of the subjects of the Russian Federation. For example, when fixing the principles of the foundations of the constitutional system in the Constitutions and Charters of the subjects of the Russian Federation. Otherwise, their legislation will not be integral, but referential and will not represent a full-fledged legal base of a particular subject of the Russian Federation (Bondareva, 2004).

In general, sharing this approach, it should still be noted that the analysis of the legislation of the subjects of the Russian Federation on food security shows that there is massive duplication of the text. Moreover, it is difficult to say unequivocally whether they copy the text from each other, or altogether (and independently of each other) from a common source—the Food Security Doctrine of the Russian Federation of 2010. However, it was discontinued in 2020 due to the adoption of a new Doctrine. Since the Doctrine is not a normative legal act the reproduction of its provisions cannot be considered in the full sense of the word duplication and evaluated negatively.

At the same time, it should be said that many laws of the subjects of the Russian Federation go far beyond the framework of the Doctrine. They create authorized bodies or lay down their indicators for assessing food security in the relevant region. The above-mentioned situation with the duplication of the text of regional laws (with each other and with the federal Doctrine) generally confirms the widespread doctrinal conclusion. More than 60% of the subjects of the Russian Federation use the practice of compiling the experience of other regions. This compilation is partial and affects only certain provisions of the legislation of the "donor region", which are transferred to the legislation of the "recipient region". Mass reception reduces the degree of exclusivity of the regional legislation of specific regions and leads to a decrease in the autonomy of this regional element of the legal system of the Russian Federation. But reception is not a panacea that saves regional legislation from the existence of regulations that have a relatively low legal quality. This vestige of regional legislation is extremely conservative, primarily due to the rather low professional level of the legislative bodies of the constituent entities of the Russian Federation (Lichichan, 2009).

28.4 Conclusions

The country's food security is an integral part of national security, which characterizes the degree of realization of the human right to food in the country and the achievement of the Sustainable Development Goals. It guarantees the physical and economic availability to citizens of high-quality and safe food in the quantities necessary for a healthy diet. The advantages of this definition are to take into account international recommendations, as well as to avoid a narrow economic perception of the goals and objectives of ensuring food security in the Russian Federation or another country.

Ensuring food security is a strategic socio-economic and environmental goal for any State. When solving this problem in Russia, it is necessary to clearly distinguish between "food security" and "food independence".

If the first category characterizes the degree of provision of the population with quantitative and qualitative food, its physical and economic accessibility, and security, then the second one focuses on "import substitution", that is, the percentage ratio between the products produced in the country and imported products. In Russia, the main focus of public authorities is on ensuring food independence at the expense of the goals and objectives of ensuring food security.

Food security is one of the guarantees of sustainable agricultural development and is located at the intersection of three types of national security: economic, social, and environmental.

As a result of the study, we consider it necessary to supplement the definition of food security in the text of the Doctrine with an indication of the social accessibility of food. We also consider it important to specify the criteria and levels of ensuring the country's food security in the Doctrine. They should be specified in separate categories of food products: "fully provided"; "mainly provided"; "partially provided"; "not provided".

This will allow for a more effective assessment of the activities of state authorities that regulate food security. In conditions when there is no law at the federal level directly devoted to ensuring food security, and the above-mentioned Doctrine is not a normative legal act, the role and importance of legislation on food security of the subjects of the Russian Federation increases. This is due to the fact, that the Russian Federation occupies a huge area, and therefore its constituent entities differ in economic, cultural, natural-climatic, and other parameters. That is why the Constitution of the Russian Federation assumes not only the subjects of the Russian Federation's jurisdiction but also the institution of joint jurisdiction on some important issues of state-building. Within the framework of which the subjects of the Russian Federation can adopt their legislative acts. One of these areas is the issue of ensuring food security. Having studied the relevant laws of the subjects of the Russian Federation, we concluded that there can be no different levels of food security in the state. Since the Russian Federation is a set of its subjects.

It follows from this that the subject of the Russian Federation cannot set food security indicators lower than it is provided for at the federal level. However, it has the right to introduce additional indicators (or indicators higher than the federal ones), which are achieved by the forces and means of such a subject of the Russian Federation. In addition, during the analysis of the legislation of the subjects of the Russian Federation on food security, we found that there is massive duplication of the text. Basically, from the Doctrine of Food Security of the Russian Federation, this is not advisable. At the same time, we found that some laws of the subjects of the Russian Federation go far beyond the framework of the Doctrine. Moreover, both in terms of creating authorized organs, and in terms of fixing their indicators for assessing food security in the relevant region.

Acknowledgements The work was supported by Russian Science Foundation (project No. 20-18-00314).

References

Belkharoev, H. U. (2003). *Legal support of food security in modern Russia: Abstract of the dissertation of the candidate of Legal Sciences*. Russian Presidential Academy of Public Administration.

Birdsong, B. C. (2013). From "food miles" to "moneyball": How we should be thinking about food and climate. *Maine Law Review, 65–2*, 410–423.

Bondareva, E. A. (2004). The legislation of the subject of the Russian Federation is an element of the legal system of Russia. In Yu. N. Starilov (Ed.), *Russian Legal State: The results of the formation and prospects of development: In 5 parts. Part 1: Constitutional and international law, theory of state and law: Materials of the All-Russian Scientific and Practical Conference*, November 14–15, 2003.

Dadalko, V. A. (2013). Food security is a component of the national and economic security of the state. *Bulletin of the Ufa State Aviation Technical University, 17*(7), 17–25.

Deineka, L. N. (2004). Food security—A systemic problem of the globalized economy. *Izvestiya TRTU, 4*, 75–78.

Eregin, A. N. (2008). *Principles of the division of powers between the state authorities of the Russian Federation and the state authorities of the subjects of the Russian Federation on the subjects of joint jurisdiction*. Ph.D. in Law thesis. Moscow New Law Institute.

Fromherz, N. A. (2012). The case for a global treaty on soil conservation, sustainable farming, and the preservation of agrarian culture. *Ecology Law Quarterly, 39*, 57–121.

Gonzalez, C. G. (2012). The global food system, environmental protection, and human rights. *National Resources & Environment, 26*, 1–15.

Inshakova, A., Goncharov, A., & Kazachenok, O. (2018). The imperatives of financial policy in the sphere of the digital economy: Impacts on increasing investment activity and tax potential of Russian regions. In *Proceedings of the international scientific conference "competitive, sustainable and secure development of the regional economy: Response to global challenges" (CSSDRE 2018)* (pp. 337–342). Atlantis Press.

Inshakova, A. O., Goncharov, A. I., Mineev, O. A., & Sevostyanov, M. V. (2017). Amendments to the civil code of the Russian Federation: Contradictions of theory and practice. Contributions to Economics. *Russia and the European Union: Development and Perspectives, XIII*, 147–153.

Lapaeva, V. V. (2002). Delineation of the legislative powers of the Federation and its subjects (based on the material of the legislation on Public Associations). *Journal of Russian Law, 12*, 10–24.

Law of Primorye Territory. (2001). *On food security of primorye territory* (Adopted 15.11.2001, No. 163-KZ).

Law of the Bryansk Region. (2017). *On ensuring food security in the Bryansk Region* (Adopted on December 4, 2017, No. 93-Z).

Law of the Kabardino-Balkarian Republic. (2012). *On food security of the Kabardino-Balkarian Republic* (Adopted 20.03.2012, No. 7-RZ).

Law of the Republic of Bashkortostan. (2000). *On food security of the Republic of Bashkortostan* (Adopted on 04.02.2000, No. 51-z).

Law of the Trans-Baikal Territory. (2011). *On food security and the market of agricultural products, raw materials, and food in the territory of the Trans-Baikal Territory* (Adopted on 01.11.2011, No. 570-ZZK).

Lichichan, O. P. (2009). Legal systems of the subjects of the Russian Federation: Current state and development. *Bulletin of Tomsk State University, 324*, 230–233.

Matytsin, D. E., & Rusakova, E. P. (2021). The strategy of quality management in industry 4.0 and formation of a cognitive economy based on industrial and manufacturing engineering in the Russian Federation and countries of the EU. *International Journal for Quality Research, 15*(4). https://doi.org/10.24874/IJQR15.04-03

McCabe, M. S., & Burke, J. (2013). The New England food system in 2060: Envisioning tomorrow's policy through today's assessments. *Maine Law Review, 65–2*, 550–579.

Parker-Flynn, J. E. (2014). The intersection of mitigation and adaptation in climate law and policy. *University of California, Davis, 38*, 1–47.

Peters, K. A. (2010). Creating a sustainable urban agriculture. *Journal Environmental Law and Litigation, 25*, 205–211.

Platonov, V. M. (2009). On the advancing role of lawmaking in the subjects of the Russian Federation. *Modern Law, 9*, 34–36.

Platonov, V. M. (2010). *Delineation of subjects of competence between the Federation and its subjects-as a principle of Russian federalism: Abstract of the dissertation of the Doctor of Legal Sciences*. Peoples' Friendship University of Russia.

Plotnikova, L. A. (2003). *Correlation of federal legislation and legislation of the subjects of the Russian Federation (questions of theory): Abstract of the dissertation of the candidate of Legal Sciences*. Moscow State University.

Reid, M. S. (2002). Agriculture and the environment—Building a sustainable and healthy future. *Environs, 25–2*, 55–58.

Safina, S. B. (2000). *Legislation of the subjects of the Russian Federation: Questions of theory and practice: Abstract of the dissertation of the Candidate of Legal Sciences*. Kazan State University.

Sternick, A. (2012). Food fight: The impending agricultural crisis and a reasonable response to price volatility. *Village Environmental Life Journal, 23–1*, 145–171.

The Law of St. Petersburg. (2015). *On ensuring food security and supporting agricultural production in St. Petersburg* (Adopted on 25.12.2015, No. 841–168).

The Law of the Kostroma Region. (2008). *On food security of the Kostroma region* (Adopted on 26.05.2008, No. 317-4-LKR).

The Law of the Kursk Region. (2009). *On food security of the Kursk region* (Adopted on 27.11.2009, No. 95-LKR).

The Law of the Nizhny Novgorod Region. (2011). *On ensuring food security in the Nizhny Novgorod region* (Adopted on 09.08.2011, No. 111-Z).

The Law of the Ryazan Region. (2005). *On food security and the market of agricultural products, raw materials, and food in the territory of the Ryazan region* (Adopted on 14.10.2005, No. 109-OZ).

The Law of the Saratov Region. (2016). *On ensuring food security in the Saratov region* (Adopted on 03.10.2016, No. 132-ZSR).

The Law of the Sverdlovsk Region. (2012). *On ensuring food security of the Sverdlovsk region* (Adopted on 31.01.2012, No. 6-OZ).

Zolotukhina, T. A. (2004). The law of the subject of the Russian Federation as a source of law. Ph.D. in Law thesis. North-Caucasian Academy of Public Service.

Chapter 29
Environmental Hazards of Nanotechnologies and Measures of Economic and Legal Incentives to Reduce Them in Russia and the EAEU Countries

Agnessa O. Inshakova and Aleksey P. Anisimov

Abstract This chapter argues for the conclusion that at the moment all the regulatory regulation of the environmental and sanitary hazards of nanotechnology in Russia is enshrined in by-laws, which are often only advisory. Therefore, it is extremely important that the Federal Law "On Environmental Protection" of 10.01.2002, as well as the Federal Law "On Sanitary and Epidemiological Welfare of the Population" of 30.03.1999, be supplemented with special articles. These articles should contain the minimum necessary amount of protective measures against real or potential threats to human health and the environment associated with the mass use of nanotechnology and nanoproducts. These articles should contain measures for mandatory labeling of nanoproducts. According to the state registration of nanoproducts of medium and high danger. For carrying out its sanitary-epidemiological and other examinations, depending on the degree of potential environmental danger of such products. To fix the obligation to research the expense of the federal budget on the presence of environmental consequences of nanotechnologies and nanoproducts for the environment and human health, which should result in changes to the existing system of environmental standards and technical regulations. To develop new methods of environmental and sanitary supervision and new types of offenses related to non-compliance with the above-mentioned sanitary and environmental measures. An alternative solution could be the adoption of a separate federal law dedicated to the development of nanotechnologies and the turnover of products obtained with their use. The law should contain a special section (chapter) on guarantees of the rights of citizens from their dangerous consequences. In addition, this chapter explains the need for a set of measures in the field of environmental education and education, as well as strengthening international cooperation. Including the adoption of some international documents regulating the creation of an international information resource containing the results of scientific research on the negative impact of nanotechnology and nanoproducts on the environment. This will allow for more rational use of the

---

A. O. Inshakova (✉) · A. P. Anisimov
Volgograd State University, Volgograd, Russia
e-mail: ainshakova@list.ru; gimchp@volsu.ru

A. O. Inshakova and E. E. Frolova (eds.), *Smart Technologies for the Digitisation of Industry: Entrepreneurial Environment*, Smart Innovation, Systems and Technologies 254, https://doi.org/10.1007/978-981-16-4621-8_29

scientific, technical, intellectual, and another potential of the leading countries of the world and will allow them to quickly exchange information about the results of such research.

Keywords Nanotechnologies · Nanoindustry · Environmental hazards · Sanitary legislation · Legal regulation · Natural resources

JEL Codes Q56 · Q57 · K32 · L26 · P28 · P48 · O44 · F64

29.1 Introduction

Nanotechnology refers to methods for producing and using substances and materials in the size range from 1 to 100 nm (nano-meter—one billionth of a meter) with specified properties and characteristics. In this size range, materials tend to exhibit different properties than at the normal scale, including greater activity and reactivity. The use of these unique properties determines the interest of the state and business in the development of new products, processes, and technologies. This gives rise to a new phenomenon—the nano industry, i.e., an intersectoral complex of organizations that provide and carry out targeted activities for the development and commercialization of nanotechnology (Matytsin and Rusakova, 2021). Developments in the field of nanotechnology have the potential to provide huge social benefits, including improvements in the diagnosis and treatment of diseases, the production of clean energy. As well as improving the efficiency of computers and electronic equipment, the production of stronger and lighter composite materials. Unfortunately, the same properties that account for many of the potential benefits of nanomaterials (especially their small size and dynamic properties) also pose risks to human health and the environment. Because nanomaterials can penetrate biological systems and react with them.

Current scientific evidence on these risks is mixed at best. Some studies of the effects of nanomaterials on experimental animals have caused toxic reactions in them. However, other studies have shown no negative consequences, which indicates that the risks of nanoproducts for human health are insignificant. Currently, there is no methodology for predicting which nanomaterials will cause a toxic reaction and which will not. Moreover, each specific nanomaterial can present significant risk variations. Extremely vague and insufficiently proven information about the side effects of nanotechnologies paralyze the activities of state management bodies in the absence of effective sanitary and environmental legislation.

At the moment, nanotechnology has not yet become widespread in Russia. At the same time, firstly, the products obtained with their use are imported to Russia from other countries. This raises the question of managing these processes and guaranteeing human health and environmental safety (Abanina et al., 2020). Secondly, the development of scientific and technological progress is inevitable, and the process of production of such products directly in Russia will sooner or later be established.

Therefore, Russian citizens and authorities will have to face the same problems that already exist in more technologically advanced countries. This should be prepared using the accumulated experience of both other countries and the Russian Federation.

29.2 Materials and Methods

The legislative base of the study was made up of Russian federal laws. Including "On the sanitary and epidemiological welfare of the population", "On Environmental Protection", "On self-regulating organizations'. As well as by-laws adopted in Russia that regulate certain guarantees of the right to health of citizens and the preservation of favorable environmental quality when using nanotechnology (GOST 54336-2011 "National Standard of the Russian Federation"). Environmental management systems in organizations that produce nanoproducts. Requirements "Methodological guidelines of MU 1.2.2636-10.1.2 "Hygiene, toxicology, sanitation. Conducting sanitary and epidemiological expertise of products obtained with the use of nanotechnologies and nanomaterials", Resolution of the Chief Sanitary Doctor of the Russian Federation of 23.07.2007 "On supervision of products obtained with the use of nanotechnologies and containing nanomaterials". Some supranational acts were also considered, including the Decision of the Customs Union Commission of 09.12.2011 "On the adoption of the Technical Regulations of the Customs Union "On Food Safety".

The purpose of this chapter was to study not only Russian legislation, but also the EEU countries (Armenia, Belarus, Kyrgyzstan, and Kazakhstan). The chapter reviewed the regulations of these countries, including the following. Fundamentals of the legislation of the Republic of Armenia "On Nature Protection" of June 9, 1991. The Law of the Republic of Armenia "On ensuring sanitary and Epidemiological safety of the population of the Republic of Armenia" of December 12, 1992, the Law of the Republic of Belarus "On Environmental Protection". Law of the Republic of Belarus "On Sanitary and epidemiological welfare of the population" of January 7, 2012, Law of the Kyrgyz Republic "On Environmental Protection" of June 16, 1999, Law of the Kyrgyz Republic "On Public Health" of July 24, 2009, Environmental Code of the Republic of Kazakhstan of January 9, 2007, Code of the Republic of Kazakhstan "On Public Health and the Health System" of July 7, 2020.

However, these regulations contain only fragmentary references to the guarantees of the right of citizens to health and a favorable environment when using nanotechnology. The chapter widely used the provisions of the scientific doctrine, including the work of Belokrylova (2014), Dana (2010), Dennis (2006), Goldstein (2010), Heselhaus (2010), Kaddour (2013), Lerer (2013), Marchant et al. (2010), Paddock (2010), Perez (2010), Petersen and Bowman (2012), Stokes (2012), Theodore and Stander (2013), Wilson (2013), Inshakova et al. (2018, 2020) and Matytsin and Rusakova (2021).

In the process of research, general scientific methods are used, such as formal-logical, dialectical, system-structural, critical cognition. Methods of synthesis, classification, and generalization were used to interpret the results of the study. The paper also uses private scientific methods: formal-legal, the principle of assessing legal processes, the method of comparative analysis, etc.

29.3 Results

29.3.1 Scope of Application of Nanotechnologies and Their Potential Danger to the Environment and Human Health

At the moment, in the scientific literature, along with the study of positive commercial and other prospects from the use of nanotechnology, attention is beginning to turn to the high probability of various negative social consequences from their use.

There is no doubt that the negative impact of nanotechnology on human health and nature must be identified before products using it enters the market. Prudent deployment of nanotechnology can have a positive impact on the environment if used as intended. The state needs to determine the most profitable way to include nanotechnology in environmental laws and regulations (Dennis, 2006). The problem is that the possible biological effects of nanomaterials entering the human body have not yet been sufficiently studied. Although there is already evidence that various substances, when converted into the form of nanoparticles, can significantly change their physical and chemical properties, which can negatively affect human health in the process of their assimilation into the body. For example, recent studies have shown that carbon nanotubes have the same carcinogenic effects as asbestos (Heselhaus, 2010). Fullerenes, tiny carbon structures that resemble soccer balls, have been found to cause brain damage in aquatic animals. Quantum pillboxes, which focus on targeted drug delivery, also pose a toxicological risk to human health and the environment. However, while carbon nanotubes may increase the risk of mesothelioma, other carbon particles of similar size do not pose a similar risk (Lerer, 2013).

The consequences of nanoparticles reacting with other substances, as well as the environmental consequences of such a combined effect, remain unexplored. Of particular concern to biologists is the impact of nanoparticles on the state of wildlife objects-animals, plants, and insects. There is a risk of the indirect impact of nanotechnologies and nanomaterials on the environment and human health. The risk lies, for example, in the effects of exposure to nano pesticides and nano agrochemicals on plants and domestic animals, as well as the person who consumes the corresponding products. The issue of the methodology for conducting ecological and hygienic studies to assess the environmental impact of waste from the production of nanomaterials, the problems of their storage, disposal, and destruction remains unexplored. There is also a threat from nanotechnology because of the potential to create incredibly destructive weapons. Such nanotechnology weapons can be more powerful than any known chemical, biological, or nuclear agent can. It will be difficult to detect, and this may lead to a new round of the arms race (Wilson, 2013).

At the same time, the environmental consequences of the use of nanotechnology cannot be considered only in a negative, negative sense. A number of scientists believe that nanotechnology can help restore a favorable quality of the environment. So, thanks to nanotechnology, solar panels are currently being exploited, which means that mass burning of coal and petroleum products is no longer possible. As well as the disappearance of the danger of the greenhouse effect and the death of the ozone layer, the consequences of oil spills and emissions of oil refining waste, air pollution by combustion products. Nanotechnology can also be an important part of cleaning up contaminated and hazardous waste. Nanotechnology can also contribute to reducing the environmental risk of toxic emissions and discharges containing arsenic, mercury, cadmium, and lead compounds. It can be achieved by developing and implementing mechanisms that convert the chemical composition of these emissions into non-toxic substances and elements. Finally, nanotechnology can be used to create more powerful sensors that can accurately detect pollutants in the environment at very low concentrations. In medicine, nanotechnology can be used for the diagnosis of diseases, in the manufacture of medicines. In the food industry—in the production of biologically active food additives and new forms of packaging that have an antimicrobial effect (or that will be gas-or moisture-proof). As well as in the production of clothing and many other areas.

These materials can eventually be used to strengthen fibers (nanowires), which will increase the safety of fabrics, including priming. Nanomaterials are used in many cosmetics to transport nutrients and other substances through the skin. This allows the body to absorb even substances that are insoluble in water. Composite materials that use nanoparticles resist heat better. They have better conductivity and exhibit a higher strength-to-weight ratio. Nanoparticles are also used for coatings that prevent the product from scratching (nanoscale ceramic particles). They exhibit antimicrobial properties in appliances that decompose organic material to make windows and roofs self-cleaning.

The military also shows interest in nanotechnology, including in such areas as optical systems, nanorobots, nanomachines, "smart" weapons, nanoelectronics, virtual reality, sensors and surveillance systems. As well as special materials for armor, nanomaterials for stopping bullets and bio-nano-devices for detecting and destroying chemical and biological agents. Much of this interest is related to protecting against attacks and minimizing the risk of military personnel (Theodore and Stander, 2013).

Despite active discussions about the negative impact of nanotechnologies and nanoproducts on the human body and its environment, there is no proper legal assessment of these consequences in both Russian and international and foreign law. It makes it difficult to protect the health of citizens and the environment from the potentially dangerous effects of nanoparticles.

29.3.2 Problems of Accounting for and Countering Threats from Nanotechnology for the Protection of the Environment and Human Health

The wide range of possibilities for using nanotechnology creates significant management problems. Since different applications of nanomaterials are regulated not within the framework of a single legal regime, such as the handling of hazardous waste, but through a wide range of regulatory rules. The fact is that no country in the world has a single regulatory framework that would cover food, chemicals, personal hygiene products, medical devices, water quality, and so on.

At the same time, in addition to the objective difficulties with managing the processes of production and turnover of nanotechnologies and nanoproducts, there is also an important subjective factor. It is associated with a lack of public knowledge and awareness in the field of nanotechnology. The population of not only Russia but also other EAEU countries is poorly informed about the benefits and possible side effects of nanotechnology. This hinders the development of this industry, and the lack of a clear management system creates a risk of public rejection of nanotechnology (Paddock, 2010). However, when planning to involve the public widely in the process of discussing the importance and possible side effects of nanotechnology, it is necessary to know the answers to the question of who is involved and for what purpose the organizers of this process plan to involve.

The fact is that the term "public participation", although currently widely used in many contexts, has many meanings and applications in practice (Petersen & Bowman, 2012). The most promising is the understanding of the "public" as all socially active citizens living in a particular locality or region (country). As well as public organizations (environmental, consumer protection, etc.). The involvement of the public in a constructive dialogue with the developers of nanotechnologies is all the more important because, in the absence of public sympathy for nanotechnologies and products produced with their help, the latter can suffer the fate of GMOs. In the case of GMOs, attention to the environmental, health and safety implications of these biotechnologies has outstripped the process of their introduction into production and the consumer market. Public concerns about this have significantly slowed the realization of the huge commercial potential of GMOs.

A negative public opinion has been formed regarding them, although the environmental or medical danger of GMOs has not yet been proven. At first, opponents of genetic engineering attributed to all GMO products any potential problems or observed negative effects from any one type of GMO. Then, as the potential benefits of GMOs became more apparent, environmentalists focused on individual GMO crops as the object of particular criticism and oversight (Goldstein, 2010).

Taking into account this negative experience of public rejection of modern technologies, when planning a nanotechnology management system, it is necessary to think through a system of measures to increase public confidence in this industry. One of the directions of this activity may be the financing of a series of toxicological examinations by several countries of the world (but not by companies producing nanoproducts). These examinations will allow you to obtain and evaluate comparable results using the same methods. However, the coordination of research methods, their implementation, and discussion of their results can take many years, while this problem is relevant now. Given the rapid market penetration of nanotechnology and the products that contain it, existing regulatory approaches cannot even be used to identify it. Not to mention the management of these processes before the results of the examinations become generally accepted. A large number of studies already suggest that many nanoparticles are not benign and can affect biological activity at the cellular, subcellular, and molecular levels. After all, nanomaterials are so small that they can be embedded in human cells and even change biological processes at the cellular level. No less significant are their potential threats to the environment, which require the development of a new generation of technical regulations, as well as new waste disposal technologies. Returning to the issue of interaction with the public, it should be noted that it seems appropriate to create special coordination bodies under the President of the Russian Federation and governors in the constituent entities of the Russian Federation. Who could participate in the discussion and solution of both modern environmental problems and be focused on a broader range of national security issues, including the tasks of ensuring the sustainable development of the Russian Federation.

The Ministry of Natural Resources and Ecology of the Russian Federation and its local authorities could coordinate the process of public discussion of the advantages and potential dangers of nanotechnology. Such a dialogue would facilitate the creation of new environmental management strategies that, together, could build public confidence, help avoid situations of unnecessary risk, and accelerate the development of technologies with greater public and business benefits. This organ could also launch a broad educational campaign, both with the help of traditional media (television, newspapers, magazines) and with the help of new Internet technologies (Inshakova et al., 2018). As well as social advertising, this is placed, for example, on billboards along transport highways. The potential for implementing educational campaigns in schools and universities is also very high. For example, recommendations on the inclusion in the curricula of the master's degree in the legal specialty of special subjects that reveal the features of legal regulation (and the dangers) of nanotechnology.

In the context of increasing computer literacy of the country's population, it would not be enough to use only the sites of environmental (or other bodies) of state power, reports, reports, or draft regulations posted there for dialogue with the public. More promising is the creation of a special Web site that regularly publishes reliable information about developments in the field of nanotechnology. This includes both the risks and benefits of nanotechnology, information on the development of state regulation of the industry, as well as information on industry standards and approaches to self-regulation.

With the development of nanotechnology, it seems inevitable that organizations that produce products containing nanomaterials will face the threat of legal liability. If exposure to nanomaterials causes harm to public health or the environment. The civil liability system plays an important role in making decisions about whether to market potentially dangerous nanoproducts. One of the key issues in tort liability is whether manufacturers of nanoproducts should be held liable only for known risks. Or they should be held responsible even if they could not reasonably have foreseen such risks (Dana, 2010). The answer to this question will also affect the development of the risk insurance market for the development and implementation of nanotechnology, including in Russia.

The following management solutions are proposed for discussion:

1. Introduce mandatory labeling of products manufactured using nanotechnology. In this regard, significant experience has already been accumulated in different countries of the world related to the labeling of GMO products. This measure will allow consumers to determine that whether nanoscale materials are contained in the products they buy, and to investigate the health and environmental impact of nanoscale products. This labeling program should also tell consumers how to properly dispose of items containing potentially hazardous nanoscale materials. Such a measure has been in place for several years in the EU (Stokes, 2012), whose experience can be used.
2. Regular supervision of nanotechnology is necessary, although problematic. The need to regulate the production and circulation of nanomaterials is caused by two interrelated considerations. First, if the use of certain nanotechnologies is not regulated, it can create very real (although not fully realized) risks of significant disruption to human health or harm to the environment. Secondly, public confidence in new technologies requires an increase in the efficiency of the regulatory authorities, which must promptly identify such potential and real threats. Thus, the environmental supervision authorities of the EAEU countries should have a new function, for which they are not yet ready. One of the problems is that "nanotechnology" is a poorly defined, insufficiently understood set of various products, processes, and technologies. Which is not easily covered by any existing regulatory definition, model, or system. At the same time, many traditional nanotechnology management tools do not work, which requires the search for new ways to regulate their use and turnover (Marchant et al., 2010).
3. Following the existing ideas in the modern theory of environmental management, conclusions about the real or potential danger of nanotechnology should be based on reliable scientific evidence. If the environmental management bodies begin to arbitrarily exercise regulatory influence on entrepreneurs engaged in the production of nanoproducts using nanotechnology. This will create, at least, excessive administrative barriers, and, at most, the ground for corruption. Thus, the risks and dangers of nanotechnology must be proven, and this requires the environmental management bodies to develop original methods for measuring the negative impact of nanotechnology on the health of citizens and the environment. However, even if such a task can be solved concerning existing types of nanotechnology, it will not save in the future. Because scientific and technological progress does not stand still, and new types of nanotechnology are introduced every year. However, even if such a task can be solved in relation to existing types of nanotechnology, it will not save in the future.

Because scientific and technological progress does not stand still, and new types of nanotechnology are introduced every year.
Nanotechnology is already progressing from using relatively simple nanoparticles to more complex and active materials, such as sensors, multifunctional drugs, and medical devices (Inshakova et al., 2020). It will be difficult for a slow-moving bureaucracy to aim at such a fast-moving object. And if excessive restrictions are created for such a business, it can move to other countries, where a more comfortable business environment is created. In addition, this will negatively affect the demand from consumers, who will reasonably consider the following. Significant efforts of the administrative apparatus to regulate and verify nanotechnologies and nanoproducts are established because of their danger to health and nature, which, as already noted, has not yet been proven. Therefore, concerning Russia and the UES countries, at the current stage of nanotechnology development, it is more appropriate to start managing these processes by creating a system. The system should include monitoring, summarizing the experience of using nanotechnology, conducting expert examinations, and only then making decisions.

4. One of the ways to solve the problem is to develop a system of self-regulation in the business related to nanotechnology. Such a measure will create additional guarantees for compensation for damage caused by nanotechnologies and nanoproducts if such is proved. Article 3 of Federal Law No. 315-FZ of 01.12.2007 (as amended on 03.08.2018) "On Self-Regulating Organizations" implies the following. Among the requirements for a self-regulating organization, among other things, are the existence of standards and rules of business or professional activity that are mandatory for all members of the self-regulating organization. As well as providing a self-regulating organization with additional property liability of each of its members to consumers of manufactured goods (works, services) and other persons. The introduction of such a mechanism in certain areas of business activity (e.g., construction) has proved quite successful, which allows us to use this experience in this area. At the same time, there is no doubt that the development of self-regulation in the production of nanoproducts does not cancel state regulation and control. Separately, it is worth highlighting the option of further development of the system of voluntary certification of products, which is already gradually developing.
5. Introduction of voluntary measures for additional reporting of nanoproduct producers. Such measures, in particular, have already been introduced by BASF Corporation in Germany, Buhler Partec in Switzerland, Du Pont Environmental Defense in the USA, some trade organizations (IG-DHS in Switzerland), as well as some other enterprises (Belokrylova, 2014). There are examples of mandatory reporting, for example, the first national rule requiring reporting on nanomaterials was established by France (Kaddour, 2013). There are cases when it was introduced at the municipal level. So, the municipal council of the city of Berkeley (USA) amended its municipal code. By requiring companies that produce or use engineered nanoparticles to disclose toxicological information, as well as information about pollution prevention. Any enterprise producing

or processing nanomaterials was subject to regulatory disclosure requirements (Lerer, 2013).

29.3.3 *Existing Achievements in the National Legal Regulation of Nanotechnology in Russia and Other EAEU Countries*

At the moment, Russia and other EAEU countries have taken certain steps to create a legal framework for the use of nanotechnology and to guarantee the safety of the resulting nanoproducts.

1. GOST R 54336-2011. National Standard of the Russian Federation. Environmental management systems in organizations that produce nanoproducts. Requirements (approved by the Order of Rosstandart No. 148-st of 27.06.2011). This standard sets out additional requirements for the environmental management system in the design and development, production, and, if applicable, installation and maintenance of nanoproducts. In particular, this GOST establishes that the organization must develop, implement, and maintain a procedure for collecting and analyzing information about the properties of nanotechnologies and nanoproducts, which are related to environmental aspects—for timely updating of the list of environmental aspects and the list of significant environmental aspects. The organization should provide stakeholders with information about its significant environmental aspects related to nanotechnology and nanoproducts. The Organization should develop, document, implement, and maintain a procedure for identifying potential emergencies and accidents, which are associated with nanotechnologies and nanoproducts that can have negative effects on the environment.
2. Decision of the Commission of the Customs Union of 09.12.2011 No. 880 (ed. of 24.12.2019) "On the adoption of the Technical Regulations of the Customs Union "On food safety". This technical regulation stipulates that food products of a new type are food products (including food additives and flavorings) that have not previously been used by humans for food in the customs territory of the Customs Union. Namely: with a new or intentionally altered primary molecular structure; consisting of or isolated from microorganisms, microscopic fungi, and algae, plants, isolated from animals, obtained from GMOs, or using them, nanomaterials and nanotechnology products. Except for food products obtained by traditional methods, which are in circulation and, due to experience, are considered safe. Thus, nanoproducts have entered a larger classification group that is subject to legal regulation that is mandatory for use in the member states of the Customs Union.

3. Resolution of the Chief Sanitary Doctor of the Russian Federation of July 23, 2007, No. 54 "On the supervision of products obtained using nanotechnology and containing nanomaterials". According to the decree, legal entities and individual entrepreneurs who produce and import products obtained using nanotechnology and/or containing nanomaterials are *recommended* to indicate the following. In the information for consumers—information about the use of nanotechnology or nanomaterials in the manufacture of products. When submitting documents for the sanitary and epidemiological examination of products—provide information about the use of nanotechnology or nanomaterials with confirmation of the safety of their use for humans. In addition, it is proposed to organize work on the objective information of the population on the use of nanotechnologies and nanomaterials.
4. Guidelines 1.2.2636-10.1.2. Hygiene, toxicology, sanitation. Conducting sanitary and epidemiological expertise of products obtained using nanotechnologies and nanomaterials. The guidelines were approved by Rospotrebnadzor on 24.05.2010. According to this document, the sanitary and epidemiological expertise of nano industry products is the activity of the Federal Service for Supervision in the Field of Consumer Rights Protection and Human Wellbeing. As well as its territorial bodies, federal-state health institutions—centers of hygiene and epidemiology, as well as other organizations accredited in accordance with the established procedure to establish compliance (non-compliance) of products with state sanitary and epidemiological rules and regulations, technical regulations. Sanitary and epidemiological examinations of products are carried out for the following purposes. Identification of nano industry products that pose a danger to human life and health, as well as the possibility of causing harm to human health during the manufacture, circulation, and use (use) of products. Assessment of compliance (non-compliance) of nano industry products, conditions of their manufacture, and turnover with the requirements of the legislation of the Russian Federation. Evaluation of the effectiveness of measures to prevent the harmful effects of nano industry products on human health during their manufacture, circulation, and use (use). As well as during the disposal or destruction of low-quality and dangerous products. All nano industry products produced in Russia or imported to the Russian Federation are subject to sanitary and epidemiological expertise. If such products: (a) are intended for use by the population as consumer products; (b) it is possible to receive significant amounts of nanoscale components that are part of the product into the human body during all stages of the product life cycle; (c) possible contamination of nanoscale components that are part of the products of the nano industry, objects of the natural environment that have a direct or indirect effect on the human body. Nanoindustry products that have passed the examination are subject to state registration with the Federal Service for Supervision of Consumer Rights Protection and Human Well-Being.

Thus, we see that certain aspects of the production of nanoproducts have received their legal regulation in fragments, which, however, is not systemic. Moreover, if we turn to the federal laws "On Sanitary and Epidemiological Welfare of the Population" of 30.03.1999 No. 52-FZ (as amended on 13.07.2020) or "On Environmental Protection" of 10.01.2002 No. 7-FZ (as amended on 08.12.2020), we will not find any mention of nanotechnologies or nanoproducts as an object of legal regulation. The situation is similar in other EAEU countries. Thus, the Law of the Republic of Belarus of November 26, 1992, No. 1982-XII (as amended. June 18, 2019) "On Environmental Protection", as well as the Law of the Republic of Belarus of January 7, 2012, No. 340-Z (ed. of June 15, 2019) "On Sanitary and Epidemiological Welfare of the population" do not contain any mention of nanotechnology, but at the subordinate level, as in Russia, certain steps have already been taken. For example, certain obligations in relation to nanotechnologies are mentioned in paragraph 38 of the Resolution of the Ministry of Health of Belarus "On Approval of Sanitary Norms and Rules" (registered in the NRPA of the Republic of Belarus on January 11, 2013, No. 8/26755). Concerning organizations engaged in the production of food products using food additives, flavorings and technological aids.

Similarly, nanotechnologies are not mentioned in the Fundamentals of the Legislation of the Republic of Armenia of June 9, 1991 "On Nature Protection" or the Law of the Republic of Armenia "On Ensuring Sanitary and Epidemiological Safety of the Population of the Republic of Armenia" of December 12, 1992 (as amended) from 04.12.2019). The study of the Law of the Kyrgyz Republic "On Public Health" of July 24, 2009, No. 248 (ed. of 07.05.2020) leads to the same results. This law replaced the law on the sanitary and epidemiological welfare of the population. As well as a study of the Law of the Kyrgyz Republic "On Environmental Protection' of June 16, 1999, No. 53 (ed. of March 23, 2020). We believe that the Environmental Code of the Republic of Kazakhstan dated January 9, 2007, No. 212-III (as amended on 09.11.2020) is the most perfect environmental legal act. Both from the point of view of legal technology and in terms of the coverage of modern threats to environmental safety within the framework of the EAEU and the CIS. It also does not mention nanotechnology as an object that poses a danger to the environment. Measures to prevent threats to the health of citizens from nanotechnology are not mentioned in the Code of the Republic of Kazakhstan on the Health of the People and the Health System of July 7, 2020, No. 360-VI. The Code replaced the law on the sanitary and epidemiological welfare of the population. At the same time, as in Russia, Kazakhstan has adopted a number of by-laws (GOST) on various aspects of the use of nanotechnology.

So, in the EEC countries, nanotechnologies are not subject to protection in environmental or sanitary legislation. Although their fragmentary references are already found in secondary sanitary (but not environmental) acts. It seems that the implementation of the international precautionary principle is impossible (reflects an attempt to find a compromise between two competing social problems—increased anxiety about the possible adverse effects of new technologies on the environment and health, and the scientific and economic desire for technological innovation) (Perez, 2010). This is impossible without the development of national environmental and health legislation, which guarantees the human right to health and a favorable environment.

In turn, to coordinate the efforts of national states, an international convention on the safety guarantees of nanotechnology and nanoproducts should be developed.

29.4 Conclusions

At the moment, all regulatory regulation of the environmental and sanitary hazards of nanotechnology in Russia is enshrined in by-laws, which are often only advisory. It seems that the Federal Law "On Environmental Protection" of 10.01.2002, as well as the Federal Law "On Sanitary and Epidemiological Welfare of the Population" of 30.03.1999, should be supplemented with special articles. These articles should contain the minimum necessary amount of protective measures against real or potential threats to human health and the environment. Which are associated with the mass use of nanotechnology and nanoproducts. These articles should contain measures for mandatory labeling of nanoproducts. According to the state registration of nanoproducts of medium and high danger. For carrying out its sanitary-epidemiological and other examinations, depending on the degree of potential environmental danger of such products. To fix the obligation to conduct at the expense of the federal one on the presence of environmental consequences from nanotechnologies and nanoproducts for the environment and human health. They should result in changes to the current system of environmental standards and technical regulations. Also, measures to develop new methods of environmental and sanitary supervision and new types of offenses that are associated with non-compliance with these sanitary and environmental measures. These articles should contain measures for mandatory labeling of nanoproducts. According to the state registration of nanoproducts of medium and high danger. For carrying out its sanitary-epidemiological and other examinations, depending on the degree of potential environmental danger of such products. To fix the obligation to conduct at the expense of the federal one on the presence of environmental consequences from nanotechnologies and nanoproducts for the environment and human health. They should result in changes to the current system of environmental standards and technical regulations. Also, measures to develop new methods of environmental and sanitary supervision and new types of offenses that are associated with non-compliance with these sanitary and environmental measures.

An alternative solution could be the adoption of a separate federal law on the development of nanotechnologies and the turnover of products obtained from their use, which would contain a special section (chapter) on guarantees of the rights of citizens from their dangerous consequences. In addition, it is necessary to carry out a set of measures in the field of environmental education and education. It is also necessary to strengthen international cooperation, including the adoption of some international documents regulating the creation of an international information resource. Such a resource should contain the results of scientific research on the negative impact on the environment of nanotechnologies and nanoproducts (with the establishment of the mode of its use). This will allow for more rational use of the scientific, technical, intellectual, and another potential of the leading countries of the world. It will also allow you to quickly exchange information about such studies.

Acknowledgements The work was supported by Russian Science Foundation (project No. 20-18-00314).

References

Abanina, E. N., Pandakov, K. G., Timofeev, L. A., Suhova, E. A., & Ustinova, A. N. (2020). Preservation of forest biological diversity as an activity to ensure environmental safety in forestry. In *IOP Conference Series: Earth and Environmental Science, 543*(1), 012022.

Belokrylova, E. A. (2014). *Legal support of environmental safety: A textbook*. Rostov-on-Don.

Code of the Republic of Kazakhstan. (2020). *On the health of the people and the health system* (Adopted on July 7, 2020, No. 360-VI) (Kazakhstan).

Dana, D. (2010). When less liability maymean more precaution: The case of nanotechnology. *UCLA Journal of Environmental Law and Policy, 28*(1), 153–199.

Decision, 2011 *On the adoption of the technical regulations of the Customs Union "On food safety"* (Adopted by the Commission of the Customs Union on 09.12.2011 No. 880 (ed. of 24.12.2019)) (Russian).

Dennis, L. V. (2006). Nanotechnology: Unique science requires unique solutions. *Temple Journal of Science, Technology and Environmental Law, XXV*, 87–119.

Environmental Code of the Republic of Kazakhstan. (2007). (Adopted on January 9, 2007, No. 212-III (ed. from 09.11.2020)) (Kazakhstan).

Federal Law. (2007). *On self-regulating organizations* (Adopted by the State Duma on November 16, 2007, No. 315-FZ) (Russian).

Federal Law No. 52-FZ of 1999 on the Sanitary and Epidemiological Welfare of the Population (Adopted by the State Duma on March 12, 1999) (Russian).

Federal Law No. 7-FZ. (2002). *On environmental protection* (Adopted by the State Duma on December 20, 2001) (Russian).

Fundamentals of the Legislation of the Republic of Armenia. (1991). *On nature protection* (Adopted on June 9, 1991) (Armenia).

Goldstein, B. D. (2010). The scientific basis for the regulation of nanoparticles: Challenging paracelsus and pare. *UCLA Journal of Environmental Law and Policy, 28*(1), 7–28.

GOSTR 54336-2011, National Standard of the Russian Federation. Environmental management systems in organizations that produce nanoproducts. Requirements (approved and put into effect by the Order of Rosstandart of 27.06.2011 No. 148-st) (Russian).

Guidelines 2010. Hygiene, toxicology, sanitation. Conducting sanitary and epidemiological expertise of products obtained using nanotechnologies and nanomaterials. Guidelines 1.2.2636-10.1.2 (Approved by Rospotrebnadzor on 24.05.2010) (Russian).

Heselhaus, S. (2010). Risk management of nanomaterials: Environmental and consumer protection under existing EC legislation on chemicals, pesticides, and biocides. *Environmental Law Review, 12*, 115–131.

Inshakova, A., Goncharov, A., & Kazachenok, O. (2018). The imperatives of financial policy in the sphere of the digital economy: Impacts on increasing investment activity and tax potential of Russian regions. *Proceedings of the international scientific conference "competitive, sustainable and secure development of the regional economy: Response to global challenges" (CSSDRE 2018)* (pp. 337–342). Atlantis Press.

Inshakova, E. I., Inshakova, A. O., & Goncharov, A. I. (2020). Engineered nanomaterials for the energy sector: Market trends, modern applications, and prospects. *IOP Conference Series: Materials Science and Engineering, 971*, 032031https://doi.org/10.1088/1757-899X/971/3/032031

Kaddour, N. (2013). No laws in nanoland: How to reverse the trend? The French example. *Pace Environmental Law Review, 30*, 486–522.

Law of the Kyrgyz Republic. (1999). *On environmental protection* (Adopted on June 16, 1999, No. 53 (ed. of March 23, 2020) (Kyrgyzstan).

Law of the Kyrgyz Republic. (2009). *On public health* (Adopted on July 24, 2009, No. 248 (ed. from 07.05.2020) (Kyrgyzstan).

Law of the Republic of Belarus. (2012). *On the sanitary and epidemiological welfare of the population* (Adopted on January 7, 2012, No. 340-Z (ed. of June 15, 2019) (Belarus).

Lerer, D. (2013). Big things in small packages: Evaluating the city of Berkeley's nanotechnology ordinance effectiveness as a model of targeted transparency. *Pace Environmental Law Review, 30*, 523–557.

Marchant, G. E., Sylvester, D. J., & Abbott, K. W. (2010). A new soft law approach to nanotechnology oversight: A voluntary product certification scheme. *UCLA Journal of Environmental Law and Policy, 28*(1), 123–152.

Matytsin, D. E., & Rusakova, E. P. (2021). The strategy of quality management in industry 4.0 and formation of a cognitive economy based on industrial and manufacturing engineering in the Russian Federation and countries of the EU. *International Journal for Quality Research, 15*(4). https://doi.org/10.24874/IJQR15.04-03

Resolution 2007, *On the supervision of products obtained with the use of nanotechnologies and containing nanomaterials* (Adopted by the Chief Sanitary Doctor of the Russian Federation on July 23, 2007, No. 54) (Russian).

Resolution of the Ministry of Health of the Republic of Belarus. (2013). *On the approval of Sanitary Standards and regulations* (Registered on January 11, 2013, No. 8/26755) (Belarus).

The Law of the Republic of Armenia. (1992). *On ensuring the sanitary and epidemiological safety of the population of the Republic of Armenia* (Adopted on December 12, 1992 (as amended) from 04.12.2019) (Armenia).

The Law of the Republic of Belarus. (1992). *On environmental protection* (Adopted on November 26, 1992, No. 1982-XII (as amended June 18, 2019)) (Belarus).

Paddock, L. (2010). An integrated approach to nanotechnology governance. *UCLA Journal of Environmental Law and Policy, 28*(1), 251–290.

Perez, O. (2010). Precautionary governance and the limits of scientific knowledge: A democratic framework for regulating nanotechnology. *UCLA Journal of Environmental Law and Policy, 28*(1), 29–76.

Petersen, A., & Bowman, D. (2012). Engaging whom, and for what ends? Australian stakeholders' constructions of public engagement concerning nanotechnologies. *Ethics in Science and Environmental Politics, 12*, 67–79.

Stokes, E. (2012). Not such a small matter after all: Nanotechnology and the European Parliament. *Environmental Law Review, 14*, 263–266.

Theodore, L., & Stander, L. H. (2013). Regulatory concerns and health/hazard risks associated with nanotechnology. *Pace Environmental Law Review, 30*, 469–485.

Wilson, G. (2013). Minimizing global catastrophic and existential risks from emerging technologies through international law. *Virginia Environmental Law Journal, 31*, 307–364.

Chapter 30
Legal Regulation of the Development of Renewable Energy Sources in Russia, the BRICS, and EAEU Countries

Agnessa O. Inshakova and Aleksey P. Anisimov

Abstract This chapter notes that in the modern world, there is a gradual greening of industrial, agricultural, and other products, which is manifested in a gradual reduction in the negative impact on the environment. Such work is being carried out to achieve several goals and objectives related to reducing greenhouse gas emissions (which will slow down the processes of global climate change), switching to sustainable development standards, and reducing the volume of production and consumption waste. One of the leading directions of this work is the greening of energy, which is manifested in the transition to environmentally friendly technologies. The world practice of recent years speaks of a rapid growth in electricity production from renewable sources, which raises the question of the role of law and legal science in understanding and stimulating this trend. The study showed that all natural resources can be classified according to different criteria. It is possible according to the criterion of origin—for biological (forests, plants, animals), mineral (minerals), and energy (solar energy, wind energy, etc.). It is possible according to the ecological content for exhaustible (renewable and non-renewable) and inexhaustible (solar, wind, tidal energy). It is possible according to the degree of actual use—real and potential, according to the criterion of substitutability of some by others-replaceable and irreplaceable. Currently, all the BRICS and EAEU countries have adopted special laws on the development of renewable energy sources. The Russian legislator went the other way, including provisions on RES in the basic Federal law "On Electric Power Industry". The authors note that all laws provide for comparable economic and organizational measures aimed at creating conditions for the development of renewable energy sources. The existing practice of developing the renewable energy system in all the BRICS and EAEU countries leaves some issues unresolved that require further discussion and resolution. Among them are questions about the role of the state in increasing the competitiveness of renewable energy sources; measures to be taken to reduce the environmental danger of renewable energy sources; measures in the field of neighborhood law aimed at guaranteeing the rights of owners of renewable energy sources, etc. A comprehensive legal solution to the issue of further development of

A. O. Inshakova (✉) · A. P. Anisimov
Volgograd State University, Volgograd, Russia
e-mail: ainshakova@list.ru; gimchp@volsu.ru

A. O. Inshakova and E. E. Frolova (eds.), *Smart Technologies for the Digitisation of Industry: Entrepreneurial Environment*, Smart Innovation, Systems and Technologies 254, https://doi.org/10.1007/978-981-16-4621-8_30

renewable energy is cross-sectoral and will require amendments and additions to civil, environmental, energy, and other legislation.

Keywords Legal regulation · Renewable energy sources · Environmental problems · Inexhaustible natural resources · Environmentally friendly technologies

JEL Codes K32 · L26 · O13 · P28 · P48 · Q20 · Q42

30.1 Introduction

In the modern world, there is a gradual greening of industrial, agricultural, and other products, which is manifested in a gradual reduction in the negative impact on the environment. Such work is being carried out to achieve several goals and objectives related to reducing greenhouse gas emissions (which will slow down the processes of global climate change), switching to sustainable development standards, and reducing the volume of production and consumption waste. One of the leading directions of this work is the greening of energy, which is manifested in the transition to environmentally friendly technologies (Inshakova et al., 2019). At the same time, different countries use their national approaches to achieve their goals, which require study and doctrinal generalization.

For the BRICS and EEU countries, these trends are of particular interest. The transition of the EU countries to new environmental standards by 2050, their gradual abandonment of the use of hydrocarbons, poses new challenges for them (and especially for Russia), which, in addition to technical aspects, have the most important economic and legal characteristics. At the moment, much has already been done in Russia for the development of renewable energy sources (hereinafter-RES) (certification of renewable energy sources, criteria for allocating subsidies for technological connection of renewable energy generating facilities, promotion of the use of renewable energy sources in the wholesale electricity market, etc.). However, the dynamic development of the events of 2020 (the decline in production due to the corona-virus) in the hydrocarbon market requires an early adjustment of the existing course of state policy, its adaptation to new challenges and threats to national security, among which an important place is given to their solution by economic and legal means.

The need to form a long-term state energy policy in the field of renewable energy sources and to create favorable conditions for the development of innovative activities in this sector is due to the exhaustion of opportunities for economic growth in Russia. This growth was based on the extensive exploitation of raw materials, against the background of the formation of the digital economy and the emergence of a limited group of leading countries with new production technologies and focused on the use of renewable resources, such a policy is necessary. A gradual transition to "green growth", in particular, in the field of renewable energy, can be achieved only through the implementation of a set of legislative and institutional measures that ensure the growth of energy efficiency and the development of renewable energy sources, the introduction of economic incentives to reduce emissions, discharges, waste generation and disposal. These are the global trends. In recent years, there has been a qualitative and structural change in the development of renewable energy sources in the world. Over the past decade of investment, RES capacity has quadrupled from 414 to 1650 GW; since 2009, solar power alone has grown more than 26-fold—from 25 GW to approximately 663 GW. In 2018, the volume of investments in the creation of new renewable energy capacity reached $272.9 billion (which is three times higher than the corresponding indicator of energy generation capacity from fossil fuels); in 2018, RES provided 12.9% of the world's electricity generation, preventing emissions of 2 billion tons of carbon dioxide (UN, 2019). Today, the search for policies that best enable a country to move from fossil fuels to renewable energy is an international competition for technological leadership in one of the fastest-growing sectors of the global economy.

International law has not yet formulated universal norms on the obligations of States to develop RES as a means of solving global energy and environmental problems, but such norms are contained in several bilateral and multilateral treaties (although as dispositive or recommendatory ones). The corresponding call is also contained in some advisory decisions of universal (UN) and regional (OAS, EU, APEC, etc.) international organizations. Thus, in paragraph 127 of the Declaration of the UN International Conference "Rio + 20", we see a call for greater use of renewable energy sources and other technologies that reduce emissions, increase energy efficiency, and increase the use of advanced energy technologies, including more environmentally friendly technologies for the use of fossil fuels. But what are the reasons for the fact that the norms of international legal regulation of renewable energy sources in the framework of international treaty law are still quite fragmented? It seems that there are several reasons for this.

As noted by Belotsky (2016), the use of solar, wind, geothermal, and tidal energy occurs mainly within the framework of sovereign state territory. At the same time, the use of such sources does not technically have a significant transboundary impact on the territories of other States. Therefore, there is no special universal international legal regime in the field of renewable energy within the framework of a single treaty, but there are some harmonized norms within the regional European legal system—in the EU legal order. The use of the recommendations of international bodies in EU law is the most obvious example of the UN's influence on the development of regional international organizations. Back in 2009, the EU adopted Directive No. 2009/28/EC on the development of the use of energy derived from renewable energy sources, which repealed the previous Directive No. 2001/77/EC on the promotion of electricity produced from renewable sources in the domestic market. The Directive established separately for each Member State the share of total energy consumption to be derived from renewable energy sources by 2020. This percentage ranged from a minimum of 10% for Malta to a maximum of 49% for Sweden and was the basis for calculating the amount of energy produced from renewable sources in 2005. To achieve this goal, each Member State was obliged to develop an appropriate development plan. The issues of calculating the energy produced and proving that it was obtained from renewable sources were specifically regulated. Special attention was paid to the problem of reducing greenhouse gas emissions, and the reporting procedure for the Member States was fixed. The 2009 Directive was revised in December 2018, when the Renewable Energy Directive 2018/2001/EC, part of the Clean Energy for All Europeans package, came into force. The Directive aims to make the EU a global leader in renewable energy and, more broadly, to help the EU meet its emissions reduction commitments under the 2015 Paris Climate Agreement.

The Directive sets a new mandatory target for renewable energy for the EU for 2030 of at least 32%, with a possible upward revision by 2023.

This experience of developing the legal regulation of renewable energy at the level of regional international organizations is of great interest to the BRICS and EAEU countries interested in the gradual transition to clean energy standards to implement their international obligations. For example, Russia, like the EU, has made several commitments to reduce greenhouse gases by signing the Paris Climate Agreement in 2015; other members of these associations have their commitments.

30.2 Materials and Methods

The legislative base of the study was made up of international and national legal acts dedicated to the development of the renewable energy system. The chapter examined the Paris Climate Agreement, adopted under the United Nations Framework Convention on Climate Change (signed in New York on April 22, 2016). As well as the EU directives (in particular, the Directive of the European Parliament and the Council of the European Union "On the promotion of the use of energy from renewable sources" (signed in Strasbourg on 11 December 2018, No. 2018/2001/EC). The authors paid special attention to the analysis of the national legislation of the BRICS and EAEU countries, in particular, the following documents were considered. The Law of the Republic of Armenia "On Energy Saving and Renewable Energy" (adopted on December 4, 2004, No. ZR-122). The Law of the Republic of Belarus "On Renewable Energy Sources "(adopted on December 27, 2010, No. 204-Z). The Law of the Republic of Kazakhstan "On Support for the use of Renewable Energy Sources" (adopted on July 4, 2009, No. 165-IV). The Law of the People's Republic of China "On Renewable Energy Sources" (adopted on December 26, 2009). Law of the Kyrgyz Republic" On Renewable Energy Sources" (adopted on December 31, 2008, No. 283). The Law of the Russian Federation "On Electric Power Industry" (adopted on 26.03.2003 No. 35-FZ). As well as some by-laws (e.g., the Order of the Government of the Russian Federation of 08.01.2009 No. 1-p (as amended). of October 24, 2020) "On the main directions of the state policy in the field of improving the energy efficiency of the electric power industry based on the use of renewable energy sources for the period up to 2035").

In the course of the study, the works of some scientists were considered. Among them are Barretto (2020), Belotsky (2016), Vankovich (2017), Vasilyeva (2011), Deynega (2020), Dmitrieva (2012), Dladla (2020), Klass (2011), Luneva (2018), Mormann (2011), Rastogi (2019), Ryzhenkov (2020), Fontana (2020), Shchepansky

(2013), Inshakova et al. (2019), Matytsin and Rusakova (2021), which allowed the authors to draw some proposed conclusions and judgments.

In the course of the research, general scientific methods were used, including formal-logical, dialectical, system-structural, and critical cognition. Methods of synthesis, classification, and generalization were used to interpret the results of the study. In the course of the work, the authors also used some private scientific methods: formal-legal, the principle of evaluating legal processes, and the method of comparative analysis.

30.3 Results

30.3.1 The Concept of Renewable Natural Resources and Their Place in the General Classification of Natural Resources

To study the legal regime of RES, it is necessary to initially formulate their definition and place it in the system of other resources.

In this regard, the scientific schools of the post-Soviet countries express several divergent points of view. Thus, in the Ukrainian ecological and legal science, it is proposed to classify natural resources into exhaustible (non-renewable natural resources). Those that are not completely restored or are restored much more slowly than they are used, for example, subsurface resources, minerals, soil resources, and renewable natural resources that are capable of full or partial self-restoration and reproduction under the influence of human activities, for example, land resources, forest resources, animal and plant life. It also offers inexhaustible natural resources—water, climate (energy, heat, moisture resources in the atmosphere and soil, light, gas), and space resources (marine energy) (Deynega, 2020).

Representatives of Russian scientific schools generally agree with this approach. Thus, I.S. Shchepansky believes that natural resources should be divided into exhaustible and inexhaustible (the latter mainly include water and climate resources). Exhaustible, in turn, must be divided into renewable (plant and animal products) and non-renewable (mineral resources of the subsurface). Renewable resources are natural resources that are capable of self-recovery after being partially withdrawn for consumption. Traditional sources of energy should include oil, natural gas, and coal, which currently dominate the global energy balance (Shchepansky, 2013).

Vasilyeva (2011) refers to alternative sources of renewable energy—the sun, wind, heat of the earth, the natural movement of water flows, the energy of biomass, which includes specially grown plants for energy production. It also includes production and consumption waste, biogas, gas released from production and consumption waste in landfills, gas generated in coal mines, and other free environmental energy. Their inexhaustibility and environmental cleanliness make it necessary to use them intensively.

Luneva (2018) believes that the rational use of renewable natural resources should necessarily lead to an increase in the sustainability of natural ecological systems, natural and natural-anthropogenic objects. Of course, this assumes such a volume of the negative impact that the environment can independently process. Under renewable natural resources, it means a part of natural resources in the redistribution of the cycle of substances in the biosphere, capable of self-recovery in terms commensurate with the terms of human economic activity (vegetation, wildlife, atmospheric oxygen, etc.). The most important characteristic of renewable resources is their self-restoration and self-purification, which should be taken into account when determining the method of their rational use.

Finally, the representative of the Belarusian scientific school Vankovich (2017) understands renewable energy sources as constantly existing or periodically appearing in the environment material objects or processes that can be continuously restored in appropriate quantities and contain or emit primary energy that can be used. Among renewable resources, it includes geothermal resources of the subsurface, which contain the energy of the earth's heat; surface water bodies, which emit the energy of the natural movement of water flows; atmospheric air, which emits wind energy; objects of the animal and plant world, which contain the energy of biomass.

Summarizing the above points of view, we note that the basic division of natural resources into exhaustible and inexhaustible is not in doubt. Concerning the goals and objectives of the use of renewable energy sources, the following should be noted. Among them, as a variety of inexhaustible natural resources, are solar energy, atmospheric air that releases wind energy; geothermal resources of the subsurface (containing the energy of the Earth's heat); the energy of the natural movement of water flows, objects of the animal and plant world that contain biomass energy.

As a temporary form of such energy, we can consider waste from production and consumption (as proposed by M. I. Vasilyeva), but with the caveat that with the development of modern technologies within the framework of the concept of circular economy, this type of resource for energy in the long term will lose its current significance.

Speaking of terminology, it should be noted that renewable energy should not be confused with clean energy. The term "clean energy" usually refers to the process of generating energy that does not pollute the environment, especially in terms of emissions of carbon dioxide and other gases. Therefore, "clean energy" has a broader meaning than "renewable energy". So, nuclear energy is a kind of environmentally friendly energy source, because it emits very little carbon dioxide. Similarly, natural gas is generally classified as a source of clean energy, provided that technologies for its extraction, transportation, and use significantly reduce or eliminate emissions of greenhouse gases and other pollutants from exhaustible energy sources (oil, coal) that are not renewable (Mormann, 2011).

So, all natural resources can be classified according to different criteria. It is possible according to the criterion of origin—for biological (forests, plants, animals), mineral (minerals), and energy (solar energy, wind energy, etc.). It is possible according to the ecological content for exhaustible (renewable and non-renewable) and inexhaustible (solar, wind, tidal energy). It is possible according to the degree of actual use—real and potential, according to the criterion of substitutability of some by others-replaceable and irreplaceable. It seems that it is also necessary to distinguish between the generic category "natural resource" and its variety "energy natural resource". The latter should be understood as a variety of natural resources that are directly used or can potentially be used by humans in the future as a source of energy.

30.3.2 Legal Regulation of the Development of Renewable Energy Sources in the BRICS and EAEU Countries: Trends and Prospects

International declarations and appeals of the UN, as well as the experience of the development of the renewable energy system in the EU countries and other regional international organizations, have had a different impact on the development of renewable energy in the BRICS and EAEU countries.

1. The Law of the People's Republic of China "On Renewable Energy Sources" of December 26, 2009, was adopted to promote the development and use of renewable energy, increase the supply of energy, improve the structure of energy, ensure energy security, protect the environment and achieve sustainable economic and social development. The term "renewable energy" is used in the law concerning non-renewable types of energy (wind and solar energy, hydropower, bio-energy, ocean energy, etc.). The Department of Energy of the State Council of the People's Republic of China, together with other departments, is developing a national plan for the development and use of renewable energy resources. The plan should contain development goals, main objectives, regional plans, key projects, the progress of work, construction of related energy networks, service systems, etc. China guarantees the purchase of electricity produced using renewable energy sources in full and supports the construction of independent electricity supply systems generated using renewable energy sources to provide electricity to local production and people in areas not covered by any networks (Ryzhenkov, 2020).
2. there is no special law on renewable energy sources in Russia, but their legal regime is defined by Federal Law No. 35-FZ of 26.03.2003 (as amended on 30.12.2020) "On Electric Power Industry". According to Article 21 of this law, the Government of the Russian Federation determines the mechanism for stimulating the use of renewable energy sources by selling electric energy produced by qualified generating facilities operating on their basis on the wholesale market at the equilibrium prices of the wholesale market, taking into account the surcharge. Such an allowance is determined following the procedure established by the Government of the Russian Federation or is stimulated by selling

> the capacity of qualified generating facilities in the volume of electricity production based on the use of renewable energy sources using the capacity trading mechanism. This is provided for by the rules of the wholesale market for the sale of the capacity of these generating facilities. At the same time, the Government of the Russian Federation also establishes a mandatory volume for buyers of electric energy in the wholesale market for the purchase of electric energy produced at qualified generating facilities operating based on the use of renewable energy.

This system of incentives is regulated in detail both in separate articles of the law itself and in the by-laws that specify its provisions. For example, according to Article 23.1 of the Federal Law "On Electric Power Industry', state regulation of prices (tariffs) in the wholesale and retail markets is manifested, in particular, in the following. A surcharge is established that is added to the equilibrium price of the wholesale market to determine the price of electric energy produced at qualified generating facilities operating based on renewable energy sources (in cases and accordance with the procedure provided by the Government).

The Order of the Government of the Russian Federation of 08.01.2009 No. 1-r (ed. of October 24, 2020) "On the main directions of state policy in the field of improving the energy efficiency of the electric power industry based on the use of renewable energy sources for the period up to 2035" notes the following. For the period up to 2024, the following target values are set for the volume of production and consumption of electric energy using renewable energy sources (except for hydroelectric power plants with an installed capacity of more than 25 MW): in 2010—1.5%; in 2015—2.5%; in 2024—4.5%.

The Order of the Government of the Russian Federation also states that no more than 8.5 billion kW is produced annually using renewable energy sources of electric energy (excluding hydroelectric power plants with an installed capacity of more than 25 MW), which is less than 1% of the total electricity production in the Russian Federation.

The main reasons for the low rate of development of electric energy based on the use of renewable energy sources are also indicated. This is the lack of competitiveness of renewable energy projects in the existing market environment compared to projects based on the use of fossil fuels. This is the presence of institutional barriers associated with the lack of necessary regulatory legal acts that encourage the use of renewable energy sources in the electric power industry, the lack of federal and regional programs to support the large-scale use of renewable energy sources. This is the lack of infrastructure required for the successful development of the electricity industry based on renewable energy sources, including the lack of the level and quality of scientific services for its development. This is a lack of human resources and the lack of mechanisms for using public resources to support the development of the electric power industry based on the use of renewable energy sources.

A similar approach to understanding the reasons for the weak level of RES development is proposed in the scientific literature. In particular, Mormann (2011) notes that any attempt to promote renewable energy sources requires a deep understanding of the obstacles that stand in the way of a timely transition to renewable energy.

The cost competitiveness of RES compared to fossil fuels may be the most obvious obstacle to their large-scale deployment. But that's not the only reason. A comparison between France and Germany demonstrates that governance structures and financial subsidies alone do not guarantee the successful promotion of renewable energy sources. Even though the promotional policies of both countries offer equally high subsidized rates for electricity from renewable energy sources, their deployment in Germany is several times greater than in France, which indicates that there are more complex reasons affecting these processes.

To correct the situation, in 2018, the Ministry of Energy of the Russian Federation selected 39 investment projects for the construction of generating facilities operating based on renewable energy sources. Projects with a capacity of 1041.5 MW of capacity, including 853.3 MW of solar generation, 148.5 MW of wind generation, and 39.7 MW of small hydro generation (Report, 2019). It follows from this that, despite the low rate of RES development, the Russian Government is making the necessary efforts to correct this situation.

3. In India, the Electricity Act (2003), the National Electricity Policy (2005), and the Tariff Policy (2016) encourage private sector participation in renewable energy development by establishing obligations to purchase renewable energy for certain organizations. Private sector organizations are present in the entire value chain in the electricity sector, including the production, transmission and distribution of electricity. The Electricity Act provides a framework for the production, transmission, distribution, trade and use of electricity. Private sector entities, including foreign investors, create renewable energy projects and supply electricity to distribution utilities, private consumers, or domestic consumption. They account for 94.92% of the installed capacity of network interactive energy in renewable energy sources (as of 31.03.2018) (Dibyanshu et al., 2019).

4. In the Republic of South Africa, the Integrated Resource Plan (IRP) was approved in October 2019, which provides that the energy balance by 2030 will consist of coal (46%), 1860 MW of nuclear energy (2%), 4696 MW of hydropower (6%), 2912 MW of pumped storage (4%), 7958 MW of solar photovoltaic (PV) energy (10%), 11,442 MW of wind (15%), 11,930 MW of gas (16%) and 600 MW of concentrated solar energy consumption (CSP) (1%). The market for "rooftop solar panels" is growing rapidly. The Mall of Africa solar photovoltaic system is the largest rooftop solar photovoltaic system in the Southern Hemisphere and the 10th largest in the world, covering an area of about 45,000 m^2. Although South African law currently does not allow the sale of excess energy back to the grid (as is the case in some parts of the USA), the sponsors, together with the national energy regulator of South Africa, are successfully conducting a lengthy legal process to ensure legal compliance and successful synchronization of the RES system with the national grid. In recent years, the Ministry of Energy of South Africa has issued regulations that establish activities that are exempt from the requirement to obtain a license to produce electricity, and activities that require registration with the Ministry of Energy. According to them, any generating facility that does not have a connection point

should not receive a license to generate regardless of the generating capacity (in megawatts). At the same time, any object that has a generating capacity of no more than 100 kW with an existing connection point must keep a register of this object, but must not receive a license for a generation. Currently, South Africa does not have significant tax incentives or other government programs similar to those in the US or the EU that would equally contribute to the growth of renewable energy in these markets (Fontana et al., 2020).

5. In Brazil, electricity production is already predominantly from renewable sources. Hydropower accounts for 60% of the country's installed capacity (a total of 105 GW is currently in operation). When measuring actual energy production, the share of hydroelectric power is even greater: almost 90% of the electricity consumed in Brazil comes from hydroelectric sources. This scenario creates a need for diversification of energy sources, since the level of dominance of hydroelectric power plants, unfortunately, also has its drawbacks. Droughts, combined with a lack of sufficient alternative energy sources, have led to spikes in spot energy prices in the recent past. Therefore, renewable sources have become more representative in recent years, with wind accounting for 11% and solar accounting for 9.7% of the installed capacity under construction, while traditional hydroelectric plants under construction account for 7%. This trend will continue in Brazil in the future (Barretto et al., 2020).

The EAEU countries have their own rather unique situation and dynamics of the development of state regulation of the renewable energy system.

1. In Kazakhstan has adopted the Law of the Republic of Kazakhstan dated July 4, 2009, No. 165-IV "On support for the use of renewable energy sources". As follows from Article 4 of this law, the following are among the main directions of state regulation in the field of support for the use of renewable energy sources. This is the creation of favorable conditions for the construction and operation of facilities for the use of renewable energy sources. This is the promotion of the production of electric and (or) thermal energy using renewable energy sources. This is the provision of investment preferences to legal entities engaged in the design, construction, and operation of facilities for the use of renewable energy sources in accordance with the Business Code of the Republic of Kazakhstan. This is the creation of favorable conditions for the effective integration of facilities for the use of renewable energy sources into a single electric power, heat system, and the market of electric and thermal energy. This is a contribution to the implementation of the international obligations of the Republic of Kazakhstan to reduce greenhouse gas emissions.
2. The Republic of Belarus has adopted the Law of the Republic of Belarus "On Renewable Energy Sources" of December 27, 2010, No. 204-Z. According to Article 18 of this law, the Republic of Belarus provides state support in the use of renewable energy sources, including through the formation of a pricing policy aimed at stimulating the use of renewable energy sources, as well as energy produced from renewable energy sources. Promotion of investment activities,

including the creation of favorable conditions for national and foreign investors. Promote the creation and application of efficient technologies in the field of renewable energy sources, as well as the production of installations for the use of renewable energy sources. Ensuring the guaranteed connection of installations for the use of renewable energy sources to the state energy networks. Establishment of tax and other benefits in accordance with legislative acts.

3. The Kyrgyz Republic has adopted the Law of the Kyrgyz Republic No. 283 of December 31, 2008 (as amended on July 24, 2019) "On Renewable Energy Sources". According to Article 7 of this Law, the Government of the Kyrgyz Republic promotes the use of renewable energy through the following measures. For example, the definition of priorities in the development of RES; guaranteed functioning of economic mechanisms and incentive measures provided for by legislation for the development and implementation of environmentally friendly technologies or technologies with low and safe waste levels in the process of RES development. And this includes wells, the evacuation of substances that pollute the environment in the production process, and the use of renewable fuels. It also supports the construction of independent renewable energy systems in cities, as well as in rural areas to provide energy services and the functioning of local production, life support for the population. It also encourages the installation and use of solar energy systems for hot water, heating, cooling, and electricity generation. The promotion of activities for the installation and establishment of a network of biogas plants for the rational use of organic waste from the agricultural production and processing industry is applied. Support is provided for the creation of service centers that ensure the stable production of renewable energy installations, repair, and maintenance of the systems being created.
4. The Republic of Armenia has adopted the Law of the Republic of Armenia No. ZR-122 of December 4, 2004 (as amended on 08.04.2020) "On Energy Saving and Renewable Energy". According to Article 1 of this law, its purpose is to establish the principles of implementation of energy conservation and state policy on the development of renewable energy and mechanisms for their implementation, aimed at the following goals. To strengthen the economic and energy independence of the Republic of Armenia, to increase the economic and energy security of the Republic of Armenia, and the reliability of the energy system. As well as the creation of new industries which promote energy conservation and the development of renewable energy and the organization of services, to reduce the man-made impact on the environment, human health. The law should create mechanisms for implementing the state policy in the field of energy conservation and promote the development of renewable energy. In particular, this law provides for the creation of a new institute of energy expertise, the activities of which will ensure the implementation of the provisions of the document.

So, all the EAEU countries have adopted special laws on the development of renewable energy sources. The Russian legislator went the other way, including provisions on RES in the basic Federal law "On Electric Power Industry". All laws provide for comparable economic and organizational measures aimed at creating conditions for the development of renewable energy sources (Inshakova and Goncharov, 2019).

30.3.3 Environmental Consequences of the Development of Renewable Energy Sources and Ways to Solve Them

Contrary to popular belief, any energy facilities always pose a certain environmental threat. Undoubtedly, such a threat from RES is incomparably less than from nuclear or thermal power plants (Matytsin & Rusakova, 2021). Nevertheless, the experience of European countries and the USA, which are massively introducing wind turbines and solar panels into practice, indicates the presence of side effects that need to be reduced. The fact is that solar panels cause shading of the land, which leads to the deterioration of the soil and the death of vegetation. A negative environmental consequence is the heating of the air as a result of the operation of such an energy facility, which leads to a change in the thermal balance of the area, humidity, and wind direction. Possible overheating and ignition of systems that use solar energy, with all the ensuing consequences.

In turn, the mass accumulation of wind turbines can affect the climate, impairing the ventilation of the area. A very important factor in the influence of wind generators on nature is the acoustic effect.

Noise effects from wind power plants are different and are divided into mechanical (noise from gearboxes, bearings, and generators) and aerodynamic effects. Noise can affect wildlife, including marine life in the area where wind farms are located. The probability of birds being affected by wind turbines is estimated at 10% if the migration routes pass through the wind park. The location of wind parks also affects the choice of bird migration routes. Interference caused by the reflection of electromagnetic waves by wind turbine blades can affect the quality of television and microwave radio broadcasts, as well as various navigation systems in the area where the wind park is located. Along with this, many citizens consider the appearance of windmills not aesthetic (it is worth noting that more and more design agencies are working on overcoming aesthetic inconveniences every year, and progress in this matter is evident). Improvements in the design of the blades and the establishment of noise limits for wind turbines by EU directives are gradually reducing the relevance of this problem. Finally, increasing the size of the wind generator and reducing the frequency of its rotation begins to reduce the number of dead birds. And although the zero levels of losses are still far away, the number of dead birds from collisions with vehicles is now definitely higher in number (Dmitrieva & Pozmogova, 2012).

A separate problem is that the construction of a wind generator or solar battery by one neighbor may violate the rights of another neighbor. This is a difficult problem for Russia and other EAEU countries because these countries have poorly developed legislation on neighborhood rights, which dates back to the times of Ancient Rome. The experience of other countries (e.g., the USA) shows that the solution to this problem should be sought in the development of a system of permits and zoning. So, in some US states (e.g., in New Mexico and Wyoming), the owner of a solar battery gets the right to access sunlight, and the right to protect his interests before neighbors or other persons who create obstacles to such access and make it unprofitable. The US state of California has passed the Solar Rights Act and the Solar Control Act. These legal acts prohibit unjustified restrictions on the installation of solar generators, provide for the creation of "solar amenities", for example, limiting the impact of vegetation on solar energy systems (in other words, the law prohibits the planting of trees that would give shade to solar panels). In some municipalities in the state of Colorado, a real "solar zoning" has been carried out, in which places are allocated where it is prohibited to build or plant trees that would violate the "solar rights" of neighbors (Klass, 2011).

Similarly, some state laws provide for "wind rights" that create conditions for the use of wind turbines and limit the possibility of creating obstacles to this. The existence of such a practice requires representatives of legal science to conduct new scientific research on the rights to the natural resources of wind and sun.

The lack of clear rules and interaction between public authorities in the post-Soviet space makes it difficult to effectively develop energy based on renewable sources. It is necessary to establish new administrative procedures for issuing permits for the construction and commissioning of wind turbines and other similar facilities. It is also necessary to adapt urban planning standards to the challenges of developing renewable energy sources and to create conditions for cost-effective and environmentally friendly heating and energy supply to cities through them.

30.4 Conclusion

The world practice of recent years speaks about the rapid growth of electricity production from renewable sources, which raises the question of the role of law and legal science in understanding and stimulating this trend. The study showed that all natural resources can be classified according to different criteria. According to the criterion of origin, for biological (forests, plants, and animals), mineral (minerals), and energy (solar, wind, etc.). According to the ecological content, for exhaustible (renewable and non-renewable) and inexhaustible (solar, wind, tidal energy). According to the degree of actual use, real and potential. According to the criterion of substitutability of some by others, replaceable and irreplaceable. Currently, all the BRICS and EAEU countries have adopted special laws on the development of renewable energy sources.

The Russian legislator went the other way, including provisions on RES in the basic Federal law "On Electric Power Industry". At the same time, all laws provide for comparable economic and organizational measures aimed at creating conditions for the development of renewable energy sources.

The existing practice of developing the renewable energy system in all the BRICS and EAEU countries leaves some issues unresolved that require further discussion and resolution. Among them are questions about the role of the state in increasing the competitiveness of renewable energy sources, as well as measures to be taken to reduce the environmental risk of renewable energy sources. As well as measures in the field of neighborhood law aimed at guaranteeing the rights of owners of renewable energy sources, etc.

A comprehensive legal solution to the issue of further development of renewable energy is cross-sectoral and will require amendments and additions to civil, environmental, energy, and other legislation.

Acknowledgements The work was supported by Russian Science Foundation (project No. 20-18-00314).

References

2015 Agreement, the Paris Agreement under the United Nations Framework Convention on Climate Change (Signed in New York on April 22, 2016).

Barretto, A. C., Figueiro, T. K., & Brasil, A. L. (2020). Brazil. In K. B. Wong (Ed.), *The renewable energy law review* (3rd ed.). Law Reviews.

Belotsky, S. D. (2016). *International legal, regulation in the field of environmentally oriented energy: Abstract of the dissertation of the Doctor of Law*. Taras Shevchenko National University of Kyiv.

Deynega, M. A. (2020). *The subject and system of natural resource law: Abstract of the dissertation of the Doctor of Law*. National University of Bioresources and Nature Management of Ukraine.

Dibyanshu, Rastogi, S., & Bhandari, P. (2019). *Renewable energy in India*. https://www.lexology.com/library/detail.aspx?g=6efdf930-f5f5-44c8-a4b3-54cf1de07b94. Accessed January 12, 2021.

Directive of the European Parliament and of the Council of the European Union 2018 on the Promotion of the Use of Energy from Renewable Sources (Signed in Strasbourg on 11 December 2018 no. 2018/2001/EC) (European).

Dmitrieva, Yu. Yu., & Pozmogova, S. B. (2012). Prospects for the use of wind power in the Ulyanovsk region. *Bulletin of the Ulyanovsk State Technical University, 2*, 37–41.

Federal Law. (2003). *On electric power industry* (Adopted on 26.03.2003 No. 35-FZ).

Fontana, L., & Dladla, Y. (2020). South Africa. In K. B. Wong (Ed.), *The renewable energy law review* (3rd ed.). Law Reviews.

Inshakova, A. O., & Goncharov, A. I. (2019). Innovation as an integral condition for the development of modern foreign trade turnover of energy resources and its legal regulation. In O. V. Inshakov, A. O. Inshakova, & E. G. Popkova (Eds.), *Energy sector: A systemic analysis of economy, foreign trade and legal regulations* (pp. 67–78). Springer Science + Business Media. https://doi.org/10.1007/978-3-319-90966-0_5

Inshakova, A. O., Goncharov, A. I., & Marchukov, I. P. (2019). Common energy policy and development paths legal regulation of foreign trade turnover of energy resources in the CIS and the EEU. In O.V. Inshakov., A. O. Inshakova, & E. G. Popkova (Eds.), *Energy sector: A systemic analysis of economy, foreign trade, and legal regulations* (pp. 207–225). Springer Science + Business Media. https://doi.org/10.1007/978-3-319-90966-0_14

Klass, A. B. (2011). Property rights on the new frontier: Climate change, natural resource development, and renewable energy. *Ecology Law Quarterly, 38*, 95–104.

Law of the Kyrgyz Republic. (2008). *On renewable energy sources* (Adopted on December 31, 2008, No. 283) (Kyrgyzstan).

Law of the People's Republic of China. (2009). *On renewable energy sources* (Adopted on December 26, 2009) (China).

Law of the Republic of Armenia. (2004). *On energy saving and renewable energy* (Adopted on December 4, 2004, No. ZR-122) (Armenia).
Law of the Republic of Belarus. (2010). *On renewable energy sources* (Adopted on December 27, 2010, No. 204-Z) (Belarus).
Law of the Republic of Kazakhstan. (2009). *On support for the use of renewable energy sources* (Adopted on July 4, 2009, No. 165-IV) (Kazakhstan).
Luneva, E. V. (2018). Legal criteria for the rational use of renewable and non-renewable natural resources. In: *Greening-the main-task of sustainable development of society: A collection of articles based on the materials of the International Scientific and Practical Conference*. Saratov State Law Academy.
Matytsin, D. E., & Rusakova, E. P. (2021). The strategy of quality management in industry 4.0 and formation of the cognitive economy based on industrial and manufacturing engineering in the Russian Federation and countries of the EU. *International Journal for Quality Research, 15*(4). https://doi.org/10.24874/IJQR15.04-03
Mormann, F. (2011). Requirements for a renewables revolution. *Ecology Law Quarterly, 38*, 903–965.
Order of the Government of the Russian Federation. (2009). On the main directions of state policy in the field of improving the energy efficiency of the electric power industry based on the use of renewable energy sources for the period up to 2035 (Adopted by the Government of the Russian Federation on 08.01.2009 No. 1-r in the ed. from October 24, 2020).
Progress report on the implementation in 2018 of the comprehensive plan for the implementation of the Climate Doctrine of the Russian Federation for the period up to 2020, 2019, Moscow. https://www.mnr.gov.ru/docs/o_vypolnenii_kompleksnogo_plana_realizatsii_klimaticheskoy_doktriny/doklad_o_khode_vypolneniya_v_2018_godu_kompleksnogo_plana_realizatsii_klimaticheskoy_doktriny_rossiy/ Accessed January 12, 2021.
Ryzhenkov, A. Y. (2020). Legal protection of the environment in China and Russia: Comparative legal aspect. *Modern Law, 7*, 123–127.
Shchepansky, I. S. (2013). *Ecological and legal problems of ensuring energy security in Russia: Abstract of the dissertation of the candidate of Legal Sciences*. Kutafin Moscow State Law University.
The UN. (2019). *Environmental program*. Official website. https://www.unenvironment.org/ru/novosti-i-istorii/press-release/desyatiletie-investiciy-v-vie-vo-glave-s-solnechnoy-energetikoy. Accessed January 12, 2021.
Vankovich, E. E. (2017). *Legal support of environmental safety in the use of renewable energy sources*. Ph. D. in Law thesis. National Center for Legislation and Legal Research, Minsk.
Vasilyeva, M. I. (2011). *Legal support of the environmental policy*. Priority areas. Institute of Sustainable Development of the Public Chamber of the Russian Federation.

Correction to: Sustainable Development and the Legal Regulation of Forced Migration in Russia

Dmitry V. Ivanov and Nikita Yu. Molchakov

Correction to:
Chapter 24 in: A. O. Inshakova and E. E. Frolova (eds.), *Smart Technologies for the Digitisation of Industry: Entrepreneurial Environment*, Smart Innovation, Systems and Technologies 254, https://doi.org/10.1007/978-981-16-4621-8_24

In the original version of the book, the misspelt co-author name "Nikita Yu. Molchkov" has been changed to read as "Nikita Yu. Molchakov" in Chapter "Sustainable Development and the Legal Regulation of Forced Migration in Russia". The erratum chapter and the book have been updated with the change.

The updated version of this chapter can be found at
https://doi.org/10.1007/978-981-16-4621-8_24

A. O. Inshakova and E. E. Frolova (eds.), *Smart Technologies for the Digitisation of Industry: Entrepreneurial Environment*, Smart Innovation, Systems and Technologies 254, https://doi.org/10.1007/978-981-16-4621-8_31

The manufacturer's authorised representative in the EU is Springer Nature Customer Service Centre GmbH, Europaplatz 3, 69115 Heidelberg, Germany. If you have any concerns regarding our products, please contact ProductSafety@springernature.com

Printed and bound by CPI Group (UK) Ltd, Croydon, CR0 4YY
30/09/2026
02225246-0019